T0140412

The Information Retrieval Series Volume 36

More information about this series at http://www.springer.com/series/6128

Peter Knees • Markus Schedl

Music Similarity and Retrieval

An Introduction to Audio- and Web-based Strategies

Peter Knees
Department of Computational Perception
Johannes Kepler University
Linz
Austria

Markus Schedl
Department of Computational Perception
Johannes Kepler University
Linz
Austria

ISSN 1387-5264
The Information Retrieval Series
ISBN 978-3-662-57031-9 ISBN 978-3-662-49722-7 (eBook)
DOI 10.1007/978-3-662-49722-7

Softcover reprint of the hardcover 1st edition 2016

Printed on acid-free paper

This Springer imprint is published by Springer Nature
The registered company is Springer-Verlag GmbH Berlin Heidelberg

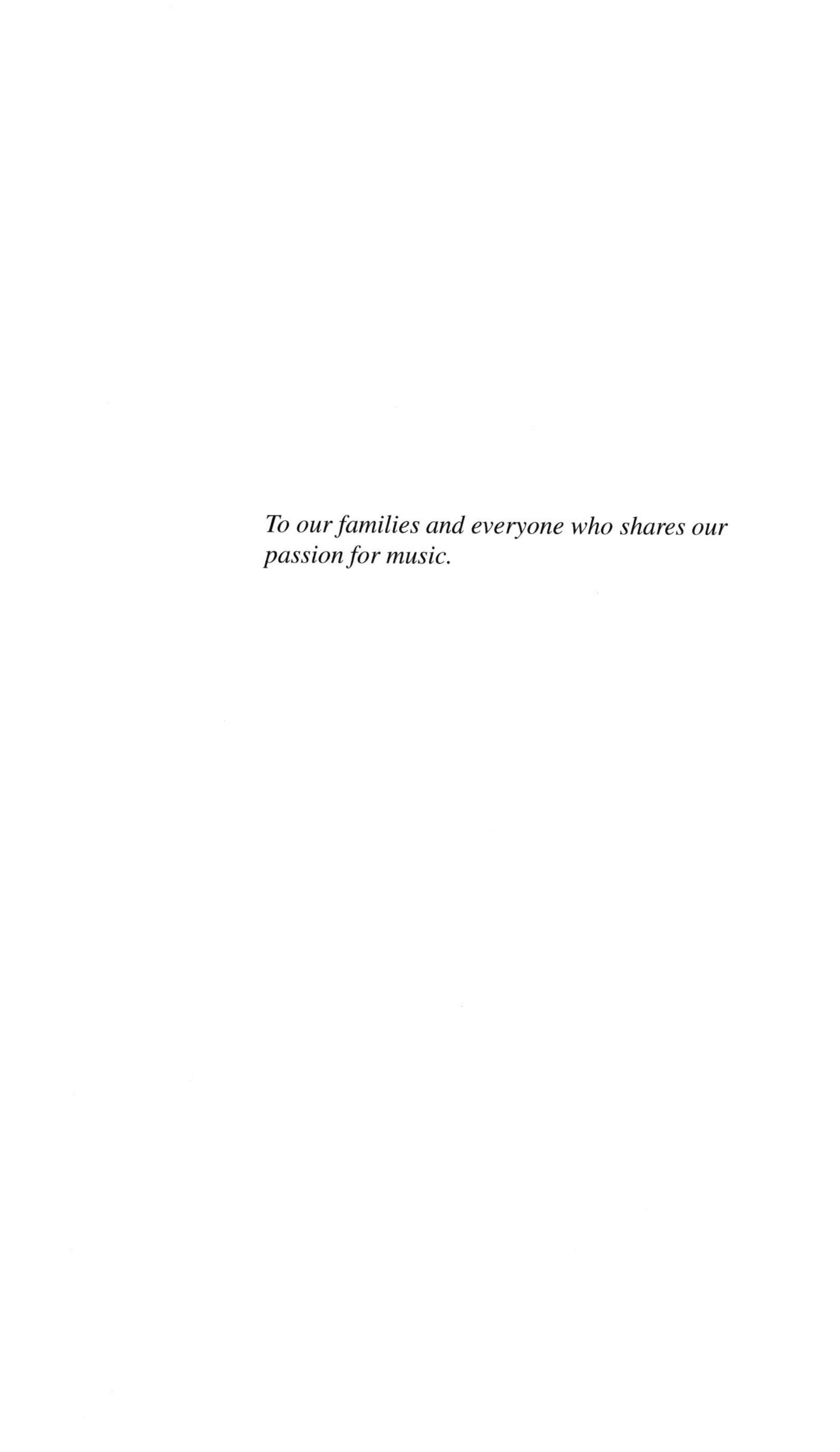

To our families and everyone who shares our passion for music.

Foreword

When Peter Knees and Markus Schedl invited me to write the foreword to their upcoming book on music information retrieval, they attached an early sketch of the contents as a copy to the email. Upon reading through their notes, I was struck by five strongly emotive thoughts in quick succession: First, Peter and Markus have done a great job capturing and explicating the many different facets that come together to make up Music Information Retrieval (MIR) research and development. Second, this is exactly the kind of text I would assign to my students and recommend to colleagues who are newcomers to MIR. Third, MIR certainly has matured and evolved in some truly interesting ways since 2000 when the first International Symposium for Music Information Retrieval was convened in Plymouth, Massachusetts. Fourth, forget 2000, how about 1988, when a much younger me was beginning my personal relationship with MIR? If it were possible to describe the state of MIR in 2016 to the 1988 version of me or any of my fellow music school students, would we be able to comprehend at all just how magically advanced music access, processing, search, and retrieval would become in just under 30 years? Fifth, and finally, if it were possible to send this book back in time to 1988, would it convince us of the marvels that lay ahead and help us to understand how they came about? Since the answer to that last question was a resounding yes, I decided to accept the kind invitation of Peter and Markus. Thus, it was my pleasure and honor to write the foreword you are currently reading.

In 1988, I was studying for my undergraduate degree in music at the University of Western Ontario in London, Ontario, Canada. I was a decent second-rate theorist. I was, however, a famously bad flautist who spent more time drinking coffee, smoking, and playing cards than I did practicing. In the spring of 1988, I became desperate to find some easy-to-learn flute music to play at my fast-approaching final flute exam. I knew vaguely of a baroque sonata in my favorite key of C major. It took me several weeks of false starts, questions to librarians, browsing CD collections, and searching the library catalogue before we were finally able to discern that the piece in question was Bach's Flute Sonata in C major, BWV 1033. The exam itself was a bit of a flop (remember, I rarely practiced) so I am now understandably a professor of library and information science, and not music. Notwithstanding that my not-so-successful flute exam was the proximate cause to a change of careers for me, the search for the music did leave me with two complementary questions that have informed my academic work ever since: (1) Why is finding music so damn hard and annoying? and (2) What can we do to make it not hard and perhaps even enjoyable? The last 30 years have shown that I am far from alone in having these questions.

After my illustrious career as a second-rate music theory student, I found a welcoming home at the Graduate School of Library and Information Science (GSLIS), University of Western Ontario. I was encouraged to explore my two questions by the Dean of the faculty, Jean Tague-Sutcliffe. Professor Sutcliffe was a leading international figure in the formal evaluation of text retrieval systems, and I was quite privileged to have studied with her. My graduate student days at GSLIS shaped my thinking about what it would mean to do MIR research. First, my graduate work dealt with music only at a symbolic level in an attempt to retrofit music information into current state-of-the-art text retrieval engines. Second, library and information science has a strong tradition of user-centered service and research. It is important to note that I paid mostly lip-service to user-centered issues while a student as all the "serious" work was being done by systems students building and evaluating new retrieval techniques. In retrospect, I was very shortsighted to have fixated on classic IR techniques and a complete idiot to have downplayed the absolute importance of the user in any successful MIR system design and evaluation. 2016 me waggles his finger at 1995 me, accompanied by a firm "tsk, tsk."

Fast forward to October of 2000 and the first ISMIR meeting. I find myself sitting in the conference room of the John Carver Inn, Plymouth, Massachusetts. In front of me, Beth Logan is presenting her paper introducing Mel frequency cepstral coefficients (MFCC) as a possible feature for building machine learning models from music audio. This paper is historic for suggesting MFCCs as an audio feature to the MIR community that has become almost ubiquitous. The presentation is also semilegendary for the spirited debate among attendees concerning the appropriateness of using MFCCs, originally designed for speech, in music applications. I must admit, the pros and cons of the debate were completely lost on me. At this point in the meeting, my head was still swimming with one overwhelming thought: "Audio! We can do real MIR things with music audio! Who knew? Too cool!" I cannot overstate how thunderstruck I was by the fact that I was in the presence of very solid researchers making very real progress working with actual music audio. Digital signal processing (DSP) seemed like alchemy to me, except this alchemy was yielding gold. Remember, my MIR world vision until this time derived from my text-based IR graduate work and my undergrad studies in manipulating and analyzing the symbols of music notation. The first major conference publication of my MIR work was, for example, at the Association for Computing Machinery, Special Interest Group on Information Retrieval (ACM SIGIR), held in June of 2000 which discussed converting melodic intervals into text for use in standard search engines.

As you can see, ISMIR 2000 was a paradigm shifting event for me. This paradigm shift involved more than my naiveté surrounding the existence of useful music audio processing techniques. Until this point, I had not thought that MIR might not be bound by the classic query-index-response model (i.e., Cranfield model) that pervades traditional conceptions of IR. I had not yet considered that MIR might be expanded to include different kinds of tasks beyond simple search and retrieval. Papers about discovering and visualizing music structures, identifying themes, classifying genres, labeling instruments, describing content, summarizing audio, and visualizing patterns illuminated new areas of possibilities that I had overlooked previously. Associated closely with these newfound MIR areas was a bundle of techniques that I had heard of but had never seen in action. For me, machine learning (ML) and its older cousin, artificial intelligence (AI), had been the stuff of science fiction and spy novels. Prior to ISMIR 2000, I had no idea how ML/AI could possibly apply to MIR research. I recall being deeply skeptical of our choice of keynote, Marvin Minksy, who was world-renowned for his groundbreaking AI work. After seeing the excellent presentations that made use of such things as neural nets and Gaussian mixture models, I became convinced.

While flying home from ISMIR, I tried my best to make sense of what I learned. I knew that I had to broaden my conception of what it meant to do MIR research. I needed to come up with better way of summarizing succinctly what the MIR project was all about. To this end, I came up with the pseudo-equation $\text{MUSIC}_{(Audio)} + \text{DSP} + \text{ML} = \text{MIR}$. I am fully aware that this model leaves out a great many important aspects of MIR such as the symbolic side of music and, of course, users in their individual and social contexts; however, it did capture quite accurately what was to become (and continues to be) the most popular, and possibly the most productive (up to now), approach to MIR research and development. While there are many different and interesting facets to MIR that do not involve the three components of my pseudo-equation—music audio, digital signal processing, and machine learning—it is an inescapable fact that one must have at least an introductory understanding of these topics, and their interactions, if one is to comprehend the accomplishments and promise of MIR research. This book you have in your hand right now provides just such an introduction, and much more.

The Music Information Retrieval Evaluation eXchange (MIREX) convened for the first time at ISMIR 2005, London, UK. A quick review of the tasks evaluated in 2005 reinforces the notion that the model represented by my pseudo-equation above had become the dominant MIR research paradigm. Most of the MIREX 2005 tasks were based upon evaluation techniques popular in the ML and DSP communities rather than those of the information retrieval domain. That is, rather than using a set of predefined test queries against which systems are evaluated, most MIREX tasks have been designed to use pre-constructed ground truth sets upon which systems are trained and then tested in n-fold cross-validation experiments. MIREX has grown substantially since 2005; however, the ML evaluation worldview still predominates. The use of ground truth sets and train-test experiments have many benefits including ease of administration, simple reproducibility, known and accepted scoring methods, and generally understandable results. There are also several rather important shortcomings to the wholesale adoption of DSP and ML evaluation methods. First, the methodology has a tendency to shape the tasks evaluated in a tail-wagging-the-dog kind of way. That is, there is a quite strong tendency for tasks to be held because there exists ground truth (or ground truth would be easy to create) rather than the inherent importance of the task topic. Second, because ground truth is so expensive to create, it is usually donated by graduate students working on specific MIR subproblems, like beat tracking or chord estimation. This means that MIREX tasks are rather narrowly scoped and thus, even when taken all together, generate a fragmented vision of the MIR problem space. Because of this, MIREX has become very good at analyzing a wide variety of interrelated trees in some detail but has failed to describe the forest in which they live. Third, MIREX is almost completely mute about the most important component of any successful MIR system: the user.

As a leader of MIREX, I bear considerable responsibility for the shortcomings listed above. As MIREX was being developed prior to 2005, I must admit a strong desire to privilege the formal evaluation of content-based approaches because they fit the model implied by my MIR pseudo-equation. When MIR researchers like Julián Urbano assert (justifiably) that MIREX, in particular, and MIR research, in general, must now focus on user-centric tasks, I can offer no counterargument. They are absolutely correct in arguing that both MIREX and MIR research are now hitting an upper limit to the useful insights it can provide absent user-inspired tasks and user-centric experiments and studies. In retrospect, I should have put more effort into developing a MIREX community to define and create user-centered evaluation tasks. One thing that remains a bit of a mystery to me about this state of affairs is the fact that my own personal MIR research evolved away from system-centric to user-focused through my marvelous interactions with such notable “user needs and uses” scholars as Sally Jo Cunningham, Jin Ha Lee, Audrey Laplante, Charlie Inskip, and Xiao Hu. These folks, and others like them, need to be applauded for always keeping the user at the heart of their MIR research thinking.

Peter and Markus, when they were not writing this fine book, have also made significant contributions to the cause of putting the user at the heart of MIR research. Their ongoing work based upon an examination of the social contexts in which all kinds of users seek and interact with music and other music lovers is very well regarded. It is Peter’s and Markus’s music-in-context line of inquiry that informs the second and third part of the text. For me, these are the most important and potentially influential contributions of the book. Parts II and III put all the technical DSP and ML pieces discussed in Part I in their proper place, that is, at service of real users doing real things with real music. If I were able to somehow send one part of the book back to 1988, it would be these sections. Armed with their knowledge and ideas about users and how they live, use and enjoy music in their everyday lives, I would have spent less time fussing about making my MIR research look like classic IR research. I would have taught myself to more joyfully embrace the noisy, contradictory, and difficult-to-comprehend data generated by users in the wild. MIREX would definitely have a large set of user-centered tasks even if they are harder to manage and run. Most significantly, my pseudo-equation would have read from the beginning, MUSIC + DSP + ML + *USER* = MIR.

Champaign, IL, USA
February 2016

J. Stephen Downie

Preface

Music is an omnipresent topic in our society—it is everywhere and for everyone. Music is more than just the pure acoustic perception. It is a pop cultural phenomenon, maybe even the most traditional and most persistent in human history. It takes a central role in most people's lives, whether they act as producers or consumers, and has the power to amplify or change its listener's emotional state. Furthermore, for many people, their musical preferences serve as a display of their personality. Given its cultural importance, it seems no wonder music was the first type of media that underwent the so-called digital revolution. Based on the technological advancements in encoding and compression of audio signals (most notably the invention of the mp3 standard) together with the establishment of the Internet as the mainstream communication medium and distribution channel and, in rapid succession, the development of high capacity portable music players, in the late 1990s, digital music has not only stirred up the IT industry but also initiated a profound change in the way people "use" music. Today, a lot more people are listening to a lot more music in many more situations than ever before. Music has become a commodity that is naturally being traded electronically, exchanged, shared (legally or not), and even used as a means for social communication. Despite all these changes in the way music is *used*, the way music collections are *organized* on computers and music players and the way consumers *search* for music within these structures have basically remained the same for a long time.

Nowadays, we are witnessing a change in this behavior. Intelligent music listening applications are on the rise and become more and more important in high-end systems for music aficionados and everyday devices for casual listeners alike. For music retrieval systems, results are often required to serve a particular purpose, e.g., as background music that fits a specific activity such as dining or working out. Moreover, the purpose and usage of music are not limited to the auditory domain in the sense that the ways that personal music collections are presented often function as displays of personality and statements of distinction. In fact, a large portion of the aura that is surrounding collecting and structuring music stems from the rich and, technically speaking, multimodal context of music. This context spans from the aesthetics of the artwork to the type of packaging and included paraphernalia to liner notes to statements made by artists in accompanying media to gossip about the members of a band. In that sense, modern music information and retrieval systems must acknowledge the long tradition of collecting analog records to gain wide acceptance. This amplifies the requirement to provide highly context-aware and personalized retrieval and listening systems. However, not all of the aforementioned context can be digitized, and therefore preserved, using today's predominant means of media delivery. Then again, this loss of detail in context is exchanged for the sheer amount of instantly accessible content.

Applications such as *Shazam* (music identification), *Pandora* (automatic personalized radio stationing), *Spotify* (music streaming), or *Last.fm* (music recommendation, information system, and social network) today are considered essential services for many users. The high acceptance of and further demand for intelligent music applications also make music information retrieval as a research topic a particularly exciting field as findings from fundamental research can find their way into commercial applications immediately. In such a setting, where many innovative approaches and applications are entering the competition and more and more developers and researchers are attracted to this area, we believe that it is important to have a book that provides a comprehensive and understandable entry point into the topics of music search, retrieval, and recommendation from an academic perspective. This entry point should not only allow novices to quickly access the field of music information retrieval (MIR) from an information retrieval (IR) point of view but also raise awareness for the developments of the music domain within the greater IR community.

To this end, the book at hand gives a summary of the manifold audio- and web-based approaches and subfields of music information retrieval research for media consumption. In contrast to books that focus only on methods for acoustic signal analysis, this book is focused on music as a specific domain, addressing additional cultural aspects and giving a more holistic view. This includes methods operating on features extracted directly from the audio signal as well as methods operating on features extracted from contextual information, either the cultural context of music pieces as represented on the web or the user and usage context of music. With the latter, we account for the paradigm shift in music information retrieval that can be seen over the last decade, in which an increasing number of published approaches focus on the contextual feature categories, or at least combine "classical" signal-based techniques with data mined from web sources or the user's context.

We hope the reader will enjoy exploring our compilation and selection of topics and keep this compendium at hand for exciting projects that might even pave the way for "the next big music thing."

Vienna, Austria/Linz, Austria
January 2016

Peter Knees
Markus Schedl

Acknowledgments

No one tells the truth about writing a book. . . . The truth is, writing is this: hard and boring and occasionally great but usually not. —Amy Poehler

Writing this book would not have been possible without the support of numerous people. We would like to express our gratitude to *Elias Pampalk*, *Tim Pohle*, *Klaus Seyerlehner*, and *Dominik Schnitzer* for sharing their audio feature extraction implementations and their support in preparing the examples in this book. For allowing us to include illustrations and screenshots, we are most grateful to *Emilia Gómez*, *Masataka Goto*, *Edith Law*, *Anita Shen Lillie*, *Irène Rotondi*, *Klaus Scherer*, *Mohamed Sordo*, and *Sebastian Stober*. We also thank *Dmitry Bogdanov* for his valuable input on hybrid recommender systems. Furthermore, we would like to thank *Gerhard Widmer* for his support and for granting us the time and freedom it takes to face the endeavor of writing a book.

We thank *Ralf Gerstner*, Executive Editor (Computer Science) for Springer, for his support, understanding, and patience. Also the constructive feedback provided by the anonymous reviewers was highly appreciated and helped a lot in sharpening the manuscript.

Finally, we appreciate the research funding organizations which provided gracious financial support, in particular, the Austrian Science Funds (FWF) and the European Commission.[1] By financing our work and projects, these organizations allowed us to gain the knowledge it takes to write such a book and provided a fruitful ground to contribute to and extend the state of the art in various tasks covered in the book.

[1]We received funding from the Austrian Science Funds (FWF) for the projects "Operational Models of Music Similarity for Music Information Retrieval" (L112-N04), "Music Retrieval Beyond Simple Audio Similarity" (L511-N15), "Personalized Music Retrieval via Music Content, Music Context, and User Context" (P22856), "Social Media Mining for Multimodal Music Retrieval" (P25655), and "Culture- and Location-aware Music Recommendation and Retrieval" (AJS3780). The European Commission supported our research within the projects "PHENICX—Performances as Highly Enriched aNd Interactive Concert eXperiences" (FP7-ICT-2011-9: 601166) and "GiantSteps—Seven League Boots for Music Creation and Performance" (FP7-ICT-2013-10: 610591).

Contents

Chapter 1
Introduction to Music Similarity and Retrieval

Traditionally, electronically searching for music, whether in collections of thousands (private collections) or millions of tracks (digital music resellers), is basically a database lookup task based on meta-data. For indexing a collection, existing music retrieval systems make use of arbitrarily assigned and subjective meta-information like *genre* or *style* in combination with objective meta-data like *artist name*, *album name*, *track name*, *record label*, or *year of release*. On top of that, often, the hierarchical scheme *(genre–) artist–album–track* is then used to allow for browsing within the collection. While this may be sufficient for small private collections, in cases where most contained pieces are not known a priori, the unmanageable amount of pieces may easily overstrain the user and impede the discovery of desired music. Thus, a person searching for music, e.g., a potential customer, must already have a very precise conception of the expected result which makes retrieval of desired pieces from existing systems impractical and unintuitive.

Obviously, the intrinsic problem of these indexing approaches is the limitation to a rather small set of meta-data, whereas neither the musical content nor the cultural context of music pieces is captured. Archival and retrieval of music is historically a librarian's task, and structure and format of databases are optimized for access by experts. Today, the majority of users are not experts—neither in database search nor in terms of musical education. When searching for music, particularly when trying to discover new music, users rarely formulate their queries using bibliographic terms but rather describe properties like emotion or usage context [207]. Therefore, different search and retrieval scenarios become more important.

Retrieval systems that neglect musical, cultural, and personal aspects are far away from the manifold ways that people organize, deal with, and interact with music collections—or expressed in information retrieval (IR) terms, these system neglect their users' *music information needs* [90, 258]. For music, information needs can be quite distinct from standard text-related information needs. Music as a media is heavily intertwined with pop culture as well as with hedonistic and recreational activities. The need to find music might not be as much one that is targeted at

P. Knees, M. Schedl, *Music Similarity and Retrieval*, The Information Retrieval Series 36, DOI 10.1007/978-3-662-49722-7_1

information but merely one targeted at pure entertainment. Thus, one could argue that for most popular and mainstream music, the average user accessing a music information system has primarily an *entertainment need* (cf. [34, 325]).

1.1 Music Information Retrieval

As a response to the challenges, specifics, and needs of retrieval in the music domain, the area of research known as *music information retrieval* (MIR) has evolved in the 1990s and emancipated itself as a dedicated field at the beginning of the millennium with the organization of the ISMIR[1] conference series [54]. Among others, MIR is researching and developing intelligent methods that aim at extracting musically meaningful descriptors either directly from the audio signal or from contextual sources. These descriptors can then be used, e.g., to build improved interfaces to music collections. In this section, we give an overview of the field of MIR. We start by looking into definitions found in the literature and proceed by describing the predominant retrieval paradigms found in MIR, illustrated by exemplary tasks and applications. We round this overview up by pointing to research areas of MIR that go beyond traditional IR tasks.

In the literature, one can find several definitions of MIR—each focusing on specific aspects. We give a selection of these in order to sketch the bigger picture. In an early definition, Futrelle and Downie emphasize the multi- and interdisciplinarity of MIR and its origins in digital library research:

> MIR is a(n) ... interdisciplinary research area encompassing computer science and information retrieval, musicology and music theory, audio engineering and digital signal processing, cognitive science, library science, publishing, and law. Its agenda, roughly, is to develop ways of managing collections of musical material for preservation, access, research, and other uses. [138]

Later on, Downie highlights research on content analysis, i.e., the automatic extraction of music descriptors from the audio signal itself, interfaces, and infrastructure:

> MIR is a multidisciplinary research endeavor that strives to develop innovative content-based searching schemes, novel interfaces, and evolving networked delivery mechanisms in an effort to make the world's vast store of music accessible to all. [110]

Finally, in our own definition, we highlight the multimodality of the field:

> MIR is concerned with the extraction, analysis, and usage of information about any kind of music entity (e.g., a song or a music artist) on any representation level (for example, audio signal, symbolic MIDI representation of a piece of music, or name of a music artist). [401]

[1] Previously: *International Symposium for Music Information Retrieval* and *International Conference on Music Information Retrieval*; since 2009: *International Society for Music Information Retrieval Conference*.

The diversity of these different definitions demonstrates the width of this field. In this book, we cannot review every aspect of music information retrieval. We focus instead on the very central part of music search and retrieval and the notion of music similarity—the underlying concept that is essential to all of the presented methods. To this end, we continue the introduction of MIR with a discussion of music retrieval tasks from an information retrieval perspective.

1.2 MIR from an Information Retrieval Perspective

Similar to other domains, we can identify three main paradigms of music information access:

Retrieval The user has a specific music information need, e.g., finding a specific musical item, and actively expresses this need using a query. This query can be represented in different modalities such as text, a symbolic music representation, or as a piece of audio. The result can be (specific parts of) audio pieces, scores, or meta-data, potentially in a ranked list.

Browsing The user has an undirected information need and wants to explore the available music collection. Browsing is an interactive and iterative task that relies on intuitive and effective user interfaces for discovering items. Like search, browsing is initiated and actively pursued by the user.

Recommendation The system filters the collection for potentially relevant items based on the user's actions or preferences. These preferences can be given explicitly by the user or derived from implicit feedback, e.g., by observing actions during retrieval and/or browsing or by tapping listening data. The user is not required to actively search for music and is presented with a personalized view.

Figure 1.1 gives a schematic overview of these access paradigms and how they connect to data structures. As can already be seen, features and similarity take a central role in this layout. To facilitate music information retrieval, naturally, a lot of research is targeted at tasks that relate to music signal processing, feature extraction, and similarity measurement, e.g., to build systems for content-based querying and retrieval; cf. Part I. As we will see, particularly in existing browsing interfaces, timbre plays an important role as a descriptor for sound similarity. Chapter 3 will therefore prominently deal with timbre-related features (as well as touching upon aspects of rhythmicity and tonality). Other musical dimensions related to pitch (key, chords, harmonies, melodies), temporal facets (rhythm, tempo), and structure, as well as retrieval tasks specific to these properties, are outside the scope of this book. For computational methods to derive this kind of information and use it for retrieval, we refer the reader to other sources (see the *further reading* Sect. 1.6). We discuss the role of features and the multifaceted notion of music similarity in Sect. 1.3. Before this, we elaborate on the three music information access paradigms *retrieval*, *browsing*, and *recommendation* by describing typical tasks in MIR and pointing to

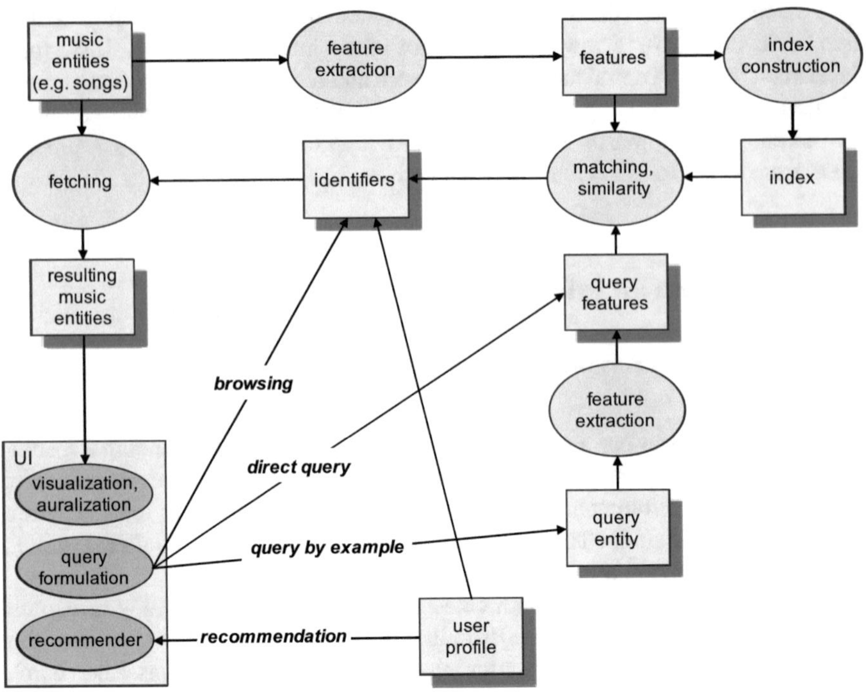

Fig. 1.1 Traditional information retrieval view on music retrieval with a focus on content-based document indexing

exemplary applications and research prototypes. Furthermore, we highlight areas of MIR beyond these three paradigms.

1.2.1 Retrieval Tasks and Applications in MIR

A retrieval approach that directly corresponds to text IR methods is searching for symbolic music using a query consisting of a symbolic representation of the same type, thus following the *query by example* retrieval scheme. The term "symbolic music" denotes digital data representations of music notation. An example of such a data representation would be the MIDI format.[2] For matching, the relevant part of the encoded information typically relates to pitch and timing information.

The *Themefinder* web search engine,[3] for instance, allows for querying a symbolic database by entering melodic sequences in proprietary text formats [240].

[2]Abbreviation for *musical instrument digital interface*.

[3]http://www.themefinder.org.

Other systems follow more intuitive approaches and are therefore also usable for less musically educated users. For example, in **query by singing/humming (QBSH)** systems, the user can hum or sing a part of the searched piece into a microphone. From that recording, musical parameters (mostly related to melody) are extracted, and the obtained sequence serves as a query to the database; cf. [143]. An example of a search engine offering exhaustive possibilities for querying is *Musipedia*.[4] *Musipedia* indexes a large number of music pieces by crawling the web for MIDI files that can then be used for identification of pieces. For indexing of pieces, the melodic contour, pitches and onset times, and a rhythm representation are extracted. To find a piece in the database, a theme (i.e., the query) can be either entered in Parsons code notation [353] or whistled into a microphone (to find matching melodies), played on a virtual piano keyboard (to find matching pitch and onset sequences), or tapped on the computer keyboard (to find matching rhythms). For a detailed explanation of the incorporated techniques, as well as a comprehensive comparison of symbolic music retrieval systems and MIR systems in general, we refer the reader to [485]. Please note that symbolic music retrieval is not within the scope of this book. Part I of this book addresses query by example systems that make use of audio similarity approaches to find the most similar recordings in a collection for a given track.

While the abovementioned systems aim at retrieving a ranked list of documents similar to the query, for the task of **audio identification** the goal is to find, i.e., to identify, the query within a large database of recordings. The query is typically a short snippet of a song, possibly recorded in low quality and in an environment with background noise, e.g., using a cellular phone. The expected result is the meta-data of the entry in the database, such as artist, album, and song name. The underlying technique is known as music fingerprinting. To this end, for each song in a music collection, a compact unique feature representation is created (the so-called fingerprint) which can be matched against the fingerprint of the query. A requirement is that fingerprints must be robust against all kinds of distortions, e.g., caused by factors such as cheap microphones or cellular phone connections; cf. [505]. Fingerprinting can be used to detect copyright infringements or in music identification services, with the most popular commercial example being *Shazam*.[5] Other examples of services that provide audio identification are *SoundHound*,[6] which also provides methods for QBSH, *Gracenote MusicID*,[7] *MusicBrainz Fingerprinting*,[8] and *Echoprint*.[9]

[4]http://www.musipedia.org.

[5]http://www.shazam.com.

[6]http://www.soundhound.com.

[7]http://www.gracenote.com/music/recognition.

[8]http://musicbrainz.org/doc/Fingerprinting.

[9]http://echoprint.me.

Conceptually related to both symbolic retrieval and audio identification is **cover song identification** or version identification. Here the goal is to find different renditions and stylistic interpretations of the query song. State-of-the-art algorithms for this task extract descriptors relating to melody, bass line, and harmonic progression to measure the similarity between two songs [391].

Another retrieval scenario is **text-based retrieval of audio and music** from the web. Some search engines that use specialized (focused) crawlers to find all types of sounds on the web exist. As with web image search, the traced audio files are then indexed using contextual information extracted from the text surrounding the links to the files. Examples of such search engines are *Aroooga* [229] and *FindSounds*.[10] Other approaches utilize text information from the web to index arbitrary music pieces. Hence, a textual context has to be constructed artificially by finding web pages that mention the meta-data of tracks. We discuss such approaches in Part II.

1.2.2 Browsing Interfaces in MIR

Next to hierarchical text-based information systems for browsing of music collections is an important access modality in MIR. Such interfaces should offer an intuitive way to sift through a music collection and to encounter serendipitous music experiences. We discuss intelligent music interfaces in detail in Sect. 9.2; however, here we want to point out some exemplary innovative interfaces that support the user in discovering music using MIR technology.

Figure 1.2 shows the *Intelligent iPod* interface that aims at providing "one-touch access" to music on mobile devices [429]. Just by using the scroll wheel of a classic *iPod*, the user can browse through the whole collection that is organized on a circular playlist according to acoustic similarity, i.e., neighboring songs are expected to sound similar and, overall, there should be smooth transitions between the different styles. Additionally, regions on the playlist are described using collaborative tags for easier navigation. After selecting a region, the same style of music continues playing. The combination of automatic, intelligent organization and the simple hardware interface resembles operating a radio dial that can be tuned to find desired music.

A combination of audio-based structuring and tag-based navigation support is also given in the *nepTune* interface [222]. Similar-sounding music is clustered and a virtual landscape is created to visualize the structure. This landscape can then

[10]http://www.findsounds.com.

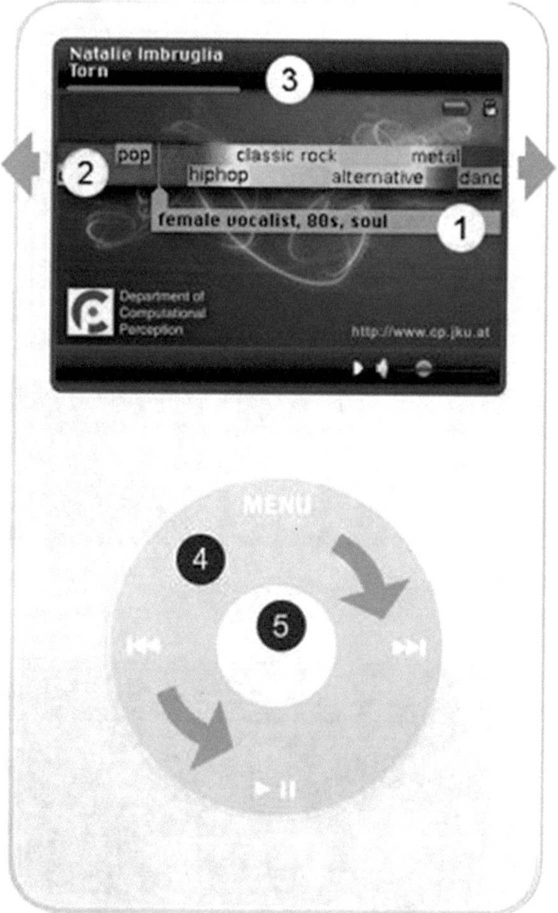

Fig. 1.2 The *Intelligent iPod* mobile browsing interface. (1) shows tags describing the music in the selected region. (2) represents the music collection as a stripe, where different styles are colored differently. The collection can be browsed by using the scroll wheel (4). The Scroll wheel can be used to browse the collection. (5), The central button to select the track. The currently playing track is shown in (3)

be navigated in the fashion of a computer game with the closest tracks auralized; cf. Fig. 1.3. Terms that describe the contents of the clusters can be displayed in order to facilitate orientation.

The final example is *Musicream*,[11] an interface that fosters unexpected, serendipitous music discoveries [153]. *Musicream* uses the metaphor of water taps that release flows of songs when opened; cf. Fig. 1.4. Different taps release music of different styles. The user can browse through the collection by grabbing songs and listen to them or create playlists by sticking songs together. To achieve consistent playlists, similar-sounding songs are easier to connect than dissimilar sounding.

[11] http://staff.aist.go.jp/m.goto/Musicream.

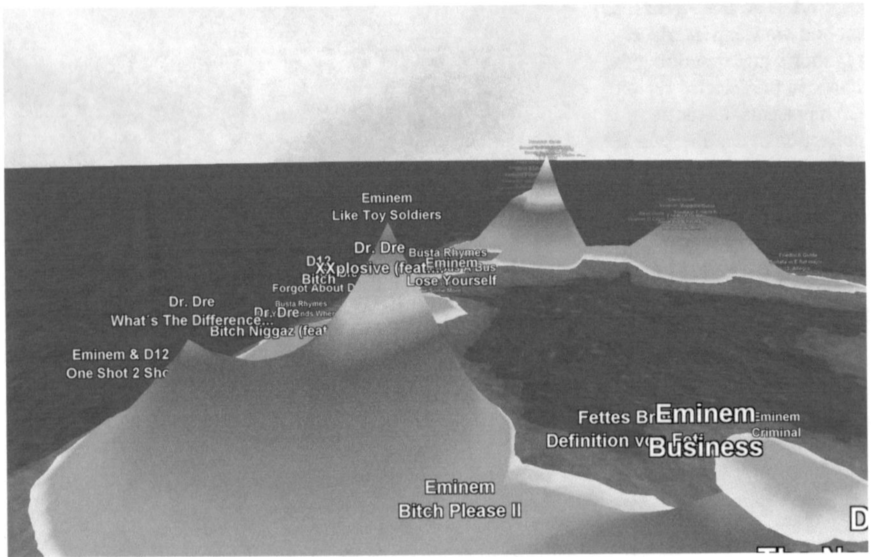

Fig. 1.3 The *nepTune* browsing interface

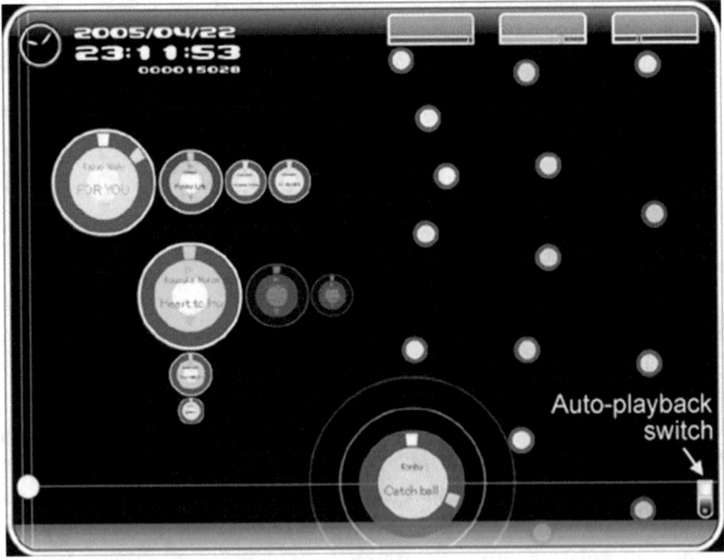

Fig. 1.4 The *Musicream* browsing interface [153] (reprinted with permission)

1.2.3 Recommendation Tasks and Applications in MIR

With virtually the complete catalog of music that has ever been commercially produced ubiquitously available and instantly streamable from the cloud, methods to provide the listener with the "right music at the right time" without requiring much (or any) interaction have become very relevant. One recommendation task that is very particular to the music domain is the **automatic generation of personalized music playlists**, i.e., recommending a sequence of songs "that is pleasing as a whole" [172]. This topic will be addressed in Sect. 9.3 of this book. Starting with a query (which can be a seed song or artist, a user profile, or the user's current context), the aim is to continuously play music that the user wants to listen to, sometimes also referred to as creating "personalized radio stations." Even though subsequent songs should sound similar, an important requirement is that the generated playlists are not boring. Hence, it is important to consider the trade-off between similarity and diversity [418, 536].

Automated playlist generation has received growing attention for about a decade and is now a standard feature of major online music retailers. Examples are *Pandora*,[12] *Last.fm Player*,[13] *Spotify Radio*,[14] *iTunes Radio*,[15] *Google Play Access All Areas*,[16] and *Xbox Music*.[17] Recommendations are typically made using (undisclosed) content-based retrieval techniques, collaborative filtering data, or a combination thereof. Moreover, most systems account for explicit user feedback on their playlists given as binary ratings to improve the personalization of recommendations.

An example of a truly content-based recommender is the *FM4 Soundpark* music player[18] that suggests other songs purely based on sound similarity [140]. The *FM4 Soundpark* is a moderated open platform for up-and-coming artists hosted by the Austrian public broadcasting station *FM4* and targets primarily alternative music. In this case, where artists are generally unknown, content-based similarity is the method of choice for recommendation. The system also offers to automatically create a "mix tape" based on a start and an end song [128]. The underlying technology makes use of content-based retrieval methods like those described in Part I.

[12] http://www.pandora.com.

[13] http://www.last.fm/listen.

[14] http://www.spotify.com.

[15] http://www.apple.com/itunes/itunes-radio/.

[16] http://play.google.com/about/music/.

[17] http://www.xbox.com/music.

[18] http://fm4.orf.at/soundpark.

1.2.4 MIR Beyond Retrieval, Browsing, and Recommendation

MIR as a research area also covers many topics that have no counterparts in IR in other domains, such as automatic real-time score following, or would not be considered a core part of IR, such as optical music recognition. The following paragraphs lay out some prominent MIR subfields that are not directly related to music similarity estimation to highlight the diversity of the field and elucidate why the acronym MIR is sometimes also expanded into the more general description "music information research."

As music is a polyphonic acoustic signal in most cases, there are several directions that aim at identifying the components and/or reducing the complexity of the signal. In the task of **instrument detection**, the goal is to identify the musical instruments present in a recording [173, 280]. One step beyond detection of sound sources is **sound source separation**. Here, the goal is to extract the different components that make up a complex audio signal, such as instrument tracks, voices, or sounds. This allows for a variety of subsequent tasks including melody, baseline, or drum track analysis on monophonic (or at least less noisy) versions, instrument and style classification, voice removal for karaoke, or reusing isolated parts of recordings in new compositions [211, 213, 361, 453, 499, 500].

Conceptually and technically strongly related is the task of **music transcription**, where a symbolic representation is estimated from a rendition [31, 66, 73, 212, 359, 361, 377]. Ideally, this results in a reconstruction of the original score. Music transcription is inherently difficult as assumptions about type and number of instruments need to be made. Thus, existing approaches are always tailored to specific use cases.

Even when the underlying score of a rendition is known a priori, it remains a challenge to synchronize the music with the score. The corresponding task is known as **audio-to-score alignment** [247, 333, 378, 432]. When performed in real time, i.e., an algorithm keeps track of the current position of an ongoing live performance in a score [9, 81, 82, 390], this enables additional tasks such as automatic page turning for performers [8], **real-time automatic accompaniment** [93, 376], and augmenting concert experiences [149]. Alternative systems provide off-line methods for automatic accompaniment of sung voice [448] or real-time methods for improvisational music accompaniment [16, 175, 337].

All of the above-described areas have in common that they try to analyze or extract the "vertical building blocks" of music, so to speak, i.e., to find or track individual components over the length of a recording [80]. In contrast, **structural segmentation and analysis** approaches focus on the "horizontal building blocks," i.e., repeating patterns or reoccurring segments over time [132]. For contemporary popular music, this mostly corresponds to identifying common musical elements such as *choruses*, *verses*, or *bridges* [151, 335, 355]; cf. Fig. 1.5. For more complex compositions, this consists in detection of *themes*, *motifs*, and *hierarchical relationships* within these [79, 315, 395]. On a meta-level, such a representation can be used to find structurally similar pieces [30, 394]. Identification of repeating segments of

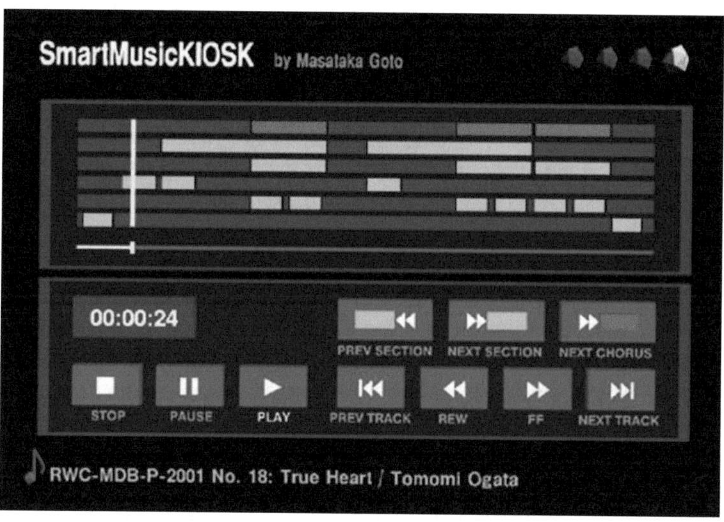

Fig. 1.5 Screenshot of the SmartMusicKIOSK [151] (reprinted with permission). *Bars* on the same line mark corresponding segments of the track. Jumping to the next section or chorus allows for fast pre-listening of the track

a piece can also be considered a step towards **music summarization** which aims at finding its most typical and representative part. Such audio thumbnailing approaches are useful whenever the essence of a piece must be graspable quickly—as is the case in digital online music stores that want to offer more than randomly extracted 30-second pre-listening snippets [67, 84, 356].

Using methods from the fields described above, the area of **computational music analysis** [314] approaches MIR technology from a musicologist's perspective. For analysis of performances, for instance, automatic extraction of musical parameters such as tempo and loudness variations allows the gaining of insights into the playing style of artists and their approaches to musical expressivity [106, 130, 367, 516]. Representational models of expressivity can be used to identify artists from their style [398, 517] or to train systems for **expressive performance rendering** [209, 476, 518].

Finally, we highlight two areas of MIR that reach out to other domains. **Optical music recognition** deals with converting printed or handwritten sheet music (or, in general, music notation) into a digital symbolic music notation [21, 55, 136, 381]. This task is analogous to optical character recognition for printed or handwritten texts and particularly relevant as an approach to *digital preservation* of cultural heritage. A subfield of *human–computer interaction* for which MIR methods become more important is **interfaces for musical expression** [107, 125, 504]. Traditionally concerned with interface design and artistic performance, automatic analysis of sound material, especially in real time, and its use in intelligent and interactive systems has been receiving growing attention.

1.3 Music Similarity

So far, the term music similarity has appeared several times and been used in different contexts and with different meanings. The notion of similarity is a central concept for retrieval systems, particularly for those applying the query by example paradigm. When assessing the relevance of a document with respect to a query, at some point in the process, the necessity to compare representations of the two arises—therefore the necessity to calculate a score reflecting their similarity (or their distance). However, as we have seen, the calculation (and definition) of similarity depends on the retrieval scenario and the user's information or entertainment need. For instance, while for retrieving cover versions calculating melodic similarity seems like the logical choice, for a task such as music recommendation, it is a priori unclear what a similarity measure should encompass. So, actually, what do we mean when we say that two music pieces "sound similar"?

Intuitively, one would characterize and describe music using concepts such as genre (e.g., "sounds like country music"; cf. [334]), instrumentation (e.g., "they only have guitar and drums"), voice characteristics (e.g., "the singer almost sounds like a woman"), melody (e.g., "the song goes something like *doo-di-doo duh-duh*"), tempo (e.g., "fast"), or rhythm (e.g., "strict four to the floor"). Undoubtedly, all these aspects have an impact on the sound and all contribute to two songs being perceived as similar or dissimilar. However, even if we were able to perfectly assess and quantify all acoustic aspects of music, i.e., everything related to *pitch*, *rhythm*, *dynamics*, *timbre*, and their evolvement over time, this signal-centric view alone would still fall short of grasping the multifaceted phenomenon of perceived similarity [519].

The question of and quest for music similarity actually goes beyond the acoustic properties of the music piece. For instance, a listener might refrain from buying music by a particular artist due to disliking couture worn by the artist on the cover of an album—despite all strictly musical aspects similar to artists favored by the listener. In this example, the personal preferences of the user in conjunction with the cultural context of the music invalidate the musical properties. In another fictitious example, a user repeatedly chooses her favorite reggae track that has female lead vocals as query for an automatic playlist service. When she is driving her car, she skips all suggested reggae songs with male lead vocals, whereas when she is running, she does not care about the type of singer and accepts all music with a similar rhythm and tempo. Here, we see a strong influence of the user's context on similarity perception and preference.

This personal preference and the perception of music are tied to the emotions conveyed by a piece. Juslin [203] argues that perceived content is related to different types of coding, i.e., specific ways in which music can carry its emotional meaning. He distinguishes between three different types of coding, namely:

1. *Iconic coding*, which relates to a formal similarity between signals, i.e., the communication of "basic emotions" through acoustic cues like loudness, tempo, or vocal expression
2. *Intrinsic coding*, which relates to "internal, syntactic relationships within the music itself" [203], i.e., more complex, structural properties
3. *Associative coding*, which conveys connotations between the music and other "arbitrary" events or objects, e.g., patriotism carried through national anthems

Thus, while iconic and intrinsic coding refer to properties that can be derived from the music content (some easier, some only to a limited extent), associative coding refers to cultural and personal attributes external to the signal. Since we cannot perfectly model all these aspects from an individual's perspective, we need pragmatic, computational approaches that incorporate as much of this information as accessible.

1.3.1 Computational Factors of Music Similarity

In lieu of a computational model of each individual listener's music perception process, we must aim at factoring in all available information related to the process of music consumption in order to derive mathematically defined measures between two representations of musical entities (e.g., songs, albums, or artists). In terms of automatically measurable factors, we can identify four categories that influence music similarity perception [420]; cf. Fig. 1.6:

1. **Music content** refers to everything that is contained in or, more pragmatically, that can be extracted from the audio signal itself, such as aspects of rhythm, timbre, melody, harmony, or even the mood of a piece. This factor refers to iconic and, on a higher level, intrinsic coding in Juslin's theory.
2. **Music context** refers to aspects of the music that cannot be inferred directly from the audio signal, such as reviews, liner notes, album artwork, country of origin,[19] decade of recording, or marketing strategies (also known as "cultural features," "community meta-data," or "context-based features"). This captures elements categorized as associative coding.

[19]While approaches to automatically detect the country of origin from the music audio exist [406, 538], up to now, they have only been shown to work on non-Western music.

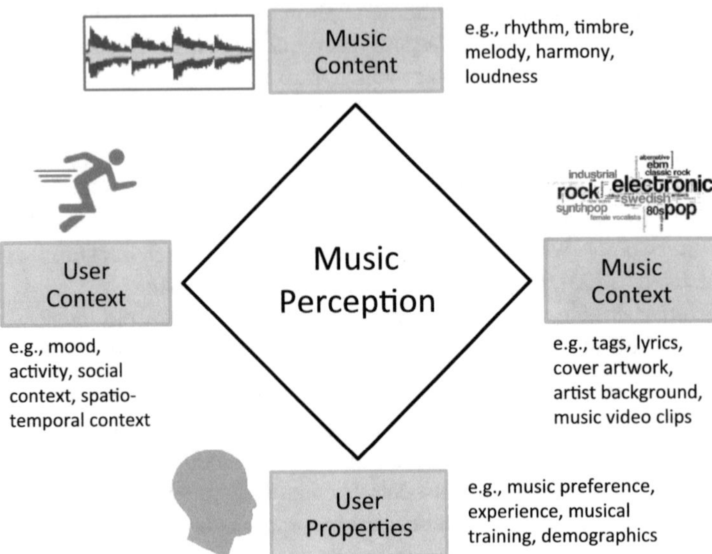

Fig. 1.6 Four different categories of factors that influence music perception

3. **User properties** refer to the listener's personality traits, such as preference and taste, as well as to musical knowledge and experience. This factor both refers to intrinsic and associative coding.
4. **User context** refers to the current situation and setting of the user, e.g., location, time, activity, peers, or mood. This factor influences the perception of all types of coding.

In order to build holistic similarity measures, in theory, we need to measure all these factors and combine them accordingly—which is nontrivial, if not elusive. However, we see that there is a strong interplay between these factors and that we need all of them to represent the different types of codings of content perception. While not every information necessary to do so is available, we aim to model these factors given the information that is accessible to us.

1.3.2 *Music Features*

To capture the abovementioned factors, we first need to identify measurable signals that are connected to or at least bear trails of the desired information. For the *music content*, this signal is the digital music representation itself. For the *music context*, a variety of signals bearing meta-data can be accessed using web technology as the web is the richest source of real-time cultural information. For the *user properties*, aspects like personality traits can be estimated using tools borrowed from psychology, such as tests and questionnaires. Questionnaires are also useful to gauge musical education, training, or preference. For the *user context*, data can be captured with sensors built into personal devices, among others. Typically, these signals are very noisy, i.e., they contain predominantly data unrelated to the factor under consideration.

Therefore, in a second step, we need to "make some sense" of the captured data, i.e., we need to derive features from the signals that contain or are suited to represent the information we are interested in. Feature definition and extraction is a crucial step in all tasks related to intelligent systems. Depending on the chosen source, features might be closer to a concept as humans understand it (*high level*) or closer to a strictly machine-interpretable representation (*low level*). For instance, features derived from the music content or the sensor data captured with a user's personal device will be mostly statistical descriptions of signals that are difficult to interpret, i.e., low-level features, whereas information extracted from web sources, such as occurrences of words in web page texts, is typically easier to interpret.

1.3.2.1 The Semantic Gap

However, for all types of features in order to be processable, we need a numeric abstraction. Between these low-level numeric representations and the factors of musical perception that we want to model, there is a discrepancy—commonly known as "semantic gap". The semantic gap has been identified in all media domains [264], and—while not necessarily justifying the definition of "semantic" [519]—pragmatically, can be seen as an umbrella term for the limitations of current intelligent systems and features in terms of user satisfaction. Ponceleón and Slaney argue that of all types of multimedia data, the semantic gap is the largest for music [369]. While they do not give an explanation for why this is the case, this can likely be connected to the strong dependency of music perception on factors external to the signal, in particular to personal factors such as musical education (see above). In fact, the presence of user factors in the model above makes clear that the term "music similarity" is per se ill defined since it is a highly subjective concept. However, this is not the only shortcoming of the term music similarity, as we will see in a bit.

In order to mitigate the problem of the semantic gap between low level and high level, a third class of features has been established: the so-called *mid-level features*. Although the boundaries are blurred and a strict categorization is difficult to make, mid-level features typically are a combination of or extension to low-level features that incorporate additional, high-level knowledge. Mid-level representations are considered closer to human concepts and should therefore allow more meaningful comparisons between entities as well as carry more relevant information for machine learning approaches to generalize from. Typical examples are psychoacoustically motivated audio features and topic cluster representations. Figure 1.7 gives a schematic overview of the relation of low-, mid-, and high-level

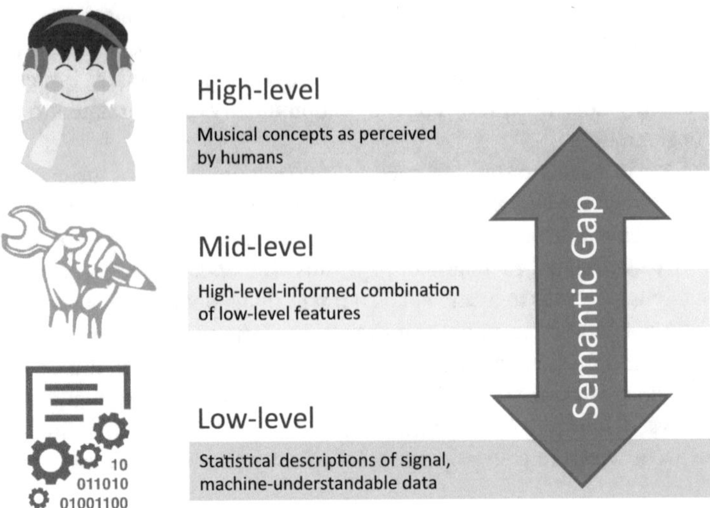

Fig. 1.7 The different levels of feature abstraction and the "semantic gap" between them

features and the semantic gap. This is a reoccurring figure throughout this book and used to exemplify the different types of features discussed for the respective sources.

1.3.2.2 A Critical View on the IR Perspective on Similarity

Even the best approximation and integration of all factors, including the user's preferences and context, will not allow us to actually study the concept of music similarity properly. One of the reasons for this is that *listening*, the main activity when it comes to music, is passive and therefore cannot be described and expressed in terms of a user's task [300]. The lack of a clearly identifiable task and goal makes evaluation as well as optimization difficult (cf. Sect. 1.5).

Apart from this intrinsic issue of retrieval for entertainment, the document-centered approach to music information retrieval adopted from information retrieval, which builds upon techniques such as statistical modeling of a chunked-up signal and discards temporal orderings, has been repeatedly subject to criticism, as it addresses music description from a thoroughly unmusical perspective; cf. [300, 470, 515]. To denounce such purely content-based approaches that are often adopted in MIR publications, foremost at the ISMIR conference, Marsden has coined the term "ISMIRality," as an anagram for similarity, to refer to "talking about music in a manner typical of ISMIR papers" [300]. He states three facts that oppose the principles habituated by "ISMIRality":

1. Music is not just a document.
2. Music is not just acoustic data.
3. Music is a trace of human behavior.

While all these points, in particular points 2 and 3, are in line with the direction of this book, these statements intend to question the concept of music similarity as we aim to exploit it. Specifically, Marsden relates to the fact that music is a process, rather than a document, and hence, musical similarity foremost depends on the outcome of such. Consequently, similarity cannot be assumed based on similar features that are extracted from recordings without musical understanding. As an alternative, he proposes to consider "two instances of music [to be] similar when there is a plausible musical process which produces identical or similar outcomes for the two instances." Thus, in order to gain a better understanding of similarity, the processes that give rise to it should be studied, such as variation, the performance of jazz standards, or the creation of cover versions. In addition, similarity (if such a concept even exists) might also stem from similarity in usage of music and overlap of musical tasks.

While most existing academic and industrial retrieval and recommendation methods indeed exhibit a lack of incorporated musical knowledge and are agnostic to most dimensions of music and their properties, to this point they have been quite successful in providing relevant results and delivering desired music [450]. This might be best explained by the fact that the information needs of musical experts and musicologists differ from the entertainment needs of music consumers and listeners. As much as being able to formulate musically meaningful queries to a music information system is desirable (e.g., *find a perfect cadence* [473] or *find a piece of classical music with a surprise at the beginning* [515]), from an everyday consumer's perspective, it seems, such a feature is not the most needed. Instead, queries that incorporate music parameters, descriptors of acoustic qualities, or metadata (e.g., *like Lucky Star by Madonna, but faster* or *What CDs did Toto make between 1975 and 1980?* [28]), query by example settings (i.e., *more like this*), and recommendations made based on the listening history of a user appear to provide satisfactory expressiveness in most cases.

Moreover, the argument that music retrieval systems make use of inadequate representations in order to index a collection equally applies to most text-based systems. Although the semantic gap is smaller in this domain, it can, rightly so, still be argued that counting word occurrences and discarding any ordering are unjust to written language in general and even more to the arts of literature and poetry, thus not capturing the essence of texts. However, in practice, such methods are widely used and generally fulfill their purpose of identifying relevant results. Therefore, we do not consider "document-centered" approaches per se inadequate for music similarity and retrieval but rather a pragmatic compromise and enabler of manifold applications. As for more expressive and adequate representations, we are confident that future developments will be able to capture more of the truly musical properties as tasks are stronger tied to specific information needs.

1.4 Contents of this Book

In this book, we address music retrieval from the document-centered perspective prevalent in information retrieval. More precisely, we address models of music similarity that extract a static set of features to describe an entity that represents music on any level, e.g., song, album, or artist, and methods to calculate similarity between entities of the same type. This excludes temporal as well as generative models that aim at modeling the progress of music over time, as well as methods that search for musical patterns and knowledge within pieces, such as musical events, musical form, or long-term dependencies (e.g., of key changes). Thus, in fact, none of the "music features" we address capture—in a strict sense—"musical" dimensions. Instead, we focus on the extraction of complementary descriptors from diverse (associated) sources. While this perspective and the chosen representations are most likely incapable of modeling the "essence" of what makes the sensation of music for a listener [515], they allow us to effectively find music of similar qualities (defined throughout the book in the respective sections) by providing abstract summarizations of musical artifacts from different modalities.

In order to cover the most relevant methods for feature extraction, indexing, and similarity calculation, the two large areas of music-content-based and music-context-based methods are essential. Furthermore, methods that combine these two domains and incorporate aspects of the user context form an important area of MIR. Finally, as music retrieval and recommendation is a highly user-targeted task, it is also necessary to take a look through current (and potential future) applications and user interfaces for retrieval and recommendation. Following this broad structure, the book is organized in four parts.

Part I deals with content-based MIR, in particular the extraction of features from the music signal and similarity calculation for content-based retrieval. In Chap. 2, we introduce the basics of audio processing, such as preprocessing and different representations of the audio signal together with some low-level audio features. Moving forward from this introduction, Chap. 3 presents methods for mid-level feature extraction describing aspects of timbre and rhythm and similarity calculation for content-based music retrieval and recommendation. Chapter 4 is devoted to methods for high-level description of music, i.e., genre and style classification, auto-tagging, and mood detection.

In Part II, we address the area of MIR methods that make use of the digital cultural context of music. Chapter 5 reviews and compares descriptive contextual data sources on music, such as tags, web pages, and lyrics as well as methods to obtain this data (web mining, games with a purpose, etc.). After discussing methods for data acquisition, Chap. 6 is concerned with the estimation of similarity and the indexing of musical entities for music retrieval from these sources.

Part III covers methods of user-aware and multimodal retrieval. Chapter 7 focuses on data stemming from music listening and the context of this activity. This includes a review of sources of music interaction and consumption traces, such as playlists, peer-to-peer networks, and ratings as well as discussions of factors of user and context awareness. Chapter 8 covers corresponding collaborative similarity and recommendation techniques, adaptive similarity measures, and aspects of diversity and serendipity.

Finally, Part IV deals with current and future applications of music search and recommendation. Chapter 9 gives a comprehensive overview of current applications of MIR including visualizations, browsing and retrieval interfaces, recommender systems, and active listening systems that are built upon the methods described in the preceding parts. The book concludes with an outlook to the possible future of MIR and the directions and applications to come (Chap. 10).

A consequence of the document-centered view on music retrieval is that it permits us to evaluate systems using the principles of IR system evaluation, in particular following the Cranfield paradigm of batch evaluation using predefined sets of documents, queries, and relevance assessments [76]. Since evaluation of music similarity algorithms is an important task that applies to all modalities alike, before discussing extraction of features from the individual domains, we want to discuss evaluation strategies and methods used in MIR and how we implement them throughout this book.

1.5 Evaluation of Music Similarity Algorithms

Following the tradition of TREC[20] in text retrieval [502, 503], establishing standards for evaluation of music retrieval algorithms has been a central concern in MIR since its beginning and, since 2005, is carried out annually through the Music Information Retrieval Evaluation eXchange (MIREX) campaign.[21] The diversity of research directions in MIR is reflected by the variety of tasks in MIREX, a selection of which is given in the following:

- Audio Music Similarity and Retrieval
- Cover Song Identification
- Classification by Genre, Mood, Composer
- Audio-to-Score Alignment / Score Following
- Melody Extraction
- Key Detection
- Onset Detection
- Tempo Estimation
- Beat Tracking
- Structural Segmentation
- Query by Tapping, Singing, Humming

[20]Text REtrieval Conference.

[21]http://www.music-ir.org/mirex.

For each of these tasks, individual evaluation criteria are relevant and corresponding procedures have been installed. In the context of this book, the *Classification* tasks and the *Audio Music Similarity and Retrieval* task are most relevant as both of these apply methods that pertain to the evaluation of music similarity measures. Another approach that can be taken to evaluate the output of music retrieval methods is to record real users' listening behavior and assess how well the actual listening history could have been predicted. This reflects the general division of evaluation strategies for music similarity into (1) methods that rely on *prelabeled data*, such as genre, as proxy for similarity and model a classification or retrieval task (Sect. 1.5.1), (2) methods that let *human experts judge* the quality of algorithms (Sect. 1.5.2), and (3) methods that measure how well *listening behavior* can be simulated by algorithms (Sect. 1.5.3). All methods have their advantages and disadvantages, as we discuss in the following sections.

1.5.1 Evaluation Using Prelabeled Data

Operating on prelabeled data and evaluating similarity tasks through classification or retrieval measures typically rely on categorical data, such as genre or tags, as a "ground truth." The concept of genre refers to a class of similar music (e.g., classical, jazz, rock, pop, blues, punk) and is frequently organized in a taxonomy (e.g., heavy metal is a subgenre of rock). Ideally, a genre describes a consistent musical style; however, as we have seen, other factors such as specific era (e.g., 1960s rock), location (e.g., Chicago blues), or arbitrary labels (such as terms invented solely for marketing purposes, e.g., love songs) can play a role in its definition. This bears a number of issues that come with the use of genre labels. While genre is a persistent and heavily used concept that is also a relevant dimension in subjects' similarity perception [334], it has been shown that genre per se is ill defined as there are cultural differences and persisting disagreement on their existence and relevance [338]. Studies have shown that humans only agree to about 80 % when asked to assign music pieces to genres [276, 442]. A fan of heavy metal music may distinguish between a power metal and a melodic metal track, whereas an aficionado of folk music may just put both in the hard rock category. Despite this, for ease of evaluation, usually, a music piece is categorized into one genre only (*1:n* relation). Keeping these limitations in mind, when using genre as evaluation criterion, there exists a natural upper bound for the performance of music similarity algorithms.

Tags, on the other hand, refer to arbitrary labels, not restricted to genre information. These can stem from a collaborative, crowdsourced tagging effort or from a pool of expert annotations and, in addition to stylistic information, can range from mood annotations to instrumentation to personal labels. In terms of ground truth annotations, one piece can have multiple labels (*n:n* relation). For a more detailed discussion on the differences of genre labels and tags, please refer to Sects. 4.1 and 4.2.

Table 1.1 Evaluating the quality of a music similarity algorithm using genre as relevance criterion

Seed artist	Seed genre	Rank	Similarity	
Beethoven	*Classical*		*1.000*	
Bach	Classical	1	0.808	✓
Chopin	Classical	2	0.805	✓
Shostakovich	Classical	3	0.745	✓
Iron Maiden	Heavy metal	4	0.412	
Kylie Minogue	Pop	5	0.404	
Black Sabbath	Heavy metal	6	0.386	
Kiss	Heavy metal	7	0.383	
Metallica	Heavy metal	8	0.381	
Chemical Brothers	Electronic	10	0.371	

The ranking is obtained by using the piano piece by Beethoven as seed

Table 1.2 Evaluating the quality of a music similarity algorithm using genre as relevance criterion

Seed artist	Seed genre	Rank	Similarity	
Radiohead	*Electronic*		*1.000*	
Kylie Minogue	Pop	1	0.556	
Snow	Rap	2	0.547	
Maroon 5	Pop	3	0.543	
Madonna	Pop	4	0.531	
Jay-Z	Rap	5	0.528	
Modeselektor	Electronic	6	0.527	✓
Kiss	Heavy metal	7	0.498	
Black Sabbath	Heavy metal	8	0.495	
Busta Rhymes	Rap	9	0.482	
Trentemøller	Electronic	10	0.478	✓

The ranking is obtained by using the Electronic track by Radiohead as seed

For reasons of simplicity, in the following, we will assume that each music piece is assigned to exactly one genre as a ground truth. To evaluate a music similarity algorithm using genre information as proxy, consider that it is given a query (seed track or artist) of known genre g for which it produces a ranked list of most similar items by sorting the contents of the collection in decreasing order based on similarity. Based on the genre information of the elements of the retrieved list, a number of evaluation measures can be calculated. In the following, we introduce the most important of these measures, which are all related to *precision* and *recall*.

To illustrate the various performance measures, consider Tables 1.1 and 1.2, where we select a piano piece by Beethoven and a song by Radiohead (see Sect. 1.5.4), respectively, as seed (query) and retrieve the ten nearest neighbors in terms of a content-based similarity measure (the details of which are explained in Sect. 3.3). The retrieved results marked with a check mark are of the same genre as the seed. The number of retrieved items is typically denoted as parameter k of the retrieval process.

Precision describes how many of the retrieved items are relevant to the query (cf. right part of Eq. (1.1), where *Rel* denotes the set of relevant items and *Ret* denotes the set of retrieved items). Using genre as proxy, this refers to the percentage of pieces in the result list that are of the same genre as the seed. In our examples, we retrieved ten pieces most similar to the seed. Considering the genre distribution within the results, precision is computed as $P = 30\,\%$ when using Beethoven as seed (three out of the ten results are from the Classical genre) and $P = 20\,\%$ for Radiohead as seed; cf. Tables 1.1 and 1.2, respectively. Note that precision does not account for the position or rank at which an item was retrieved. However, it is common to compute precision at certain levels of retrieved items k, in which case it is referred to as **precision@k** (P@k). Hence, precision@k is computed only on the top k retrieved items.

$$R = \frac{|Rel \cap Ret|}{|Rel|} \qquad P = \frac{|Rel \cap Ret|}{|Ret|} \tag{1.1}$$

Recall measures how many of the relevant items are retrieved [cf. left part of Eq. (1.1)]. In our case, relevant items are those with the same genre as the seed. Since the toy music data set we use in this example (see below) shows a uniform distribution of genres, namely, four pieces per genre, we can retrieve at most three pieces of the same genre as the seed. In the example of Beethoven, recall is 100 %, since all other Classical pieces are in the results list. In contrast, using Radiohead as seed yields 67 % recall, since only two out of three other Electronic songs are retrieved. Analogous to precision@k, we can also define **recall@k** (R@k) as recall computed on the top k retrieved items.
Sometimes precision is computed using the number of relevant items in the data set as k, in which case precision and recall values are identical. This can be easily seen from the formulas for precision and recall given in Eq. (1.1).

Considering precision and recall, it is noteworthy to mention that there are frequently natural limits for both precision and recall. If we retrieve the k most similar pieces for a seed, but there as less than $k-1$ pieces of the same genre in the collections, precision can never reach 100 %. Likewise, if the number of pieces of the seed's genre in the collection (excluding the seed itself) is larger than k, recall can never reach 100 %. In general, precision and recall are inversely related; the larger the number of retrieved items, the more likely to retrieve more items of the seed's genre (higher recall) but also the more likely to have a higher number of unrelated items in the results (lower precision).

F-measure is defined as the harmonic mean of precision and recall, optionally also computed only on the top k items, in which case it is abbreviated as F@k; cf. Eq. (1.2). It is only possible to score high on F-measure if both precision and recall are high. F-measure further facilitates comparing different retrieval algorithms, because it is an aggregate. In our examples, $F = 46\,\%$ for Beethoven and $F = 31\,\%$ for Radiohead.

$$F = 2 \cdot \frac{P \cdot R}{P + R} \qquad F@k = 2 \cdot \frac{P@k \cdot R@k}{P@k + R@k} \tag{1.2}$$

Average precision is the arithmetic mean of all precision values up to the number of retrieved items but only considering relevant items. The individual precision values are computed from the first ranked item up to the number of retrieved items, ignoring positions of irrelevant items; cf. Eq. (1.3), where $rel(i)$ is 1 if the item retrieved at position i is relevant and 0 otherwise. Please note that average precision (AP) implicitly models recall because it accounts for relevant items that are not in the results list. Again, AP@k can be defined, considering precision values only up to k retrieved items. When using Beethoven as seed for our similarity algorithm, $AP = \frac{1}{3} \cdot (P@1 + P@2 + P@3) = 100\,\%$. For Radiohead, $AP = \frac{1}{3} \cdot (P@6 + P@10) = \frac{1}{3} \cdot \left(\frac{1}{6} + \frac{2}{10}\right) = 12\,\%$.

$$AP = \frac{1}{|Rel|} \cdot \sum_{i=1}^{|Ret|} rel(i) \cdot P@i \qquad AP@k = \frac{1}{|Rel|} \cdot \sum_{i=1}^{k} rel(i) \cdot P@i \tag{1.3}$$

Average precision is defined on a single seed only. In practice, however, we are usually interested in the performance of a similarity or retrieval algorithm on a whole collection. To this end, **mean average precision** (MAP) computes the arithmetic mean of the average precision values over all possible seeds; cf. Eq. (1.4), where *Items* denote the set of items in the music collection under consideration. MAP hence assigns a single score to an algorithm's performance on a given data set.

$$MAP = \frac{\sum_{i \in Items} AP(i)}{|Items|} \tag{1.4}$$

In a classical information retrieval task, the two classes of interest are whether a document is relevant or irrelevant to a query. Each document retrieved for a query belongs to one of these classes. In contrast, in our setting, as is the usual case, we typically have more than two genres in the collection of interest. This fact opens up more ways to measure performance of a music retrieval system in terms of genre, for instance, the average intra-/inter-genre similarity ratio (see Sect. 1.5.4).

Another common way to investigate the quality of a music similarity algorithm is analyzing genre confusions, typically via a **confusion matrix**. To this end, the genre of the seed is assumed to be unknown and is predicted by a classification algorithm based on the retrieved items. The confusion matrix then displays, for the entire music collection, which genres are typically confused, by contrasting all pairs of predicted genres and actual genres. Given this kind of information, algorithms can be tuned towards improving similarity estimation for particular genres. For instance, if an audio-based algorithm often confuses rap and electronica, feature designers could try to incorporate into their algorithm a component that detects the typical vocal parts in rap music. Note that for quickly assessing which types of errors occur when using different similarity algorithms, throughout this book, we make use of confusion matrices in combination with the average intra-/inter-genre similarity ratio mentioned above; cf. Sect. 1.5.4.

It was already mentioned that genre-based evaluation techniques for music similarity and retrieval algorithms, although being easy to implement and widely adopted, might not be the best way to assess performance. This is further elaborated in detail by Sturm [470, 471] and Urbano et al. [490], among others. In addition to the previously mentioned arguments of a problematic definition of genre, there is also another factor to which the above evaluation strategies do not pay attention, namely, there is a perceptual difference between the closeness and proximity of genres. For instance, most music listeners will agree that music of the genres hip-hop and rap is more similar than music of the genres hip-hop and classical. However, this fact is not accounted for in the evaluation protocols that model a retrieval or classification task, which both assume genres to form constant and hard classes. Even though there are algorithms that predict genre assignment in a probabilistic way, it is unclear how to properly measure and incorporate genre similarity.

1.5.2 Evaluation Using Human Judgments

A presumably better, albeit more expensive, evaluation strategy is to ask humans to judge the quality of music similarity algorithms. While this yields more significant evaluation results, which are obviously closer to what humans perceive as similar, relying on human judgments of similarity is costly and labor intensive. Also this evaluation strategy is prone to biases, given the subjectivity in perception of music similarity. While one person may judge two songs as similar because of similar melodies or rhythm, another may perceive the same songs very different because the lyrics of one relate to war, whereas the other is a love song. This subjective perception of music similarity is further investigated by Novello et al. [334] and Wolff and Weyde [520]. Furthermore, even for small collections, considerable effort, including the detection of intentionally erroneous or irrelevant input, is required to obtain a complete assessment of similarity from human judgments [121].

One way of dealing with this is to only acquire judgments on the items retrieved by the algorithms under evaluation, a strategy employed in the MIREX *Audio Music Similarity and Retrieval* task. Given a number of algorithms and a list of seed songs, first, all songs retrieved by any algorithm for a given seed are merged into one list, a procedure referred to as *pooling*. Subsequent shuffling of these pooled results ensures that participants cannot identify patterns in subsequent items and in turn may be able to make out a particular algorithm. The pooled and shuffled results for a seed, together with the seed itself, are then presented to the judge(s), who have to rate the similarity of each result to the seed. In the case of MIREX, two rating scales are used: a broad scale and a fine scale. The former requires judges to rate each retrieved song as either "very similar," "somewhat similar," or "not similar" to the seed. The latter asks judges to indicate a level of similarity in the range 0 ("not similar") to 100 ("perfectly similar"). Following this procedure for a number of randomly selected seed songs (50 in the case of MIREX), we can compute different performance scores for each algorithm.

As performance measure, one option is to calculate the **average similarity score** of all songs retrieved by the algorithm under consideration, as indicated by the judges. When computing these scores on the broad scale, the judgments need to be mapped using the range 0 to 2 (0 = "not similar"; 1 = "somewhat similar"; 2 = "very similar"). For the fine scale, the top performing content-based algorithms achieve an average fine score of above 50 %.[22]

Another category of performance measures is based on **rank correlation**. They assess the performance of an algorithm by comparing the song rankings produced by it with the rankings given by the judges. Doing this for each (seed, pooled results) pair gives the average rank correlation coefficient for an algorithm. There are several statistical methods to compute correlation between two rankings. The most common ones are *Spearman's rank correlation coefficient* (or *Spearman's* ρ) [462] and *Kendall's rank correlation coefficient* (or *Kendall's* τ) [205]. To identify whether differences found in the rankings obtained by different approaches are significant, *Friedman's test* [134] can be applied to the ranking data. To give a detailed description of these methods would be beyond the scope of this book but can be found, for instance, in [447].

A frequently raised criticism of the evaluation procedure employed in the MIREX *Audio Music Similarity and Retrieval* task is that each pair of seed and pooled result is evaluated by a single judge only. On the one hand, it is of course costly to have each pair rated by several judges. On the other hand, assigning more, e.g., three, judges to each pair allows identification of agreement or disagreement between them. Utilizing, for instance, three judges would also make an unambiguous majority voting possible. In fact, in a meta-study, Flexer has shown that inter-rater agreement in the *Audio Music Similarity and Retrieval* task is indeed low, resulting in an upper bound of algorithm performance with respect to subjective gradings [127]. Moreover, evaluated algorithms have reached this upper bound already in 2009 and subsequent developments were not able to surpass it since then.[23] This shows that neither prelabeled data nor human judgments are per se the best choice to satisfactorily evaluate music similarity and retrieval systems.

[22]http://www.music-ir.org/mirex/wiki/2013:Audio_Music_Similarity_and_Retrieval_Results.

[23]Note that these algorithms are described in Chap. 3, specifically in Sects. 3.2 and 3.3.

1.5.3 *Evaluation Using Listening Histories*

A third method that is more targeted at evaluating music recommender systems in general (cf. Part III) can also be used to assess the quality of similarity-based retrieval systems. Based on a collection of real-world listening activity logs, i.e., chronological lists of music pieces listened to by a person, a retrieval task is constructed by splitting the collection into a training and a test set. The splitting criteria can be a ratio of the sizes of training and test set or defined by a point in time, i.e., all listening activity before such a point in time belongs to the training set and can be used to learn a model of music preference (or simply serve as queries for finding similar pieces), whereas all activity after this point is used for evaluating the predictive capabilities of the learned model. This evaluation method was, for instance, applied in the MSD challenge [304], where, as main evaluation criteria, recall of positively-associated items for each user and mean average precision (MAP) were used.

The advantages of this type of evaluation are its user centricity and the operation on real-world data. It also elegantly circumvents the requirements of explicitly labeling music or quantifying the degree of similarity between music pieces. However, despite the need to collect substantial amounts of actual listening data, also this approach entails several drawbacks. First, using the outlined strategy, the ground truth consists only of music pieces that have been listened to. This does not mean that alternative pieces would have been irrelevant at the time. Hence, we can identify true positives only, i.e., pieces that both the user has listened to and the algorithm has predicted, which is not enough information to meaningfully calculate recall. Second, we have no knowing of the intention behind the listening events (nor of context or engagement of the user), thus, we do not know which task we are simulating and evaluating. Third, as a consequence of this, the recorded listening logs could already be the result of an automatic music recommender. Therefore, a retrieval system would in fact not be tested for imitating user behavior but rather for simulating another automatic system.

As could be seen from the last three sections, there are different approaches to evaluate music similarity algorithms and retrieval with different advantages and disadvantages. The choice of methods that should be applied and the expressiveness of the obtained results therefore strongly depend on the given scenario and evaluated task. Therefore, results need to be interpreted with the respective limitations in mind.

1.5.4 *Music Collection and Evaluation in this Book*

In order to demonstrate the various methods presented, throughout this book, we will refer to a small music collection (which we call the "toy music data set"). The goal is to show the results, thus highlighting advantages and disadvantages, of the individual approaches using well-known artists and music pieces. This should allow the reader to get a realistic and graspable view of the state of the art in content- and web-based MIR. We select genres, artists, and tracks in the data set such that a reader with a common knowledge of popular Western music can relate to them. To allow for this, as a necessary evil, we have to refer to commercial music that we cannot distribute due to legal restrictions.[24]

The toy music data set in total comprises of 20 pieces attributed to five conceptually distinguishable genres, namely, Electronic, Classical, Metal, Rap, and Pop. While genre labels cannot provide a perfect ground truth [338, 469], they should just serve as a reference point for following the described approaches. In particular, this is apparent for the genre Classical that, strictly defined, contains only one Classical composer, i.e., Beethoven, whereas Bach belongs to the Baroque era, Chopin to the Romantic, and Shostakovich to the twentieth century, being often classified as neoclassical, post-romanticism, and modernist. However, as an umbrella term, the word classical is often used. What all of them have in common is that the chosen examples are piano pieces.

For each genre, we select four pieces, each by a different artist or composer. One of the members in each genre can be considered "less obvious," in order to demonstrate the breadth of the genres, as well as to avoid obtaining only trivial results. For Electronic, the less obvious example is Radiohead, who are typically considered alternative or indie rock despite their heavy influences from electronic music. The piece chosen is one of their more electronic-sounding ones. For Classical music, Shostakovich sticks out as he falls in a completely different style of Classical (see above) and has a distinct cultural background. For Metal, Kiss as a disco and glam influenced rock band are not as typical to metal as the others, namely, Black Sabbath, Iron Maiden, and Metallica. The artist Snow, who is assigned to Rap, is also frequently classified as reggae and dancehall, despite the fact that his chosen (only) hit Informer features strong rap passages. In the genre Pop, finally, three of the artists are female singers (Kylie Minogue, Madonna, Lady Gaga) and one is a band with a male singer (Maroon 5), who is, however, known for his rather high pitch of voice. Additionally, the chosen track by Maroon 5 also features the voice of Christina Aguilera. Despite these variations, artists and tracks within the same genre are considered more similar to each other than to artists and tracks from other genres. Thus, we assume artists and tracks from the same genre to be most relevant when retrieving similar items.

A complete list of music pieces in the toy music data set is shown in Table 1.3. Note that for the genre classical, instead of the performing artist, we refer to the composer of the piece. For more background information on the individual pieces, we refer the reader to Appendix A.

In order to show the effects of the similarity measures defined throughout the book, we visualize the obtained similarity values for the complete toy music data set in a 20×20 confusion matrix, where similarity values are normalized and mapped

[24]However, previews of the pieces are available for free through major online music stores. We selected the examples from the *7digital* catalog (http://www.7digital.com).

Table 1.3 Composition of the toy music data set used throughout the book

Artist/composer	Music piece
Electronic	
Chemical Brothers	Burst Generator
Modeselektor	Blue Clouds
Radiohead	Idioteque
Trentemøller	Moan
Classical	
Ludwig van Beethoven	Piano Sonata No. 14 in C Sharp Minor, Op. 27/2: "Moonlight Sonata"
Johann Sebastian Bach	Das Wohltemperierte Klavier: Prelude in C Major, BWV 846
Frédéric Chopin	Nocturne for Piano, No. 1 in B flat minor, Op. 9/1
Dmitri Shostakovich	Prelude and Fugue No. 2 in A Minor
Metal	
Black Sabbath	Paranoid
Iron Maiden	The Wicker Man
Kiss	I Was Made For Lovin' You
Metallica	Master of Puppets
Rap	
Busta Rhymes	Put Your Hands Where My Eyes Could See
Jay-Z	Big Pimpin'
Snoop Dogg	Drop It Like It's Hot
Snow	Informer
Pop	
Kylie Minogue	Get Outta My Way
Madonna	Hung Up
Maroon 5	Moves Like Jagger
Lady Gaga	Bad Romance

to a gray scale with black indicating the highest similarity and white the lowest. Figure 1.8 shows an example of a similarity matrix on the toy music data set that has been obtained from human judgment by asking music lovers of various genres to rate the similarity of each pair of tracks on a scale from 0 to 100 %.

To quantitatively assess the effectiveness of the similarity measure, we analyze how well it separates songs from different genres. To this end, we compute the **intra-/inter-genre similarity ratio**, which is displayed for each genre on the right-hand side of the figure (the values in parenthesis next to the genre names). This intra-/inter-genre similarity ratio for genre g is defined as the average similarity between all songs within g (excluding self-similarity which is always the maximum) divided by the average similarity between the songs in g and the songs in other genres. More formally, it is defined as shown in Eq. (1.5), where T_g is the set of tracks in genre g, $T_{G\setminus g}$ is the set of tracks not in genre g, t_1 and t_2 denote the tracks themselves, and $sim(t_1, t_2)$ refers to the similarity between t_1 and t_2 as given by the approach under investigation. Therefore, a value of 20.192 for the Electronic genre means that the

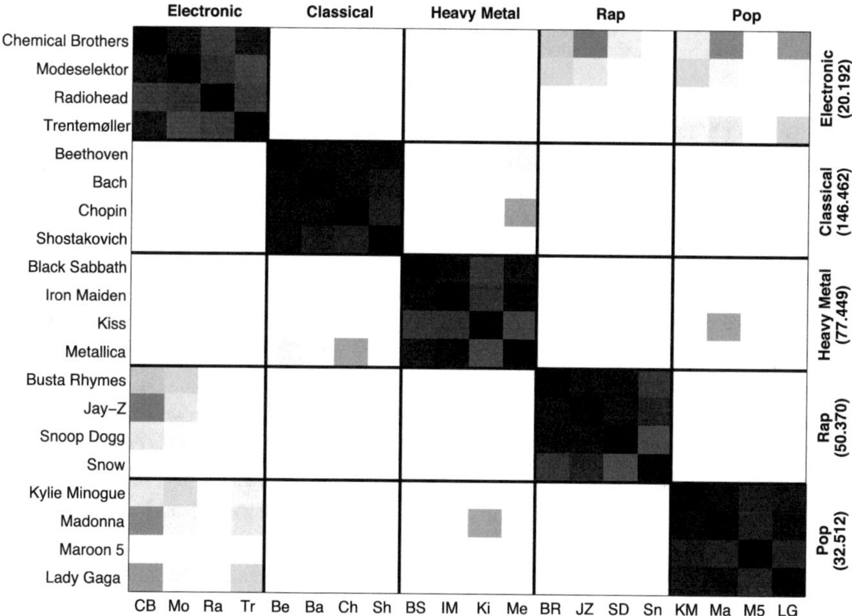

Fig. 1.8 Visualization of track similarity as assessed by a human annotator

similarity between two randomly chosen Electronic pieces is, on average, about 20 times higher than the similarity between a randomly chosen Electronic piece and a randomly chosen piece from any other genre.

$$ii_g = \frac{\left(|T_g|^2 - |T_g|\right)^{-1} \cdot \sum_{t_1 \in T_g} \sum_{t_2 \in T_g,\, t_1 \neq t_2} sim(t_1, t_2)}{\left(|T_{G\setminus g}| \cdot |T_g|\right)^{-1} \cdot \sum_{t_1 \in T_g} \sum_{t_2 \in T_{G\setminus g}} sim(t_1, t_2)} \tag{1.5}$$

To obtain an overall effectiveness measure for an approach's ability to distinguish music of different genres, we compute the average intra-/inter-genre similarity ratio over all genres, by taking the arithmetic mean or the median over all genre-specific ratios. In our example of similarities judged by humans, the mean amounts to 65.397, while the median, which suppresses the otherwise high influence of outliers, is 50.37. Note that none of the computational approaches we will discuss in the course of this book is capable of rivaling these values, as this can be seen as the upper bound, defined by human judgment.

1.6 Further Reading

As pointed out before, the breadth of the field of MIR does not allow us to cover every aspect extensively. For the topics not covered in this book, we recommend the following books, all with a different focus.

With respect to *melodic and symbolic retrieval*, Typke gives a comprehensive overview [485]. Besides covering symbolic indexing methods, different distance functions for melodic retrieval, and evaluation methods of such systems, an extensive survey of early MIR systems is provided.

Focusing on symbolic MIDI and audio signal representations of music, topics of *computational music analysis*, which can be employed in music retrieval and similarity tasks, are covered in a very recent book edited by Meredith [314]. The contributions include articles about analyzing harmonic surface and musical form, chord- and note-based voice separation, interactive melodic analysis, pattern discovery and recognition in melody and polyphonic music, detection and analysis of motifs, composer classification, and analysis of non-Western music (e.g., Ethiopian begena songs).

Müller provides in his textbook an introduction to essential topics of *music processing and retrieval* [326]. He covers a wide range of content-based MIR methods, with a focus on Fourier analysis of audio signals, content-based retrieval, audio decomposition, synchronization, structural analysis, chord recognition, and tempo and beat tracking.

In their edited book, Li et al. cover the broad topic of *music data mining*, including contributions on audio feature extraction (timbre, rhythm, pitch, and harmony), music classification (instrument, mood, and emotion), automatic as well as human music tagging and indexing, and a few advanced topics, such as hit song detection and musicological symbolic data mining [267]. Another recent book titled "Advances in Music Information Retrieval," edited by Ras and Wieczorkowska, reports on current developments in chord recognition and analysis, instrument detection in polyphonic music, genre classification, sound quality assessment, emotion-based retrieval of MIDI files and of vocal music, cover song identification and similarity, multimodal retrieval with audio and lyrics, and MIREX [379].

For a more detailed elaboration on general purpose IR methods and performance measures used to evaluate retrieval systems, we recommend the books by Baeza-Yates and Ribeiro-Neto [20] or Büttcher et al. [53]. With respect to the statistical significance of MIR system evaluation, Urbano's PhD thesis provides an excellent in-depth analysis [489].

Celma's book is a highly recommended read for the topic of *music recommendation* [62]. He provides a compact introduction to the main recommendation strategies, collaborative filtering, content-based filtering, context-based filtering, and hybrid systems, before digging deeper into the tasks of analyzing and recommending the "long tail" [5] of music items. Celma further discusses network-centric and user-centric evaluation and presents two of his applications for music discovery and recommendation in the long tail.

Part I
Content-Based MIR

Part I of this book provides an introduction to audio signal processing methods and, building upon these, an overview of methods to extract musical features. In Chap. 2, we introduce basics in audio processing, such as preprocessing and different representations of the audio signal. Furthermore, we discuss some low-level features to model music via audio. Moving forward from this introduction, Chap. 3 presents methods for feature extraction on the frame and block level as well as state-of-the-art similarity measures. Among others, we deal with the questions of how to describe rhythm and timbre and how to use these descriptions to compute a similarity score between music pieces, which is a key ingredient to content-based music retrieval and recommendation systems. Additionally, Chap. 4 is devoted to related methods for high-level description of music via semantic labeling.

Chapter 2
Basic Methods of Audio Signal Processing

A central task in content-based MIR is to extract music features from the audio signal, which is encoded in a digital music file. We use the term "feature" to refer to a numerical (scalar, vector, or matrix) or nominal description of the music item under consideration, typically being the result of a feature extraction process. Feature extractors aim at transforming the raw data representing the music item into more descriptive representations, ideally describing musical aspects as perceived by humans, for instance, related to instrumentation, rhythmic structure, melody, or harmony. In general, a music item can be, for example, a song, an album, a performer, or a composer.

In this chapter, we lay the foundation for content-based MIR techniques by introducing the basics in audio signal processing. The music items under consideration are therefore music pieces or songs. In particular, we first give a categorization of different music audio features according to various aspects (level of abstraction from low level to high level, temporal scope, musical meaning, and frequency vs. time domain). We then show the steps involved in a typical music audio feature extraction process. Subsequently, we give some examples of frequently used low-level music features.

2.1 Categorization of Acoustic Music Features

Computational features that aim at modeling aspects of human music perception can be categorized according to various dimensions—among others, according to (1) their level of abstraction, (2) their temporal scope, (3) the musical aspect they describe, and (4) the signal domain they are computed in.

The *level of abstraction* of a feature is typically described on a scale of three levels: low level, mid-level, and high level. An illustration of this categorization is depicted in Fig. 2.1. From low level to mid-level to high level, the features increase

P. Knees, M. Schedl, *Music Similarity and Retrieval*, The Information Retrieval Series 36, DOI 10.1007/978-3-662-49722-7_2

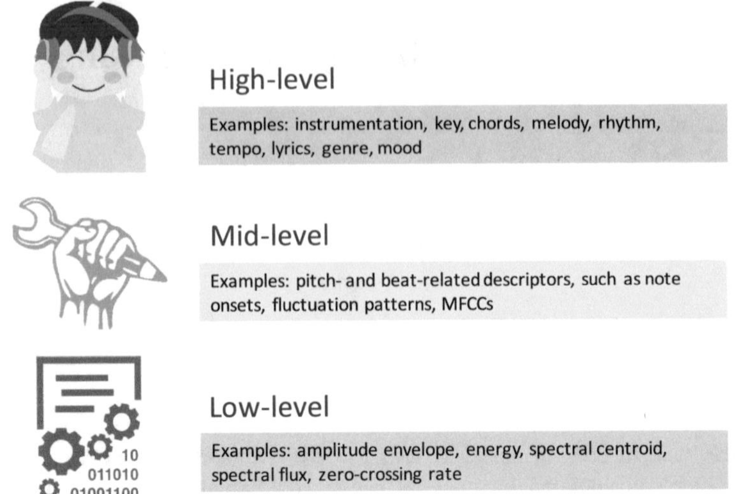

Fig. 2.1 Examples of content-based features on the different levels of abstraction

in terms of their semantic meaning for the user and decrease in terms of closeness to the audio signal. While low-level features are typically computed directly from the raw audio waveform and are frequently simple statistical summaries of the waveform, mid-level features capture aspects that are already musically more meaningful, such as note- or beat-related properties. Corresponding mid-level feature extractors frequently combine low-level features or apply some psychoacoustic model. On the highest level, we find features that describe music in terms of how humans perceive it. Such aspects include instrumentation, rhythm, melody, and song lyrics. To give a hint of the interpretation of the three levels of abstraction, one could say that low-level features can only be deeply understood by a scientist working in signal processing, mid-level descriptors are understandable by the knowledgeable music expert, whereas high-level features are comprehensible also by the average human listeners.

From the point of view of the features' *temporal scope*, we can distinguish instantaneous, segment-level, and global features. Instantaneous features are computed for a particular point in time. Given the temporal resolution of the human ear, which is around 10 ms depending on the person [290], such features typically cover at most a few tens of milliseconds. Segment-level features represent a predefined length of consecutive audio. They are either computed on a fixed length window (e.g., of length 5 s) or using a semantically more meaningful definition of segment, for example, a musical phrase or the chorus of a song. Global features describe the entire music item of interest, such as a song, a movement, or an audio excerpt.

Of course, music audio features can also be categorized in terms of the *musical aspect* they describe. Such aspects include beat, rhythm, timbre, instrumentation, pitch, melody, chords, and harmony.

Another categorization of music features is according to the domain in which they are computed. Basically, an audio signal can be represented in the *time domain* or in the *frequency domain*. The time domain representation indicates the amplitude of the signal at each point in time, or more precisely at each time when the signal was observed, i.e., a sample was taken. The frequency domain representation is typically the result of a Fourier transform (see Sect. 2.2.3), which decomposes the signal into a number of waves oscillating at different frequencies. This process allows the description of the signal by its magnitude (power) at various frequencies.

For an illustration of the two representations, consider Figs. 2.2 and 2.3. The audio signals in these cases represent a violin sound and a synthesized drum pattern sample, in Figs. 2.2 and 2.3, respectively.[1] Both are encoded with 16 bits and a sampling rate of 44,100 Hz; cf. Sect. 2.2.1. The left parts of the figures show the representation of the respective audio signal in the time domain, plotting amplitude values over time. In case of the violin sound, the covered time period equals 5 s, corresponding to 220,500 samples at a sampling rate of 44,100 Hz. In case of the drum sample, the duration of the audio is 81,627 samples, which corresponds to 1.851 s using again a sampling rate of 44,100 Hz. The right part of the figures depicts the Fourier-transformed signal, plotting magnitude information over frequencies. Looking at the time domain representations, it already becomes obvious that the drum sound shows repetitive patterns, indicating a rather rhythmic, less melodic sound, in contrast to the violin. Comparing the frequency domain representations of the two sounds, we observe that the frequency range covered by the violin is much larger than that covered by the drums. We see that most of the energy in the violin sound is concentrated between 1000 and 2500 Hz, while the main frequency

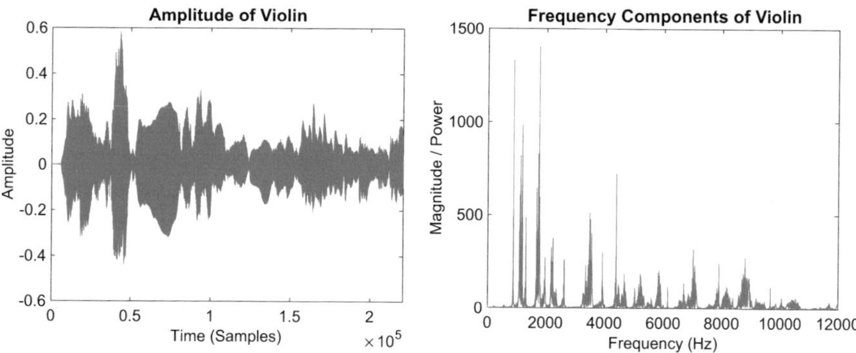

Fig. 2.2 Time domain representation (*left*) and frequency domain representation (*right*) of a violin sound

[1]Both examples taken from *Freesound* (https://freesound.org) a collaborative database of freely available audio snippets: violin sound sample (https://freesound.org/people/jcveliz/sounds/92002) licensed under Creative Commons Sampling Plus 1.0; drum pattern sample (https://freesound.org/people/Goup_1/sounds/190613) licensed under Creative Commons Attribution 3.0 Unported.

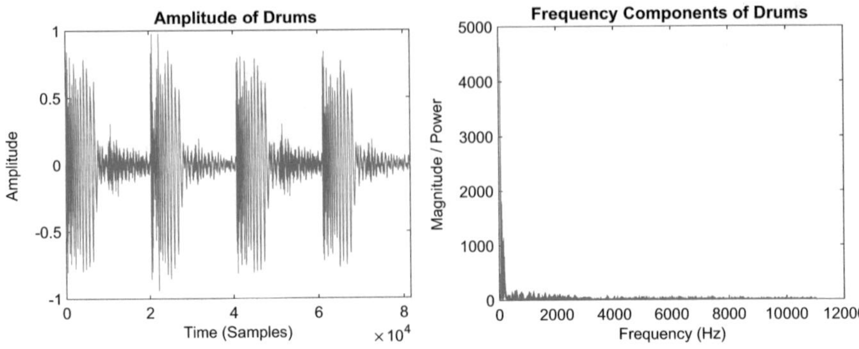

Fig. 2.3 Time domain representation (*left*) and frequency domain representation (*right*) of a synthesized drum pattern sound

component in the drum sound appears up to approximately 300 Hz. Comparing the magnitude values of the two sounds further reveals that the low frequency component of the drums is much more powerful than any component in the violin sound. From the distribution of power over frequencies present in the violin sound, we can infer that this sound is more harmonic.

2.2 Simplified Scheme of a Music Content Feature Extractor

Let us now take a look at the entire process of computing some kind of acoustic music feature from an audio signal. To this end, consider Fig. 2.4, which shows the pipeline of a primitive feature extractor. The input to the system is a sound-producing device, for instance, a trumpet as shown in the top left of the figure. We first have to record and convert the analog audio signal produced by this instrument into a digital representation which can be easily processed by a computer. This results in a so-called *pulse code modulation* (PCM) representation of the audio signal, in which an amplitude value is assigned to each sample, i.e., measurement taken at fixed time interval. Since these samples represent a point in time and are hence too fine-grained to be used as meaningful audio features, the next step is *framing*, which means that consecutive samples are concatenated and referred to as frame. The entire audio signal is thus represented by a number of frames. From this frame representation, it is already possible to extract various features in the time domain. However, if we aim at computing features in the frequency domain, we have to further process each frame. To this end, it is common to first perform *windowing*, which means applying a windowing function to the samples of each frame. This is done for the following reason: If the last sample and the first sample of a given frame show a discontinuity, computing the Fourier transform will produce artifacts in the spectrum, a problem known as *spectral leakage*. To alleviate this problem,

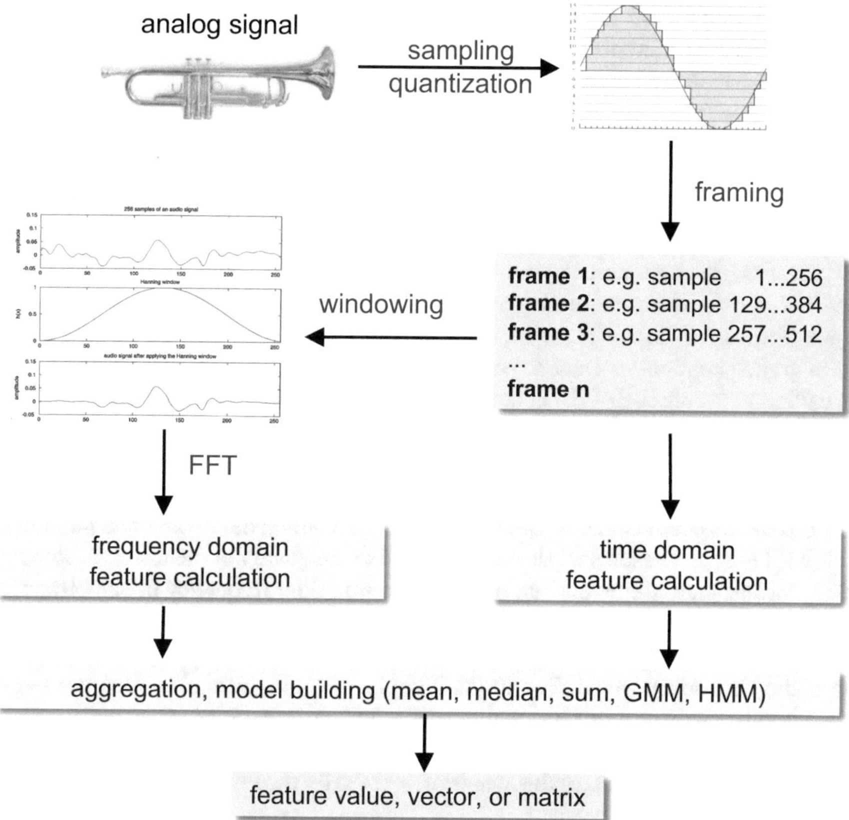

Fig. 2.4 Simplified workflow of a typical audio feature extractor

applying a windowing function such as the *Hann* or *Hanning function* smooths out such discontinuities, thus can be interpreted as making the signal periodic. From the frequency domain representation resulting from the FFT, we can now compute a variety of simple audio features, which are discussed below. As each of these features still describes only one frame, in a final step, they need to be combined in order to eventually obtain a feature that describes the entire audio segment or music piece under consideration. Each of the steps briefly outlined above is explained in detail in the following sections.

2.2.1 Analog–Digital Conversion

Real-world musical instruments, in comparison to synthesized sounds, produce a continuous, analog audio wave by making the air oscillate at different frequencies.

Since recording devices to capture sound do not have an arbitrarily high time resolution, the continuous signal needs to be sampled at uniform time intervals and quantized, i.e., encoded with a certain limited number of bits. This twofold process is referred to as analog–digital conversion (ADC).

Sampling refers to measuring the signal's amplitude, typically using a microphone. Note that this still gives an analog value. In order to be further processed, the signal needs to be quantized, which means the analog signal is transformed into a digital representation using a smaller, finite set of values. This is typically performed by rounding to some unit of precision. Figure 2.5 shows an example, in which the red curve represents the analog signal over time and the gray bars illustrate the quantized values. In this example, 4 bits are used to encode the signal, which allows to map each continuous value to one of $2^4 = 16$ discrete values. This value of 4 bits is called the resolution of the analog–digital conversion.

However, this quantization process introduced a quantization error, which is the difference between the value of the continuous signal and the discrete value the continuous value is mapped to. This quantization error can be observed in Fig. 2.5 as the difference between the red and the black curve, along the y-axis. The number of bits used to encode each sample is thus crucial to the quality of the result. A compact disc, for instance, stores the audio signal at a sampling frequency or sampling rate of 44,100 Hz, i.e., samples per second, and a resolution of 16 bits per channel.

Besides the resolution, another important factor to consider in ADC is the sampling frequency. According to the Nyquist–Shannon sampling theorem [443], the sampling frequency of a signal needs to be $2 \cdot f$ to capture frequencies up to f Hertz. If the sampling rate is chosen too small, it becomes impossible to account for the full range of frequencies that can be perceived by the human ear (approximately 20–20,000 Hz).

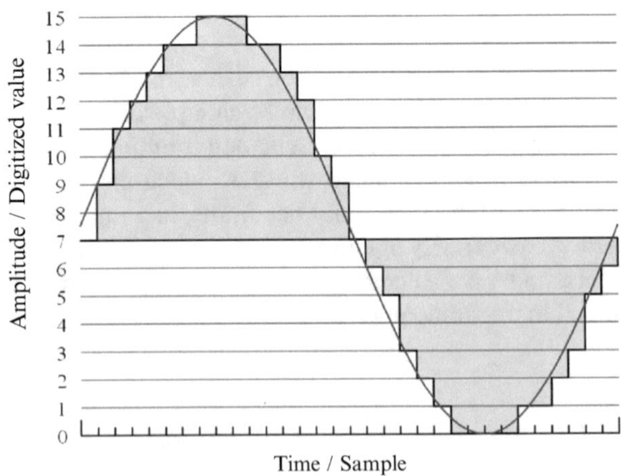

Fig. 2.5 Quantization of an audio signal. The analog signal is represented by the *red curve*. Measurements are taken at fixed points in time (x-axis) and encoded using a range of 4 bits (y-axis)

2.2.2 Framing and Windowing

After having converted the analog signal to a sampled, discrete amplitude representation, each resulting amplitude value covers only a tiny fraction of a second. Consider again the sampling frequency of a recording on a compact disc at 44,100 Hz. Each amplitude value is valid for only 0.0227 ms.

In order to compute meaningful music features that describe a chunk of the audio signal long enough to be perceivable by the human ear, we thus need to aggregate consecutive samples into frames. A frame is typically composed of a number of samples that equal a power of two, because conversion to the frequency domain via a Fourier transform is much faster in this case. Common choices for the frame size are 256–8192 samples. When deciding on a proper frame size, the sampling frequency needs to be considered though. As the time resolution of the human ear is around 10 ms, choosing a frame size that covers a much smaller time range does not make sense.

The resulting frame representation of a music piece can already be used as input to feature extractors that operate in the time domain. For features computed in the frequency domain, we still need to convert each frame into a representation of magnitudes over different frequencies, using a Fourier transform as explained below. Before we can compute this transform, we first have to perform *windowing* of each frame. Windowing refers to applying a windowing function to each frame. A common choice is the *Hann function*, named after the Austrian meteorologist Julius von Hann. It is defined in Eq. (2.1), where K is the frame size and $k = 1 \ldots K$.

$$w(k) = 0.5 \cdot \left(1 - \cos\left(\frac{2 \cdot \pi \cdot k}{K - 1}\right)\right) \tag{2.1}$$

By multiplying each frame sample-by-sample with the respective value of the Hann function, we eventually obtain a periodic signal. This is important because the Fourier transform might otherwise produce artifacts in the spectrum, a problem known as *spectral leakage*. The windowing process using a Hann window is illustrated in Fig. 2.6. The upper plot shows the input signal, i.e., a frame composed of $K = 256$ samples. The middle plot depicts the Hann windowing function over the K samples. The lower plot finally shows the windowed output signal of the frame under consideration.

As can be seen in the figure, windowing suppresses samples towards both ends of the frame. In order to avoid information loss resulting from this suppression, consecutive frames typically overlap by a certain amount, which is referred to as *hop size*. To give an example, using a frame size of 256 and a hop size of 128 results in frame 1 covering samples 1–256, frame 2 covering samples 129–385, frame 3 covering 257–513, and so on. Thus, consecutive samples show an overlap of 50 %, which is a common choice in music signal processing.

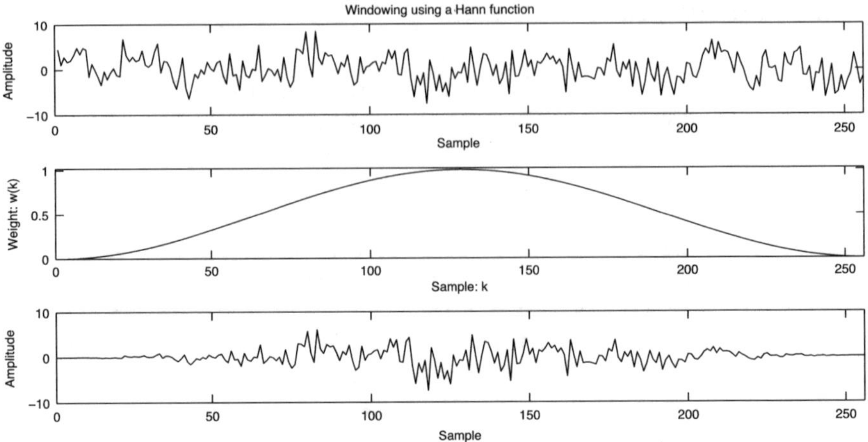

Fig. 2.6 Windowing of a 256-sample frame using a Hann function

2.2.3 *Fourier Transform*

The Fourier transform was named after the French mathematician and physicist Jean Baptiste Joseph Fourier. As already pointed out, it is a transformation of a signal from the time domain into the frequency domain. Fourier's theorem underlying the respective transform states that any continuous and periodic function can be represented as the sum of sine and cosine waves oscillating at different frequencies. The Fourier transform results in a discrete set of complex values, which describe the frequency spectrum of the input signal.

Since the Fourier transform represents a fundamental tool in signal processing, quite a few approaches and algorithms have been developed. The most important ones in audio and music processing are detailed in the following.

A computationally inexpensive variant of the Fourier transform, the fast Fourier transform (FFT), can be used if the input signal is discrete, which is the case in our signal representation as samples and frames. Hence, the FFT is a discrete Fourier transform (DFT), as defined in Eq. (2.2), where m is an integer ranging from 0 to $K-1$, K is the number of samples in the frame, and x_k represents the kth input amplitude, i.e., the kth sample in the frame.[2]

$$X_m = \sum_{k=0}^{K-1} x_k \cdot e^{-\frac{2\cdot\pi\cdot i}{K}\cdot k\cdot m} \tag{2.2}$$

Each output value X_m of the DFT computation is a complex number, in which the real-valued part $Re(X_m)$ and the imaginary part $Im(X_m)$ illustrate the importance of the cosine and the sine waves, respectively. From these, we obtain the *magnitude*

[2]We assume a signal that has already been windowed.

and the *phase* information as follows: $|X_m| = \sqrt{Re(X_m)^2 + Im(X_m)^2}$ gives the magnitude, which represents the amplitude of the combined sine and cosine waves; $\phi(X_m) = \tan^{-1} \frac{Im(X_m)}{Re(X_m)}$ defines the phase, which indicates the relative proportion of sine and cosine.

Computing X_m over different frequencies from the time domain input signal x_k directly from Formula (2.2) has quadratic computational complexity—for each of the K outputs, we need to sum over K terms. In order to significantly reduce the computational complexity, Cooley and Tukey proposed in 1965 an algorithm that runs in $O(K \cdot \log K)$ [83]. As additional prerequisite to the input being a discrete signal, their algorithm requires K to be a power of two, which is typically not a limiting restriction in real-world usage.

Among other related variants of the Fourier transform is the discrete cosine transform (DCT), which uses only real values instead of complex ones as in the DFT. While the DCT models the input signal only by cosine functions, the DFT models it by both cosine and sine functions. Considering again Eq. (2.2), this becomes obvious when expressing the complex exponential factor as $e^{i \cdot x} = \cos(x) + i \cdot \sin(x)$.

Of high practical importance in the context of audio signal processing is the short-time Fourier transform (STFT), which computes several DFTs over different segments of the input signal. These segments correspond to the windowed frames in temporal order. Concatenating the resulting frequency domain representations of the individual DFTs and depicting them along the time axis yields a *spectrogram*.

To exemplify the result of the STFT, Fig. 2.7 shows one piece of the Classical genre and one of the Electronic genre from the toy music data set, both in time domain and frequency domain. The time domain representation depicted in the upper plots shows the amplitude values of the first 384,000 samples of each piece, which correspond to roughly 8.7 s at the used sampling frequency of 44,100 Hz. These 384,000 samples further correspond to the first 1500 frames of each piece, using a frame size of 256 samples and the hop size being the full length of a frame, i.e., no overlap between consecutive frames. Taking a closer look, we can see a few distinguishing characteristics of the genres. For instance, the STFT representation of the Classical piano piece by Chopin reveals individual notes and overtones—the horizontal lines in the spectrogram. The Electronic piece by Modeselektor, in contrast, shows a rather vertical characteristic of the spectrogram, where distinctive repeating patterns emerge. Please also note the different range of the amplitude values. While the piano piece's amplitude values do not exceed 0.25, the Electronic piece exhausts the whole range of $[-1, 1]$.

2.3 Common Low-Level Features

Low-level features are directly computed from the representation of the audio signal in either the time domain or the frequency domain. Some of the most common low-level features in music retrieval are summarized below. An overview and categorization is given in Table 2.1. All of the features are computed on the frame level and are thus instantaneous according to temporal scope. Although each

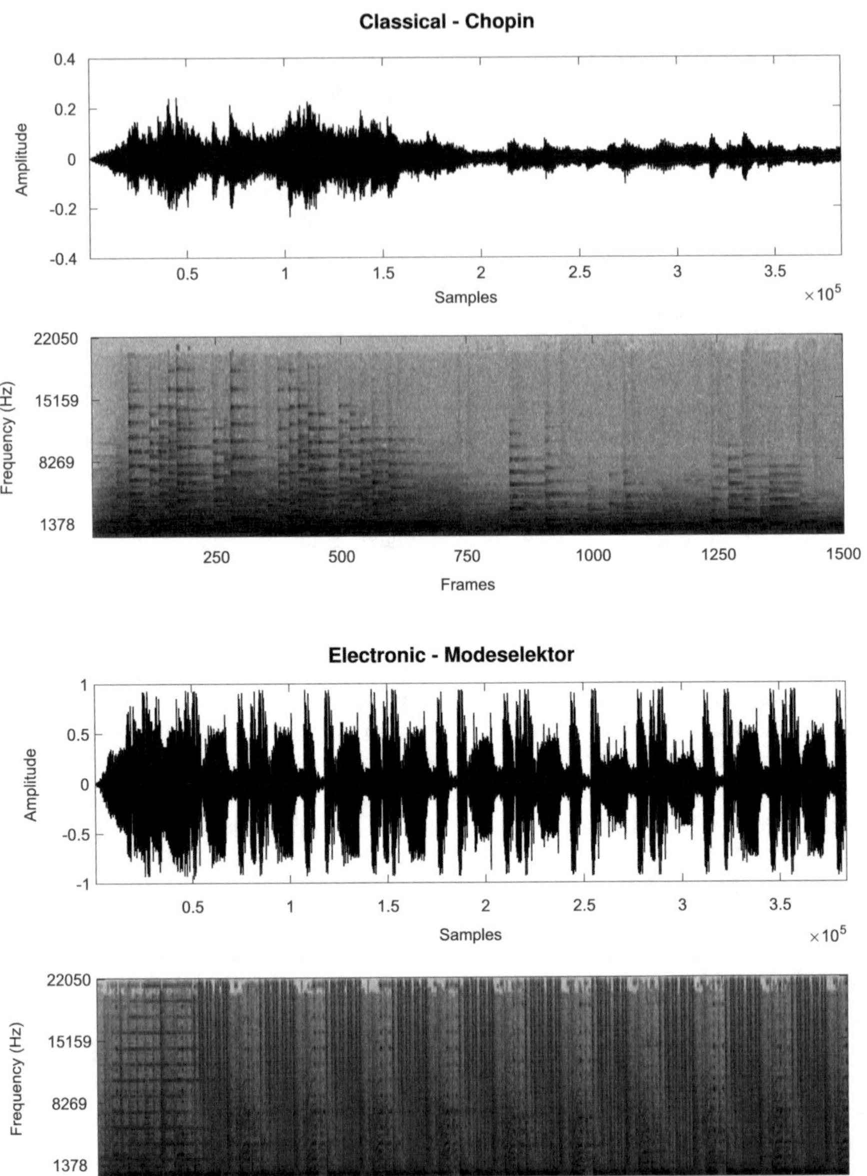

Fig. 2.7 Time domain and frequency domain representations of Classical and Electronic pieces from the toy music data set

Table 2.1 Overview of various low-level audio features

Feature	Abbreviation	Domain
Amplitude envelope	AE	Time
Root-mean-square energy	RMS	Time
Zero-crossing rate	ZCR	Time
Band energy ratio	BER	Frequency
Spectral centroid	SC	Frequency
Bandwidth (spectral spread)	BW	Frequency
Spectral flux	SF	Frequency

Table 2.2 Terms used in the definitions of the low-level features

Term	Description
$s(k)$	Amplitude of the kth sample
$m_t(n)$	Magnitude of the signal in the frequency domain at frame t in frequency band n
K	Frame size, i.e., the number of samples in each frame
N	Number of frequency bins, i.e., the number of the highest frequency band

feature value hence describes only one frame, it is possible to aggregate all values of a particular feature that belong to the same piece of music. This can be done by simply computing some statistical summarization function such as mean, median, or maximum. However, quite a lot of information is lost when doing so. Alternative aggregation methods, for instance, modeling the distribution of the individual features belonging to the same piece via a Gaussian mixture model (GMM) [528], typically yield a much better description of the piece.

To define the subsequently discussed low-level features, we will use the common terms defined in Table 2.2. In order to compare the characteristics of the different features, Figs. 2.8, 2.9, and 2.10 depict time domain and frequency domain low-level features for selected pieces of the toy music data set, chosen from the three very distinct genres Electronic, Classical, and Metal. The top plot in each figure shows the waveform (amplitude values over time or samples). The subsequent three plots depict the discussed time domain features. Below them, the spectrogram is shown, followed by illustrations of the four frequency domain features.

2.3.1 *Time Domain Features*

Amplitude Envelope (AE) is simply the maximum amplitude value of all samples in a given frame t. It is hence computed in the time domain. The formal definition is given in Eq. (2.3).

$$AE_t = \max_{k=t\cdot K}^{(t+1)\cdot K-1} s(k) \tag{2.3}$$

AE may be used, for instance, as a very simple beat-related feature for onset detection [210] but is highly sensitive to outliers in the amplitude.

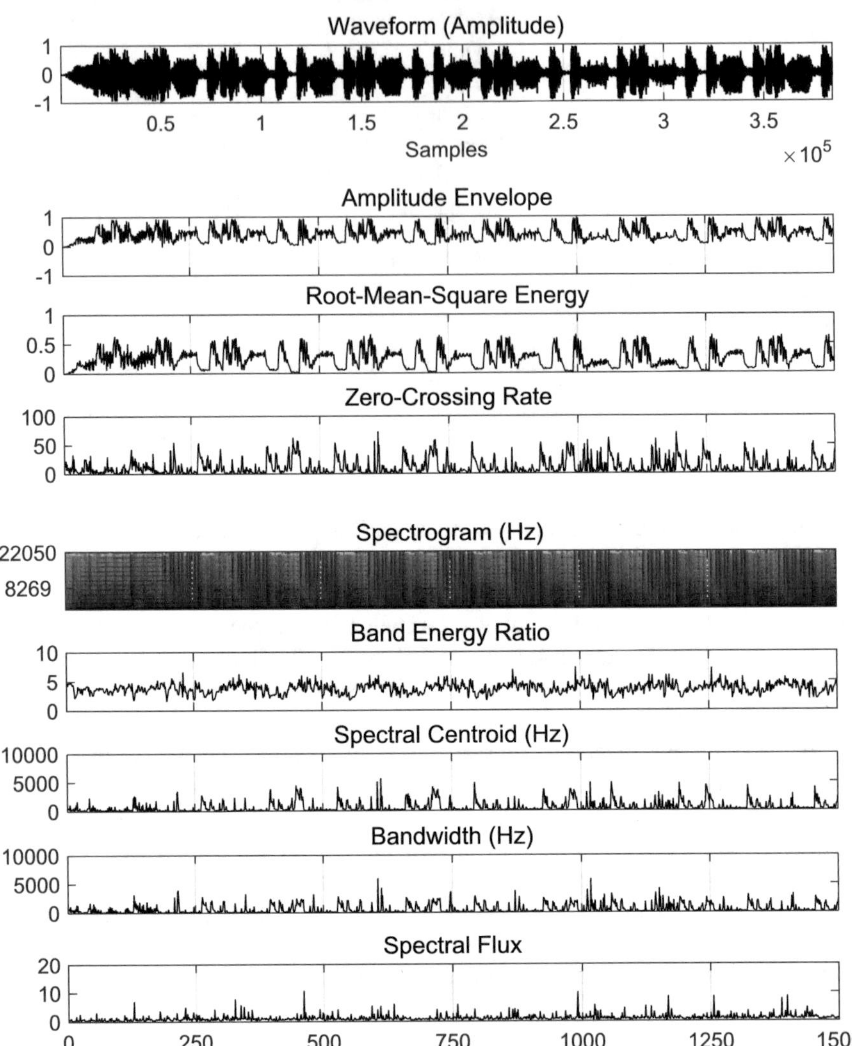

Fig. 2.8 Low-level time domain and frequency domain features of the Electronic piece by Modeselektor

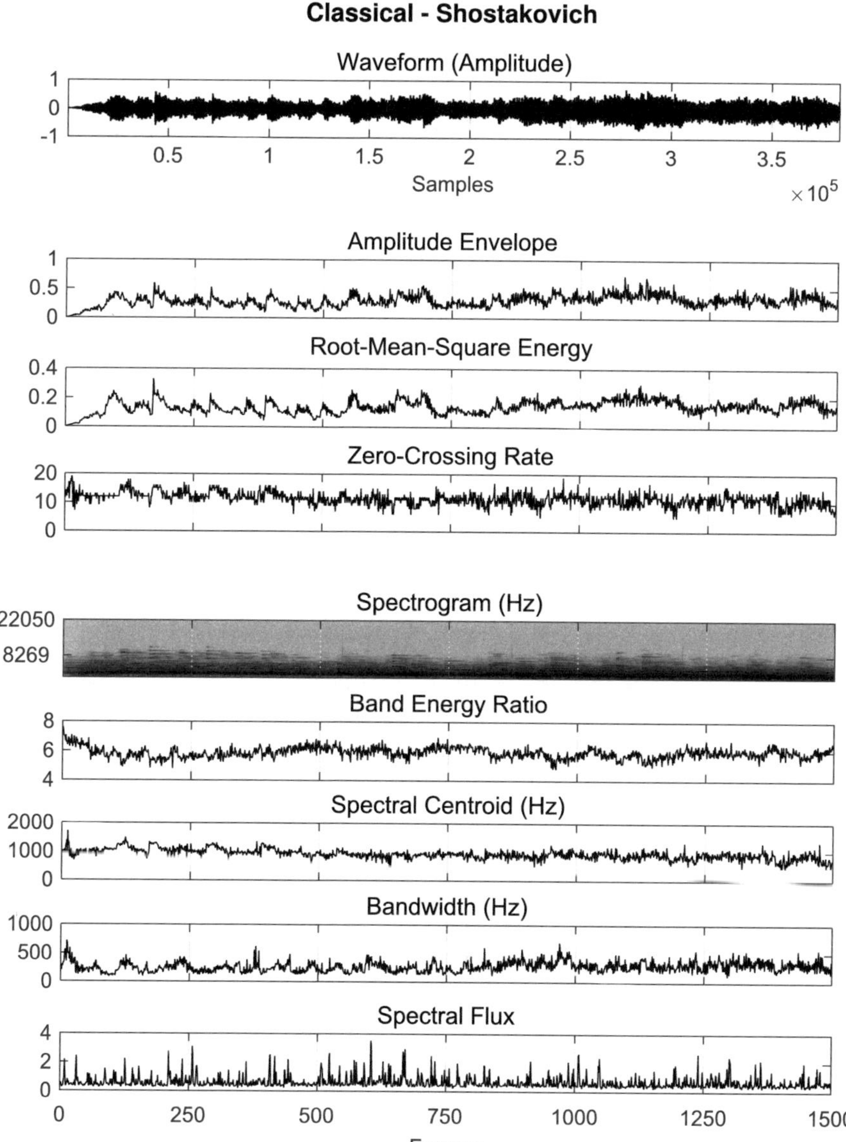

Fig. 2.9 Low-level time domain and frequency domain features of the Classical piece by Shostakovich

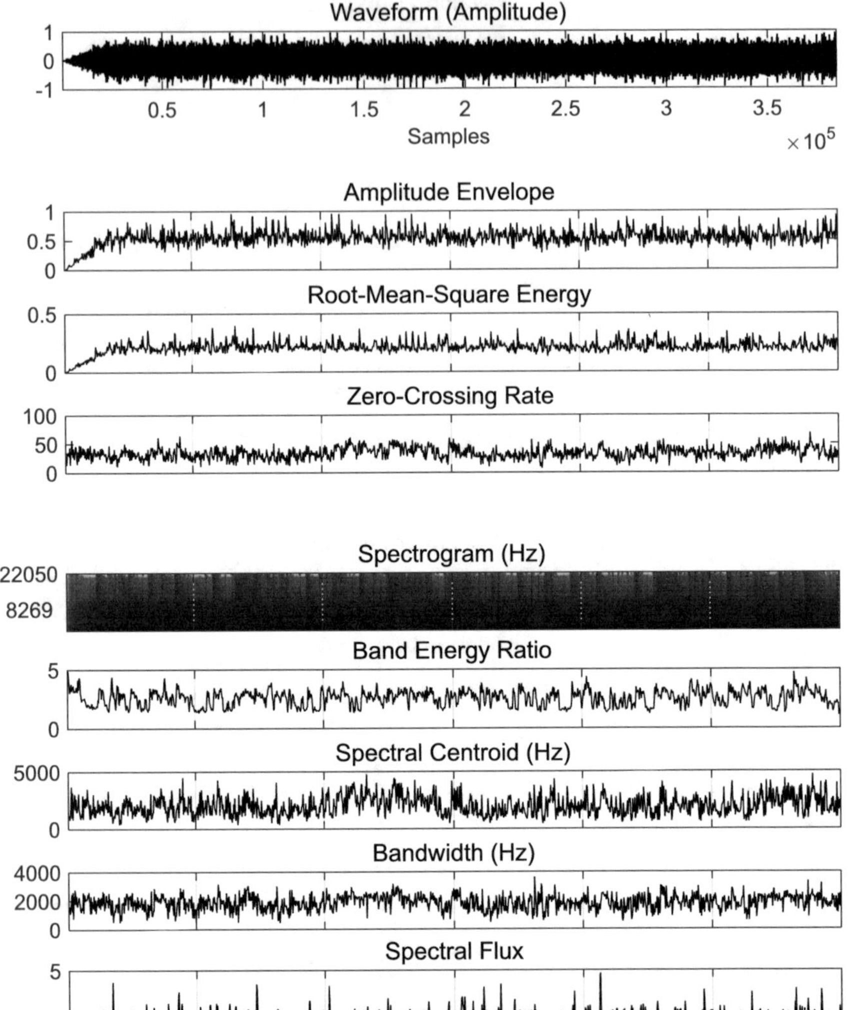

Fig. 2.10 Low-level time domain and frequency domain features of the Metal piece by Iron Maiden

Root-Mean-Square Energy is typically denoted as RMS energy, RMS level, or RMS power and is another time domain feature, computed as shown in Eq. (2.4).

$$RMS_t = \sqrt{\frac{1}{K} \cdot \sum_{k=t\cdot K}^{(t+1)\cdot K-1} s(k)^2} \tag{2.4}$$

As it relates to perceived sound intensity, RMS energy can be used for loudness estimation and as an indicator for new events in audio segmentation [56]. It is a feature similar to AE but less sensitive to outliers. Note the lower energy and AE values for the Classical music piece in comparison to the songs in other genres; cf. Fig. 2.9.

Zero-Crossing Rate (ZCR) measures the number of times the amplitude value changes its sign within the frame t under consideration; cf. Eq. (2.5).

$$ZCR_t = \frac{1}{2} \cdot \sum_{k=t\cdot K}^{(t+1)\cdot K-1} |\operatorname{sgn}(s(k)) - \operatorname{sgn}(s(k+1))| \tag{2.5}$$

Although this simple time domain feature is foremost used in speech recognition, it is also applied in MIR to detect percussive sounds and noise [156]. It can further be used as an indicator of pitch for monophonic music signals as higher ZCR values typically correspond to higher frequencies. Again, note the lower ZCRs and their much higher stability for Classical music (cf. Fig. 2.9), which can be explained by the absence of distorted and percussive instruments in the piano piece by Shostakovich.

2.3.2 *Frequency Domain Features*

Band Energy Ratio (BER) is a feature computed in the frequency domain as shown in Eq. (2.6). It is used, among others, for speech/music discrimination [254] and, together with other features, music (genre) classification [312]. BER relates the energy in the lower frequency bands to the energy in the higher bands and in this way measures how dominant low frequencies are.

$$BER_t = \frac{\sum_{n=1}^{F-1} m_t(n)^2}{\sum_{n=F}^{N} m_t(n)^2} \tag{2.6}$$

F denotes the split frequency band, i.e., the band that separates low from high frequency bands. The choice of F highly influences the resulting range of values.

Since the energy in low bands is often much higher than that in high bands, it is common practice to either apply a cutoff for BER values or use the inverse BER, i.e., numerator and denominator in Eq. (2.6) are switched. The BER plots depicted in Figs. 2.8, 2.9, and 2.10 use the standard formulation [Eq. (2.6)] and a split frequency of 2000 Hz.

Spectral Centroid (SC) represents the center of gravity of the magnitude spectrum, i.e., the frequency band where most of the energy is concentrated. This feature is used as a measure of "brightness" of a sound, thus relates to music timbre [160, 430]. However, SC is highly sensitive to low-pass filtering as high frequency bands are given more weight than low ones; cf. Eq. (2.7). This is particularly problematic when the audio signal is down-sampled, i.e., the sampling rate is reduced (e.g., from 44,100 to 11,025 Hz). Since such a down-sampling will cut off or distort frequencies higher than half of the sampling rate (due to the Nyquist–Shannon theorem), the SC value will change accordingly.

$$SC_t = \frac{\sum_{n=1}^{N} m_t(n) \cdot n}{\sum_{n=1}^{N} m_t(n)} \tag{2.7}$$

The Classical piece shows the lowest SC values and by far the most stable ones, whereas the SC fluctuates much more the Electronic and Metal songs. This evidences the "clearness" or "brightness" of the sound in the (monophonic) piano piece by Shostakovich, in contrast to the pieces from other genres.

Bandwidth (BW), also known as **Spectral Spread** (SS), is derived from the spectral centroid, thus also a feature in the frequency domain. Spectral bandwidth indicates the spectral range of the interesting parts in the signal, i.e., the parts around the centroid. It can be interpreted as variance from the mean frequency in the signal. The definition is given in Eq. (2.8). The average bandwidth of a music piece may serve to describe its perceived timbre [262].

$$BW_t = \frac{\sum_{n=1}^{N} |n - SC_t| \cdot m_t(n)}{\sum_{n=1}^{N} m_t(n)} \tag{2.8}$$

Comparing Figs. 2.8, 2.9, and 2.10, we observe that Classical music shows much smaller bandwidths than Electronic and Metal. This indicates a lower spread of energy among different frequency bands.

Spectral Flux (SF) describes the change in the power spectrum between consecutive frames. It is therefore a frequency domain feature and computed according to Eq. (2.9), where D_t is the frame-by-frame normalized frequency distribution in frame t.

$$SF_t = \sum_{n=1}^{N} (D_t(n) - D_{t-1}(n))^2 \tag{2.9}$$

As can be seen, SF sums the differences of consecutive frames between the magnitudes over all frequency bands. While SF is often used as speech detector [389], it can also be used to describe the timbre of an audio signal and in onset detection [105].

2.4 Summary

This chapter gave a short introduction to basics in audio signal processing, which are fundamental to understand the more advanced content-based techniques in the next chapter. We first characterized different music audio features according to their level of abstraction into low level, mid-level, and high level (from close to the signal to understandable by the average listener) and according to their temporal scope into instantaneous, segment-level, and global features. We further distinguished features computed in the time domain (amplitude over time) from those computed in the frequency domain (magnitude over frequency) and presented the Fourier transform which transfers the time domain representation of the audio signal into its frequency domain representation.

Introducing a typical pipeline of a simple audio feature extractor, we discussed analog–digital conversion of an analog audio signal by sampling and quantization, which yields a sequence of samples (e.g., 44,100 per second), each one described by an amplitude value (e.g., encoded by 16 bits). Given the digitized time domain representation, we showed how sequences of samples form larger segments called frames, a process referred to as framing. Each frame typically corresponds to a few tens of milliseconds of audio. From such frames, low-level time domain features like amplitude envelope, root-mean-square energy, or zero-crossing rate can be computed. Alternatively, frames can be windowed by multiplying all samples in each frame with a Hann(ing) function in a point-wise manner. Subsequent computation of the Fourier transform yields a frequency domain representation of each frame. From this, the low-level features band energy ratio, spectral centroid, bandwidth (also known as spectral spread), and spectral flux can be calculated, among others.

While these low-level features have been shown to be applicable to some extent in certain tasks, e.g., onset detection (amplitude envelope), modeling timbre (spectral flux and spread), or audio segmentation (root-mean-square energy), their discriminative power and semantic meaning are limited. This makes them largely unsuited for tasks such as genre classification or music emotion recognition, where better representations are required.

2.5 Further Reading

Unlike many other textbooks about MIR, this book does not focus purely on audio- or MIDI-based approaches, but takes a broad perspective by discussing audio, web, and user-centric strategies. Due to this broad view and the availability of excellent textbooks on music audio processing, audio features are only dealt with in brief in the book at hand. For readers who want to deepen their knowledge about audio signal processing and analysis, a selection of books we recommend is given below. While most books on the topic not only discuss low-level features but also higher-level concepts and are therefore presented in Chap. 3, the selection given here includes texts which devote a major part to audio signal processing and basic content analysis.

Lerch [262] provides a very good and detailed introduction to low-level audio processing. He starts by characterizing different types of audio signals (random and periodic signals), before digging deeper into their processing (including sampling, convolution, Fourier transform, and filter banks). A major part of the book is then devoted to low-level music features, providing a more detailed treatment of the features discussed here and many more. Lerch further categorizes music audio features into intensity-related (intensity and loudness), tonal (pitch, chords, harmony, etc.), and temporal (onsets, tempo, beat, rhythm, etc.) descriptors.

While most of his textbook is devoted to tasks involving high-level representations or descriptors of music signals (e.g., tempo, harmony, chords), Müller [326] also gives a comprehensive introduction to signal processing basics, including Fourier transform and analysis.

In addition to textbooks specifically focusing on music applications of audio processing, a very comprehensive elaboration on all aspects of digital signal processing (DSP) is provided by Smith [454]. Among the many topics covered in his book, the ones most relevant to the book at hand are detailed discussions of analog–digital conversion, convolution, Fourier transform, digital filters, and applications of DSP. In particular, his treatment of the Fourier transform and related techniques is very comprehensive.

Chapter 3
Audio Feature Extraction for Similarity Measurement

In this chapter, we will introduce techniques to extract audio features for the purpose of modeling similarity between music pieces. As similarity is a crucial concept for any retrieval and recommendation task, having at hand accurate audio features and corresponding similarity functions is key to building music retrieval systems. Derived from low-level audio features, but not yet representing semantically meaningful music concepts, this information can be categorized as mid-level features, according to Fig. 1.7. This chapter builds upon the previous one in that some of the audio features introduced here rely on the low-level features discussed earlier. In the following section, we will first briefly touch upon psychoacoustic models of music perception, which can be used to ameliorate music feature extraction. Subsequently, we will present the two main categories of audio features used for music similarity: frame-level and block-level features. We will describe state-of-the-art feature extractors in both categories and detail how the resulting feature vectors for a whole piece of music can be aggregated to form a global representation. Techniques covered range from simple statistical summarization to vector quantization to Gaussian mixture models. From the resulting global acoustic feature representation, we show how similarity between music pieces can be computed. The chapter is concluded with a discussion of (and an approach to solve) a problem known as "hubness," which occurs when computing similarities in high-dimensional feature spaces.

3.1 Psychoacoustic Processing

There is an obvious discrepancy between physical characteristics of sound and its human perception. The frequently used Mel frequency cepstral coefficients, for instance, employ a nonlinear scale to model feature values according to how humans perceive pitch intervals; cf. Sect. 3.2.1. We cannot give a comprehensive discussion

P. Knees, M. Schedl, *Music Similarity and Retrieval*, The Information Retrieval Series 36, DOI 10.1007/978-3-662-49722-7_3

of the field of psychoacoustics here but will outline some of the most important psychoacoustic facts to consider when elaborating audio feature extractors. In particular, we focus on psychoacoustic aspects of loudness and frequency. The reader interested in more details should refer to [124] or [360].

3.1.1 Physical Measurement of Sound Intensity

On a physical level, the intensity of a sound is typically described as sound pressure level and measured in decibel (dB). Intensity in dB is defined as shown in Eq. (3.1), where i refers to sound power per unit area, measured in watts per m^2, and i_0 is the hearing threshold of 10^{-12} W/m^2.

$$i_{dB} = 10 \cdot \log_{10}\left(\frac{i}{i_0}\right) \tag{3.1}$$

Decibel is a logarithmic unit. An increase of 10 dB thus corresponds to an increase of ten times in sound pressure level.

3.1.2 Perceptual Measurement of Loudness

Human perception of loudness generally varies with sensitivity of the ear, which again varies between different people. In addition, perception of loudness also varies between different frequencies of sound. The human ear is most sensitive to frequencies between 2000 and 5000 Hz. Tones with the same physical sound pressure, measured in decibel (dB), but with frequencies out of this range are thus perceived as softer. To describe sound pressure levels that are perceived as equally loud, irrespective of sound frequency, several listening experiments have been conducted in research groups around the world. The results have eventually been integrated into the ISO standard 226:2003. Figure 3.1 shows the resulting equal loudness curves. Each curve describes the relationship between sound pressure (dB) and a particular level of loudness perception, varying over frequencies. The unit for loudness perception is phon. Phon is defined as having the equal dB value at a frequency of 1000 Hz. Thus, for a pure tone of 1000 Hz, the dB value equals the phon value, which can also be seen in the figure.

To account for human perception of loudness in music signal processing, the transformation from dB to phon is a vital step. As illustrated in Fig. 3.1, the phon scale reflects the nonlinear relationship between sound pressure and human sensation of loudness. Please note, however, that the phon scale is still logarithmic.

For easier interpretation due to its linearity, the sone scale defines 1 sone as being equivalent to 40 phons in terms of loudness, which again amounts to a physical sound pressure level of 40 dB for a signal with 1000 Hz. Doubling the sone value

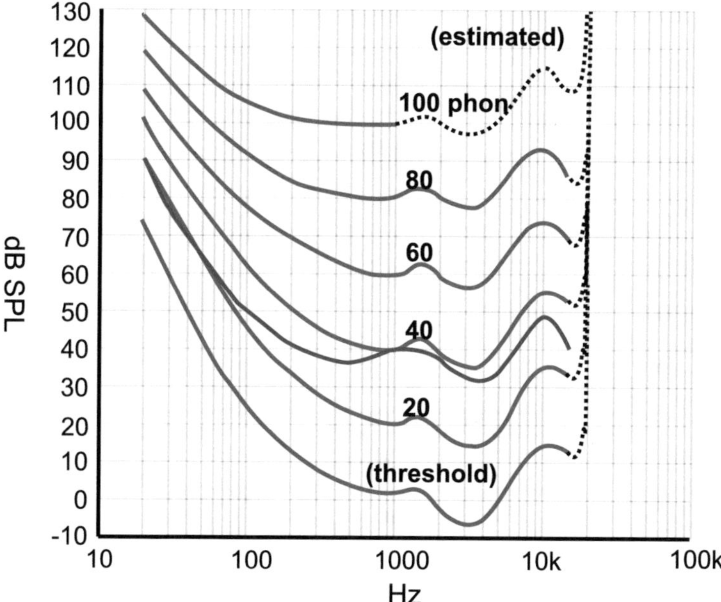

Fig. 3.1 Equal loudness curves as defined in ISO 226:2003 (*red lines*) and an earlier standard (*blue line*). Adapted from Wikimedia Commons [275]

roughly corresponds to a doubling of perceived loudness. An increase of loudness by 10 phons leads to roughly a doubling of the corresponding sone value for loudness $l_{sone} > 1$, as shown in Eq. (3.2), which describes the approximate relationship between loudness in phon (l_{phone}) and in sone (l_{sone}).

$$l_{phon} = 40 + 10 \cdot \log_2 (l_{sone}) \tag{3.2}$$

To give an example, while loudness of a quiet conversation is around 1 sone (corresponding to 40 phons), a jackhammer is perceived about 64 times as loud (64 sones = 100 phons), and the threshold of pain is reached at about 512 sones = 130 phons.

3.1.3 Perception of Frequency

A signal's frequency is measured in Hertz (Hz), the number of cycles per second. Humans can perceive frequencies between approximately 20 and 20,000 Hz. However, the upper limit decreases with age. Any audio signal out of this range is thus not informative for music audio feature extraction.

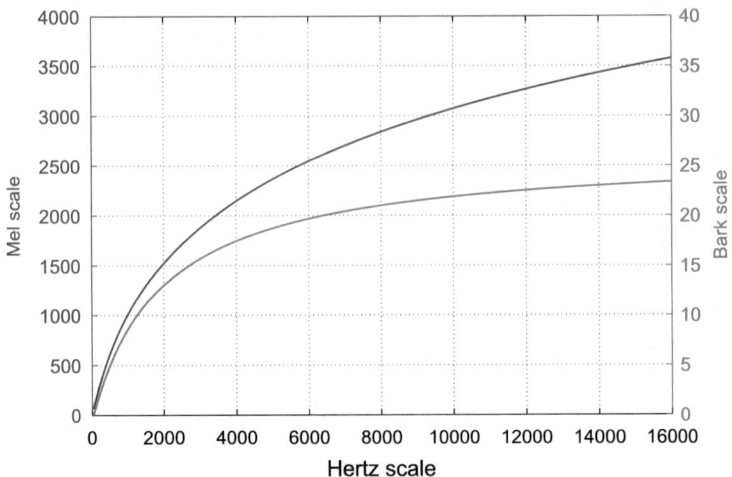

Fig. 3.2 Relationship between Hertz–Bark and Hertz–Mel

Perception of pitch intervals is nonlinearly correlated with intervals in frequency. Two frequently used scales that reflect this nonlinear relationship are the Mel and the Bark scales. Both scales are based on pitch ranges that are judged as equal by humans. Several formulas exist to describe the relationship between frequency in Hz and Bark or Mel. Due to their origin in listening experiments, these formulas are only approximations, however. Two popular versions are shown in Eqs. (3.3) and (3.4), respectively, for Bark and Mel. The corresponding plots illustrating the relationship between Hz and Mel (blue curve) and between Hz and Bark (green curve) are shown in Fig. 3.2.

$$b = \left(\frac{26.81 \cdot f}{1960 + f} \right) - 0.53 \tag{3.3}$$

$$m = 1127 \cdot \log \left(1 + \frac{f}{700} \right) \tag{3.4}$$

According to the Mel scale, the relationship between Hertz and Mel is approximately linear for frequencies below 500 Hz. From about 500 Hz upwards, in contrast, frequency ranges that are judged to yield equal pitch intervals become larger and larger. The scale's reference point is 1000, which means that 1000 Hz equals 1000 Mel.

For what concerns the Bark scale, it is common to quantize music signals into critical bands, whose range corresponds to human auditory perception. The range between highest and lowest frequency within each critical band is thus judged as equal in pitch interval, irrespective of the actual band. To give an example, the range [10, 11] Bark corresponds to [1250, 1460] Hz, whereas the interval [18, 19] Bark equals [4370, 5230] Hz. However, even though the first range only covers a frequency interval of 210 Hz, the perceived pitch difference between 1250 and 1460 Hz equals the perceived pitch difference between 4370 and 5230 Hz, which covers 860 Hz.

Another important psychoacoustic concept for modeling frequency perception is spectral masking. Spectral masking effects refer to the occlusion of a sound by a different sound in a different frequency band. This typically happens when the first sound is considerably louder than the occluded one. Audio feature extractors can detect such spectral masking effects and suppress the imperceptible sound. On a side note, this effect is also used in audio compression, such as in MPEG Audio Layer III, also known as *mp3*.

3.2 Frame-Level Features and Similarity

Methods for audio feature extraction that aim at estimating music similarity between pieces can be categorized into approaches that extract features at the level of individual frames and approaches that consider longer chunks of the audio signal—typically a few seconds. The former are referred to as frame-level features and the latter as block-level features. Both have in common that they eventually model similarity at the level of the entire music piece, which means that feature values or vectors need to be aggregated, regardless of whether they are valid for a frame or a block. For what concerns frame-level features, by far the most prominent of this type of features are variants of the Mel frequency cepstral coefficients. Before we elaborate on these, please note that we define the "similarity" between two music tracks t_1 and t_2 (given by a function $s(t_1, t_2)$) and the "distance" (given by $d(t_1, t_2)$) to have an inverse relationship, which means for similarities normalized to the range [0, 1], the equation $s(t_1, t_2) = 1 - d(t_1, t_2)$ holds. Furthermore, similarity and distance measures in the context of music similarity are almost always commutative or symmetric, i.e., $s(t_1, t_2) = s(t_2, t_1)$ and $d(t_1, t_2) = d(t_2, t_1)$.

3.2.1 Mel Frequency Cepstral Coefficients

Mel frequency cepstral coefficients (MFCCs) have their origin in speech processing but were also found to be suited to model timbre in music. The MFCC feature is calculated in the frequency domain, derived from the signal's spectrogram.

Given the magnitudes over frequencies for each frame, the frequency scale in Hertz is first converted into the Mel scale. From the Mel representation of the magnitudes of a given frame, the logarithm is subsequently taken. The resulting logarithmic magnitude values over Mel bands are then fed into a DCT (cf. Sect. 2.2.3), which yields a spectrum computed over Mel frequencies, rather than over time. This output spectrum eventually represents the MFCCs for the frame under consideration. Performing the steps detailed above for all frames in the music

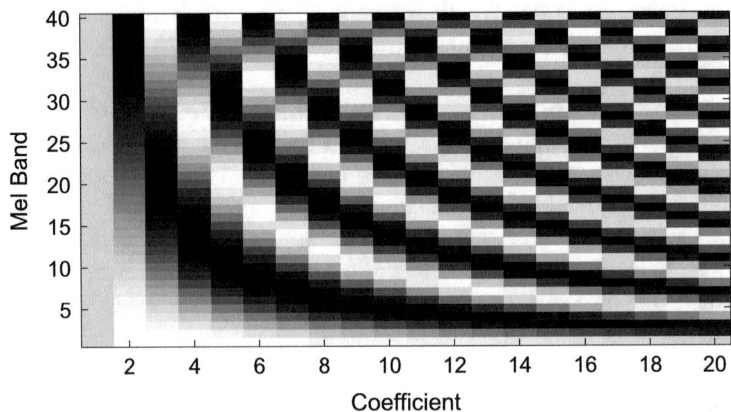

Fig. 3.3 Mel frequency cepstral coefficients bands

piece results in temporally ordered vectors of MFCCs, similar to the spectrogram produced by an STFT.

One MFCC vector thus describes the periodicities found in the magnitude values over the frequency distribution of a given frame. In music signal processing, between 13 and 25 MFCCs are typically computed for each frame. The first MFCC corresponds to the average energy of the signal in the frame under consideration. It is frequently omitted in feature modeling for music retrieval. Increasing MFCC positions correspond to higher periodicities; cf. Fig. 3.3. To give some examples, Fig. 3.4 visualizes waveform, spectrogram, and the corresponding 20 MFCCs over time, for a Classical and a Metal piece of our toy collection.

Computing MFCC vectors or other frame-level features for an entire music piece typically yields several tens of thousands individual feature vectors. Consider, for instance, a 3-min song sampled at 44,100 Hz. Applying to this song a frame-level feature extractor with a frame size of 512 samples and a hop size of 50 % produces more than 20,000 feature vectors to describe it; cf. Eq. (3.5), where the factor of 1.5 accounts for the hop size increasing the number of frames by 50 %.

$$N_{fv} = 1.5 \cdot \left(\frac{44{,}100 \text{ samples/s}}{512 \text{ samples/frame}} \right) \cdot 180\,\text{s} = 23{,}256 \text{ frames} \tag{3.5}$$

Aggregating this large number of feature vectors is typically performed either by statistical summarization of all resulting vectors, by applying vector quantization, or by fitting probabilistic models to the data. All of these methods discard the temporal ordering of the frames, thus providing representations of the distribution of MFCC vectors over the whole piece ("bag of frames" approaches; cf. "bag of words" methods in text, Sect. 5.4). From the resulting representation of each song, pairwise similarities between songs can be computed in various ways, depending on the aggregation technique.

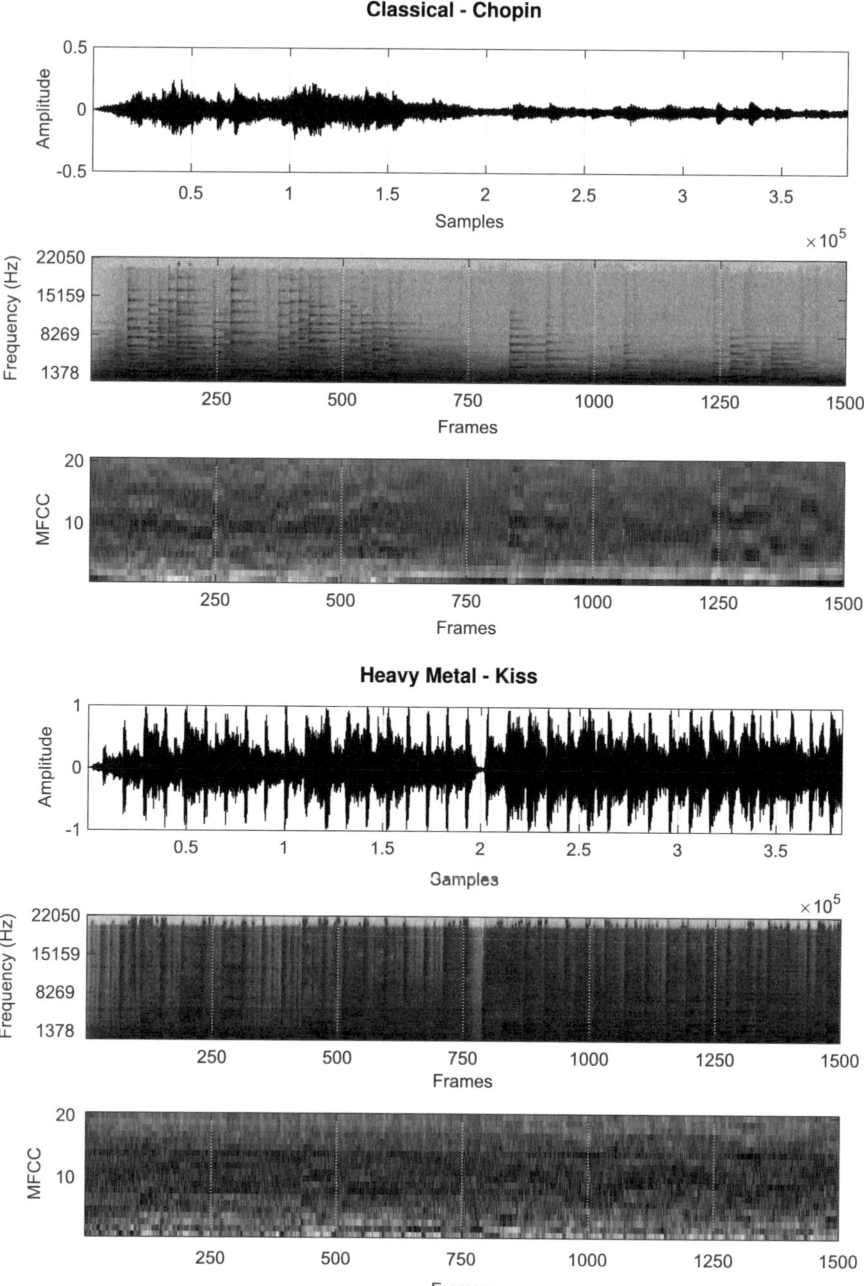

Fig. 3.4 Amplitude, spectrogram, and MFCCs of the Classical piece by Chopin (*top*) and the Metal piece by Kiss (*bottom*)

3.2.2 Statistical Summarization of Feature Vectors

One approach to create a unified description of the MFCC vectors of a music piece is to use statistics of the distribution as features. Mandel [292] and Tzanetakis [486], for instance, propose to create such a statistics vector by concatenating (1) the mean of each of the MFCCs over all frames and (2) the respective flattened covariance matrix. When using 22 MFCCs, this results in a feature vector of dimensionality 275 (22 means, 22 variances, and 231 covariances[1]), regardless of the length of the music piece.

To compute a similarity estimate between two songs represented by feature vectors $\mathbf{x}$ and $\mathbf{y}$, L_p-norm distance can be used; cf. Eq. (3.6), where $\mathbf{x}$ and $\mathbf{y}$ denote the n-dimensional feature vectors of the two pieces to compare, respectively. The most prominent ones are the Manhattan and Euclidean distance, respectively, L_1-norm and L_2-norm distance.

$$d(\mathbf{x},\mathbf{y}) = \|\mathbf{x}-\mathbf{y}\|_p = \left(\sum_{i=1}^{n} |x_i - y_i|^p\right)^{\frac{1}{p}} \tag{3.6}$$

Figure 3.5 shows a visualization of pairwise similarities computed on our toy music data set. Euclidean distances were calculated on statistical summaries as described above. As can be seen from the figure, this simple approach to computing similarities from frame-level features is barely suited to separate different genres. The sole exception is the track by Snoop Dogg, whose high dissimilarity to most other pieces can be explained by its minimalist performance; cf. Appendix A.4.

Looking at the intra-/inter-genre similarity ratio of the Classical genre, we see that the similarity between two randomly chosen Classical pieces is, on average, about 1.5 times higher than the similarity between a randomly chosen Classical piece and a randomly chosen piece from any other genre. Using the statistical summarization approach to music similarity measurement discussed in this section, the intra-/inter-genre similarities are highest for Rap and Classical. The global values for the statistical summarization approach amount to 1.386 (mean) and 1.390 (median).

3.2.3 Vector Quantization

Another way of deriving a representation with a fixed size of dimensions is vector quantization (VQ), also known as codebook approach. In contrast to statistical summarization methods and the probabilistic models presented in the following,

[1]Given the 22×22 covariance matrix, only the upper triangle contain unique covariances, while the diagonal contains the variances, hence we have $\frac{22 \cdot 21}{2}$ covariance values.

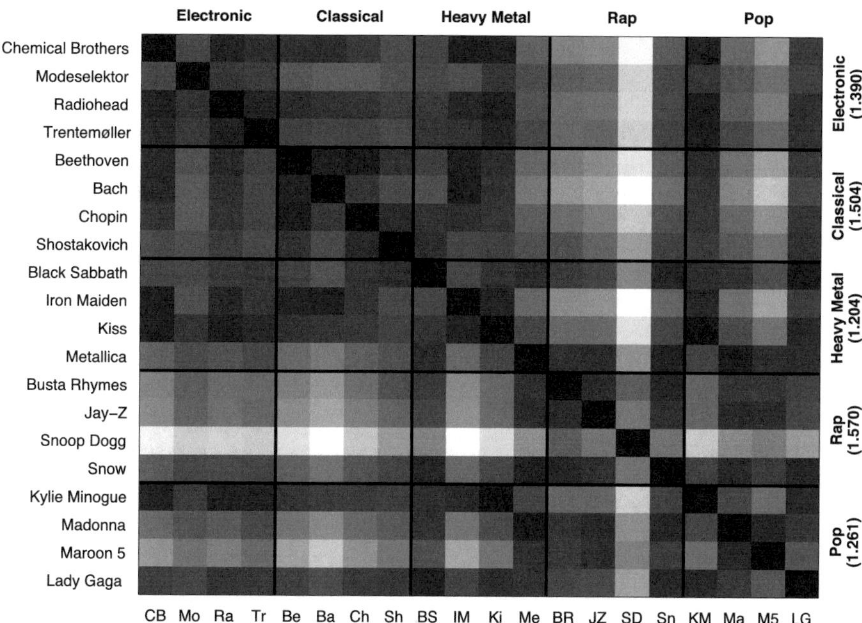

Fig. 3.5 Visualization of similarity matrix computed on L_2-norm distances (Euclidean distances) on statistical summaries over MFCC vectors

which both derive piece representations solely on the basis of the individual pieces, VQ is best performed simultaneously on all music pieces in a collection (e.g., a training collection) to cover the space of possible feature values.

After extracting all MFCC vectors of all music pieces in the training collection, the resulting feature distribution over all pieces is modeled by finding clusters, e.g., through k-means clustering. The k identified cluster centers are the entries ("words") of the so-called codebook. Each word of the codebook corresponds to a dimension of the resulting feature space; thus, the choice of k defines the dimensionality of the feature vector space. To obtain a codebook representation for a music piece (which can be either a previously unseen piece or already contained in the training set), for each of its MFCC vectors, the closest word in the codebook is identified. Subsequently, the relative frequency of each word is calculated, i.e., it is counted how often each word is identified as being closest in relation to the total number of MFCC vectors of the piece. The relative frequency of each word in a given music piece serves as corresponding feature value of the k-dimensional global feature vector of the piece. These values therefore model the affinity of the piece to prototypes of feature clusters in the whole collection. Pairwise similarity of music pieces can then again be obtained by calculating, for instance, L_p-norm distances between their feature vectors; cf. Eq. (3.6).

To illustrate these steps, consider the simplified example given in Fig. 3.6. Here, we assume the distribution of frames (modeled in two dimensions representing

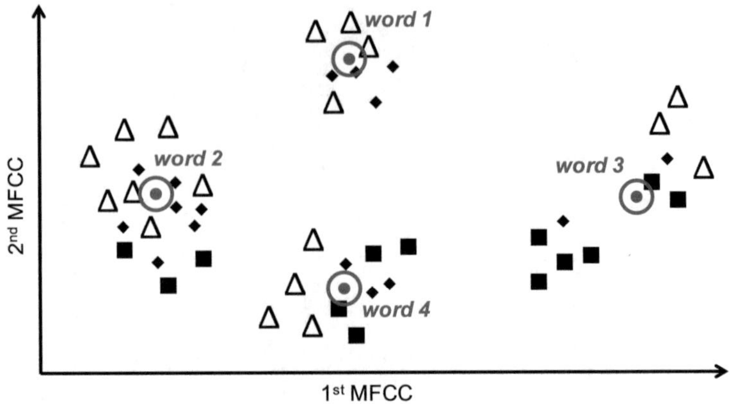

Fig. 3.6 Schematic example of vector quantization in two dimensions trained on a collection of three music pieces. The codebook is chosen to contain $k = 4$ words (resulting in four-dimensional feature vectors for the music pieces)

the first two MFCC values) of a collection of three different music pieces (the frames of which are represented by diamond, square, and triangle symbols, respectively). Learning a four-dimensional codebook using k-means clustering results in words 1–4 (depicted as circles). Calculating the relative frequencies of word occurrences per music piece, in this toy example, would lead to the feature vector $[0.25, 0.44, 0.13, 0.19]$ for the piece represented by the diamond which overall consists of 16 frames, to vector $[0.00, 0.23, 0.46, 0.31]$ for the piece represented by the square (13 frames total), and to vector $[0.22, 0.39, 0.17, 0.22]$ for the piece represented by the triangle (18 frames total). Comparing these vectors by means of, e.g., Euclidean distance then indicates that the diamond- and the triangle-represented pieces are most similar in the collection.

Despite the shortcoming of requiring a training collection to define the resulting feature space, the codebook approach has several favorable properties that stem from the use of a vector space representation, foremost simple similarity calculation and efficient indexing; cf. Sect. 6.1.1. An extension to this principle is applied by Foote [131], and similarly by Seyerlehner et al. [438], who describes the distribution of MFCC-transformed frames via a global supervised tree-based quantization.

3.2.4 Gaussian Mixture Models

Integrating elements from both methods discussed so far, Gaussian mixture models (GMMs) [38] aggregate the MFCC vectors of an individual piece by means of a probabilistic model. GMMs are characterized by a number K of Gaussian distributions $\mathcal{N}_{k=1...K}$, each of which is defined by their common parameters: means μ_k and covariances Σ_k. As the Gaussian components model the original data points given by the MFCCs, μ_k is of dimensionality d, which equals the number of MFCCs considered in the feature extraction process. Likewise, the covariance matrix Σ_k is of dimensionality $d \times d$ and its elements $\sigma_{ii} = \sigma_i^2$ contain the variance of dimension i, whereas elements $\sigma_{ij\ i \neq j}$ contain the covariance of dimensions i and j. Furthermore, each Gaussian $\mathcal{N}(\mu_k, \Sigma_k)$ in the mixture model is assigned a weight w_k that indicates how many data items, i.e., MFCC vectors, are represented by it. These weights must sum up to 1, i.e., $\sum_{k=1}^{K} w_k = 1$, since they indicate the probability that a randomly chosen data item belongs to component k.

To illustrate how a GMM describes or summarizes a set of several thousand MFCC vectors extracted from a music piece, Fig. 3.8 depicts GMMs learned from the first two MFCCs for different pieces in the toy music data set. As in Sect. 3.2.3, for illustration purposes, we only consider the first and second coefficients and fit a two-dimensional GMM with three Gaussians to the MFCC vectors. The original MFCC vectors are shown as cyan points. The center or mean μ_k of each Gaussian is depicted as a red dot, whose size corresponds to the mixture weight w_k. The covariances Σ_k are visualized as ellipsoids drawn at a distance of one standard deviation from the respective mean, i.e., the area within the ellipsoid covers approximately 39 % of the total probability mass.

In order to create a GMM for a given set of MFCC vectors, we first have to determine the number of components K. It has been shown that $K = 3$ seems a suitable number in case of MFCC inputs [12, 283]. The next step is to fit the mixture model to the data items, which is performed by a technique called maximum likelihood estimation (MLE). The purpose of MLE is to estimate the parameters of a statistical model, a GMM in our case, by maximizing the probability of the data items under the model. In our case, it hence tries to determine parameters w_k, μ_k, and Σ_k that best describe the input data, meaning that the GMM corresponding to these parameters is the most likely to explain the data. In practice, MLE can be realized by the *expectation maximization* (EM) algorithm, which iteratively performs two steps: (1) assessing the quality of fit of the model to the data ("expectation" step) and (2) adapting the parameters of the GMM to improve the fit ("maximization" step). EM can thus be thought of as a probabilistic version of k-means clustering, which is an iterative process in which each cluster center is moved towards the centroid of its nearest data points. However, in GMM-EM it is not only the centers or means but also the covariances that are adapted iteratively. For more details, please consider, for instance, Bishop's book on pattern recognition [38].

To give some real-world examples in our music domain, Fig. 3.7 illustrates GMMs with three components that have been fitted via the EM algorithm to 12-dimensional MFCC vectors. The MFCCs themselves were computed on the 30-second snippets of our toy music data set. The index of the MFCC is depicted on the x-axis, while the actual corresponding MFCC values are indicated on the y-axis. The means for each Gaussian component are illustrated as a black curve over the MFCC indices, while the variances are visualized by shades of turquoise (larger) and blue (smaller). As can be seen in the figure, for some pieces or genres, all Gaussian components have very similar parameters. This is particularly the case for the Heavy Metal and Pop songs.

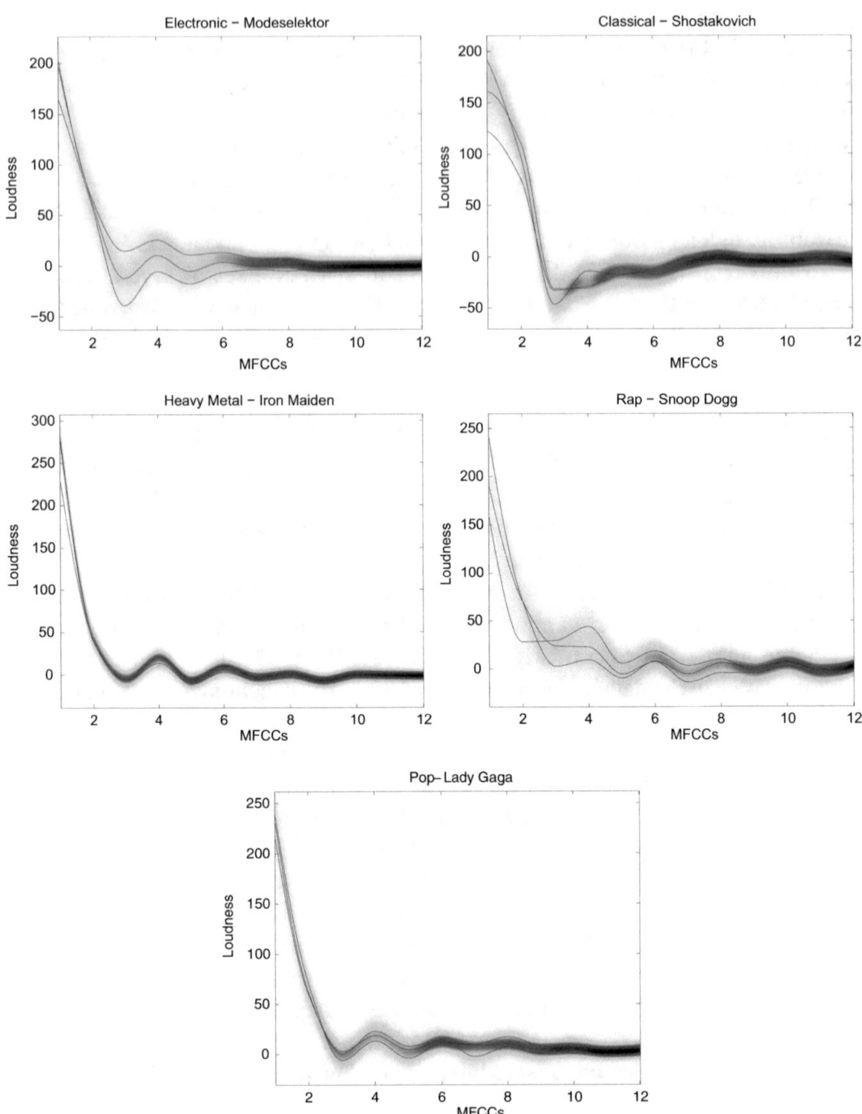

Fig. 3.7 Visualization of GMMs fitted to MFCC vectors of selected songs in the toy music data set

Equipped with the techniques to represent each music piece in a collection as a GMM that aggregates the piece's frame-level features, we can now compute similarities between two song's GMMs. Two frequently employed methods are Earth Mover's Distance (EMD) and Markov chain Monte Carlo sampling. Regarding the former, consider again Fig. 3.8 and two songs between which we want to estimate similarity. EMD then aims to transform one song's distribution P into the other

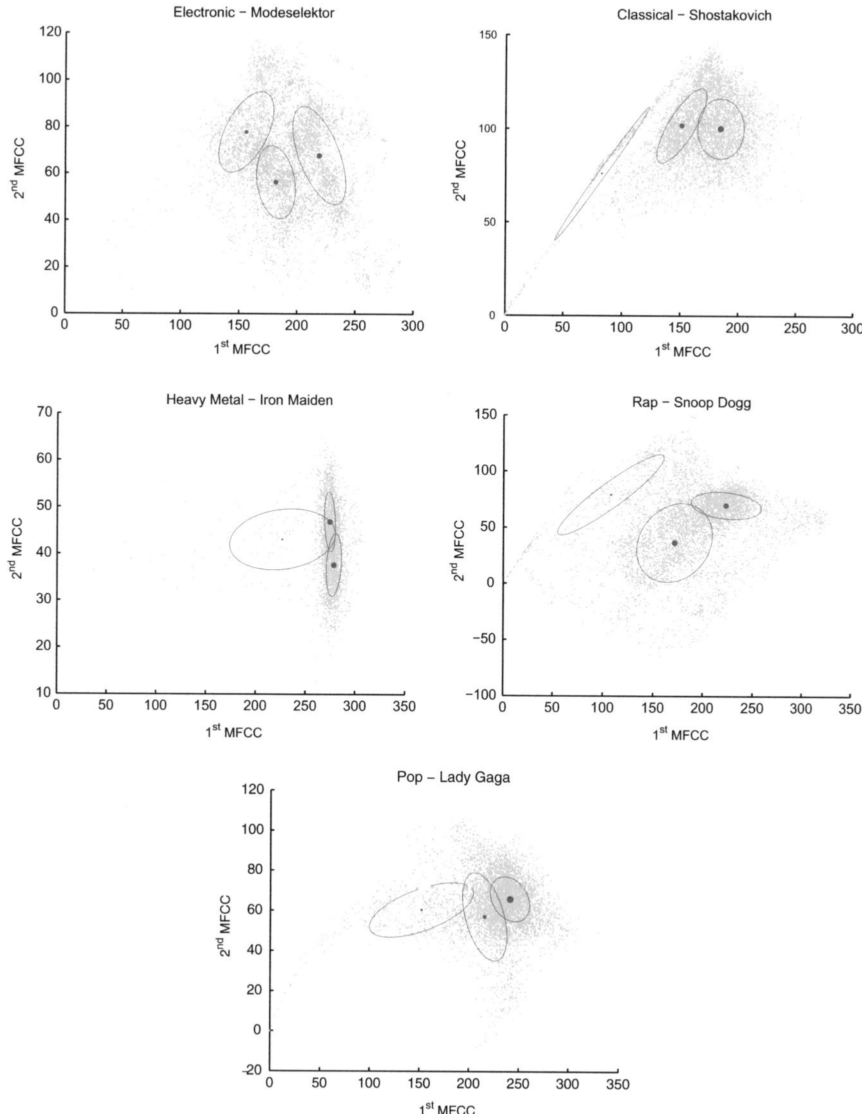

Fig. 3.8 Visualization of three-component GMMs fitted to the top two MFCCs of selected songs in the toy music data set

ones, Q. To this end, EMD defines an optimization problem that can be thought of as minimizing the amount of "probability mass" that is required to transform the GMM P into the GMM Q. An alternative, less artificial interpretation is a transportation problem, in which P represents supply (or a number m of manufacturers) and Q demand (or a number n of customers). Given distances in an $m \times n$ distance matrix D between the individual components of P and Q, for instance, road distances between each pair of manufacturer and customer, the rationale of EMD is then to find the minimum flow F that is required to transport all goods provided by the m suppliers to the n customers. Another way of expressing the problem is to interpret P as representation of piles of earth and Q as representation of holes. We then want to find the minimum amount of work required to fill the holes with earth.

Formally, the optimization problem is given in Eq. (3.7), where F are the flows to be determined and D is the distance matrix in which d_{ij} represents the distance between the ith component of P and the jth component of Q. The Earth Mover's Distance between P and Q is then computed as shown in Eq. (3.8), which normalizes the distance-weighted flows by the total flow.

$$\min_F \sum_{i=1}^{m} \sum_{j=1}^{n} f_{ij} \cdot d_{ij} \tag{3.7}$$

$$d_{EMD}(P,Q) = \frac{\sum_{i=1}^{m} \sum_{j=1}^{n} f_{ij} \cdot d_{ij}}{\sum_{i=1}^{m} \sum_{j=1}^{n} f_{ij}} \tag{3.8}$$

To fully specify the optimization problem, we still need to formally define P and Q as well as some constraints that ensure determination of the optimization algorithm. Concerning the distributions P and Q, each of them is defined by a signature containing each component's centroids and weights, more formally, $P = \left[(p_1, w_{p_1}), \ldots, (p_m, w_{p_m})\right]$, where p_i represents the centroids, or means in the case of GMMs, and w_{p_i} denotes the corresponding weight. Again, the weights of each distribution's components must sum up to 1, which enables the direct use of the GMM mixture weights for w_{p_i}. The signature of Q is defined analogously.

Given the signatures P and Q, the optimization problem requires four additional constraints. The first constraint shown in Eq. (3.9) ensures that all flows are positive, meaning that no probability mass is moved in the wrong direction. Only movements from P to Q are allowed. Equation (3.10) ensures that no more probability mass than available in each of P's components is moved to components in Q. Analogously, Eq. (3.11) ensures that no more earth than each target component in Q can accommodate is transported from P. Finally, Eq. (3.12) forces the maximum amount of mass in P to be moved to Q but no more than the maximum supply provided or demand requested by P and Q, respectively.

$$f_{ij} \geq 0 \qquad 1 \leq i \leq m,\ 1 \leq j \leq n \tag{3.9}$$

$$\sum_{j=1}^{n} f_{ij} \leq w_{p_i} \qquad 1 \leq i \leq m \tag{3.10}$$

$$\sum_{i=1}^{m} f_{ij} \leq w_{q_j} \qquad 1 \leq i \leq n \tag{3.11}$$

$$\sum_{i=1}^{m} \sum_{j=1}^{n} f_{ij} = \min\left(\sum_{i=1}^{m} w_{p_i}, \sum_{i=1}^{n} w_{p_j}\right) \tag{3.12}$$

To solve this optimization problem, there are highly efficient linear programming methods. Another advantage of EMD is that it is a metric, provided the underlying distance matrix D is metric and the summed weights of both P and Q are identical. More details on the EMD can be found in [386].

Another popular method to estimate similarity between two song's GMMs P and Q is Markov chain Monte Carlo sampling. The idea is to sample points from P and compute the likelihood that these points were in fact sampled from Q. Obviously, the higher this likelihood, the more similar the two songs. However, sampling a large number of points from P and computing the likelihood of the samples given Q does not yield the same result as performing the same computation with switched roles of P and Q, without loss of generality. This means that the result is no metric and thus needs to be symmetrized and normalized, as proposed, for instance, by Aucouturier and Pachet [15] via Eq. (3.13), where s denotes the number of drawn samples and $\mathscr{S}_i^P$ represents the ith sample drawn from distribution P.

$$d_{MC}(P,Q) = \sum_{i=1}^{s} \log p(\mathscr{S}_i^P|P) + \sum_{i=1}^{s} \log p(\mathscr{S}_i^Q|Q) - \sum_{i=1}^{s} \log p(\mathscr{S}_i^P|Q) - \sum_{i=1}^{s} \log p(\mathscr{S}_i^Q|P) \tag{3.13}$$

3.2.5 *Single Gaussian Model*

Other work on music audio feature modeling for similarity has shown that aggregating the MFCC vectors of each song via a single Gaussian may work almost as well as using a GMM [343]. This approach was first proposed by Mandel and Ellis [293]. When employing such a single Gaussian model, an analytical solution can be computed with the Kullback–Leibler divergence (KL divergence). Doing so decreases computational complexity by several magnitudes, in comparison to GMM-based similarity computations. Given two distributions P and Q, the KL divergence is defined as shown in Eq. (3.14). It can be interpreted as an estimate of how well distribution P can be represented using distribution Q as codebook.

$$KL_{(P||Q)} = \int p(x) \cdot \log \frac{p(x)}{q(x)}\, dx \tag{3.14}$$

In our case, the single Gaussian models for two pieces P and Q to compare are given by their means μ_P (μ_Q) and covariances Σ_P (Σ_Q). The KL divergence has then a closed-form solution as shown by Penny [357], which is given in Eq. (3.14), where $|\Sigma_P|$ denotes the determinant of Σ_P; $Tr(\cdot)$ denotes the trace, i.e., the sum of the diagonal of a matrix; and d is the dimensionality of the input data, i.e., the number of MFCCs considered.

$$KL_{(P||Q)} = \frac{1}{2}\left[\log\frac{|\Sigma_P|}{|\Sigma_Q|} + Tr\left(\Sigma_P^{-1}\Sigma_Q\right) + \left(\mu_P - \mu_Q\right)^{\mathsf{T}} \Sigma_P^{-1}\left(\mu_Q - \mu_P\right) - d\right] \tag{3.15}$$

In order to be used as a metric similarity measure between music pieces, the results of the KL divergence need to be symmetrized, for instance, as shown in Eq. (3.16).

$$d_{KL}(P,Q) = \frac{1}{2}\left(KL_{(P||Q)} + KL_{(Q||P)}\right) \tag{3.16}$$

Similar to Fig. 3.5, Fig. 3.9 visualizes the similarity matrix computed by applying the single Gaussian model to our toy data set. Classical music is pretty well separated from music of other genres, which is underlined by an intra-/inter-genre similarity ratio of almost 3. In contrast, no strong separation between other genres can be made out, although all intra-/inter-genre similarity rations are well above 1. As for the average intra-/inter-genre similarity ratios that describe the overall performance over all genres, the variant using the mean yields a value of 1.761; the variant using the median yields 1.464. The correction for outliers (the Classical genre) can be nicely made out when using the median variant.

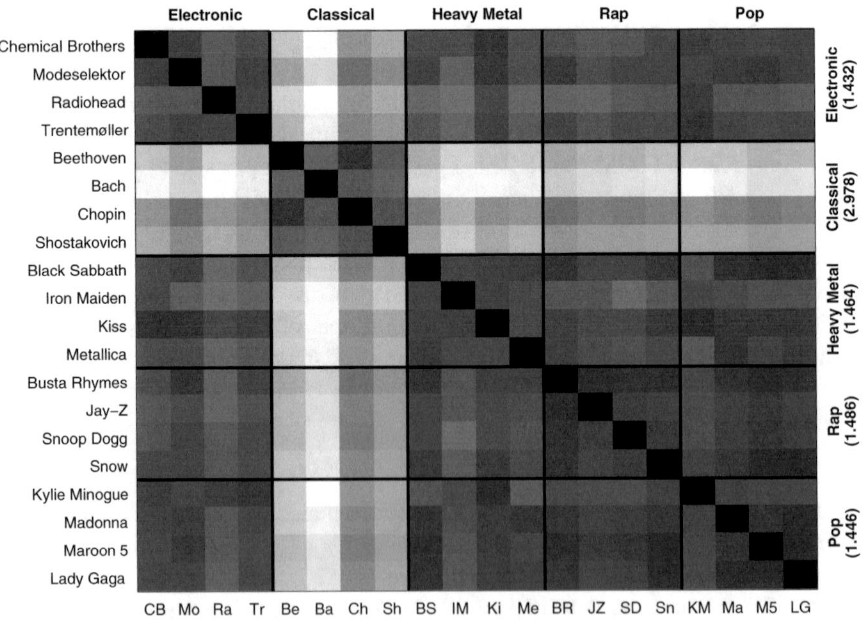

Fig. 3.9 Visualization of similarity matrix computed on symmetric KL-divergences using a single Gaussian model

3.3 Block-Level Features and Similarity

In contrast to frame-level features, where short segments of a music piece are considered one at a time and summarized via some statistical or probabilistic model, block-level features describe larger segments of a piece, typically a few seconds. Due to this property, block-level features can incorporate temporal aspects to some extent. Figure 3.10 illustrates the feature extraction process on the block level. Based on the spectrogram, blocks of fixed length are extracted and processed one at a time. The block width defines how many temporally ordered feature vectors from the STFT are in a block. Also a hop size is used to account for possible information loss due to windowing or similar methods. After having computed the feature vectors for each block, a global representation is created by aggregating the feature values along each dimension of the individual feature vectors via a summarization function, which is typical the median or another percentile. Please note that the summarization functions (different percentiles) reported in the following treatment of block-level features are the result of optimization on different data sets, performed by Seyerlehner [437, 439]. They might not generalize to arbitrary data sets.

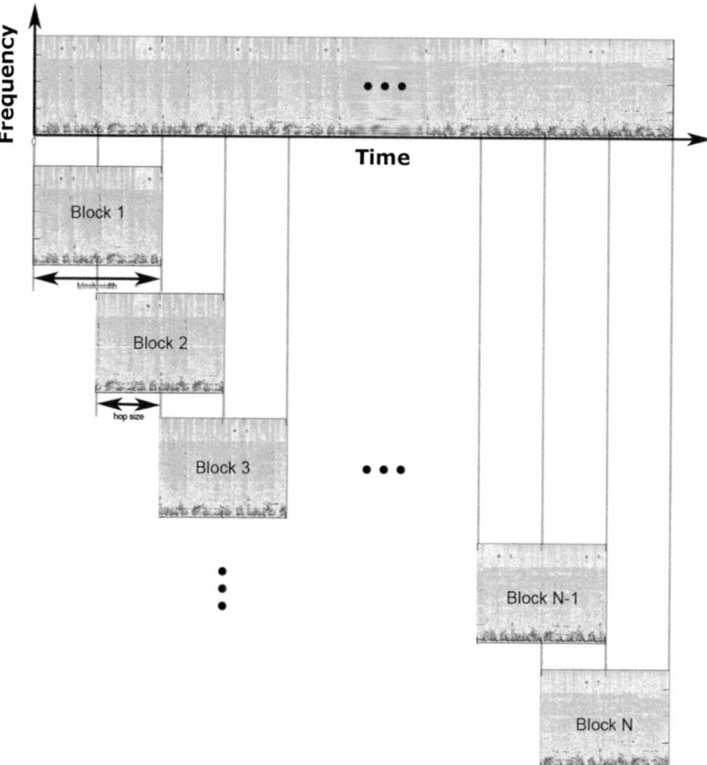

Fig. 3.10 Overview of a block-level feature extractor

3.3.1 Fluctuation Pattern

An early example of a block-level feature is the *fluctuation pattern (FP)*, proposed by Pampalk et al. [346]. Figure 3.11 shows an overview of the feature extraction process, which is comprised of ten steps. As you can see, the input music track is first cut into 6-second segments, before these snippets are fed into the actual feature extraction process. Next, the FFT is computed to obtain the power spectrum. In the subsequent steps 3–6, several psychoacoustic preprocessing techniques are applied, which have already been detailed in Sect. 3.1. Step 7 is comprised of applying a second FFT over time to the critical bands in order to reveal information about the frequency structure. This transforms the time–frequency band representation into a periodicity–frequency band representation. These periodicities show frequency patterns repeated over time, which are also referred to as fluctuations. They reveal which frequencies reoccur at certain intervals within the 6-second segment under consideration. The result is a 20×60 matrix that holds energy values for 20 critical

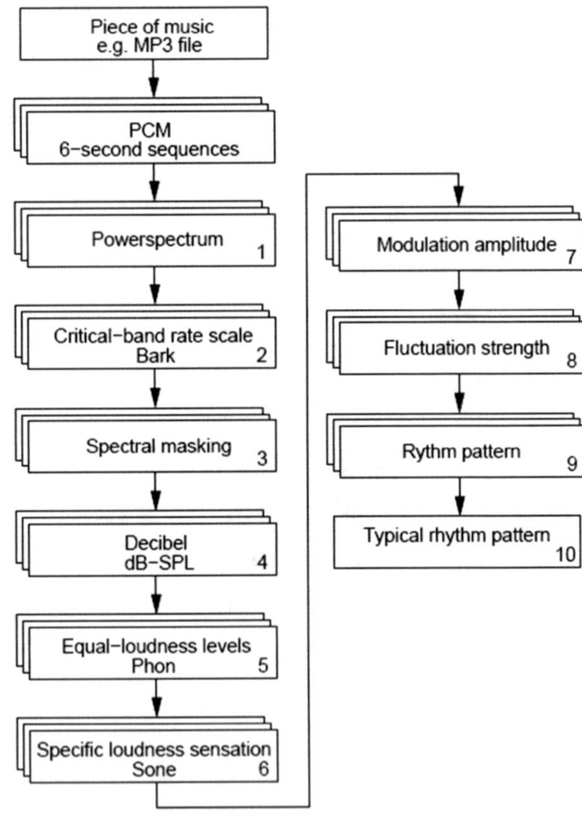

Fig. 3.11 Pipeline of the extraction process to obtain fluctuation patterns

frequency bands over 60 bins of periodicities, ranging from 0 to 10 Hz, i.e., 0 to 600 beats per minute (bpm). In step 8, a psychoacoustic model of fluctuation strength is applied, since perception of fluctuations depends on their periodicity. For instance, reoccurring beats at 4 Hz are perceived most intensely. Performing these steps eventually yields a rhythm pattern for the music segment under consideration. In order to create a global feature model, Pampalk et al. propose to aggregate the rhythm patterns of all segments in a track by computing the median over each dimension in the 1200-dimensional fluctuation pattern matrix for all segments. Similarities between tracks are finally computed via L_1 or L_2-norm distance between the corresponding vector representations of the fluctuation patterns. To obtain this representation, the fluctuation pattern matrix is flattened, similar to the approach described in Sect. 3.2.2 to create a statistical summary from the mean vector and covariance matrix.

To illustrate what fluctuation patterns look like, Fig. 3.12 shows visualizations of the fluctuation pattern matrix for selected pieces from the toy music data set. Please note that we display 24 critical bands, measured in Bark, along the y-axis, while Pampalk et al. use 20. This is due to the fact that we employ a sampling frequency of 44,100 Hz, whereas Pampalk et al. use music sampled at only 11,025 Hz. Along the x-axis, we depict periodicities between 0 and 300 beats per minute. The figure shows that fluctuation patterns are quite different for different music styles, both in energy values (cf. color bars on the right-hand side of the figure) and actual pattern. The Classical piano piece by Shostakovich shows a clearly horizontal characteristic, i.e., activations over a wide range of periodicities, but in a limited number of frequency bands. The main energy is concentrated between Bark bands 7 and 12 and periodicities in the range $[30, 100]$ bpm. The fluctuation patterns of the Electronic and Pop songs, in contrast, reveal a rather vertical characteristic, which can be explained by the excessive use of synthetic elements such as drum loops and similar. In the Lady Gaga song, the main energy is concentrated in a short periodicity range between 100 and 120 bpm, whereas the Modeselektor song shows a slow beat at low frequencies (from 25 to 40 bpm and below 5 Bark) and activations in the upper bands at periodicities around 130 bpm. The fluctuation pattern of the Heavy Metal piece nicely shows the effects of excessive use of distorted guitars and drums, covering a very wide range in both frequency and periodicity. Please also note the different color bars for different pieces, which correspond to the mapping of colors to energy values. The energy in the Pop and Rap songs is highly concentrated at the main beat, around 110 bpm and 80 bpm, respectively. In contrast, energies in the other pieces are more spread out and less focused on a few "hot spots."

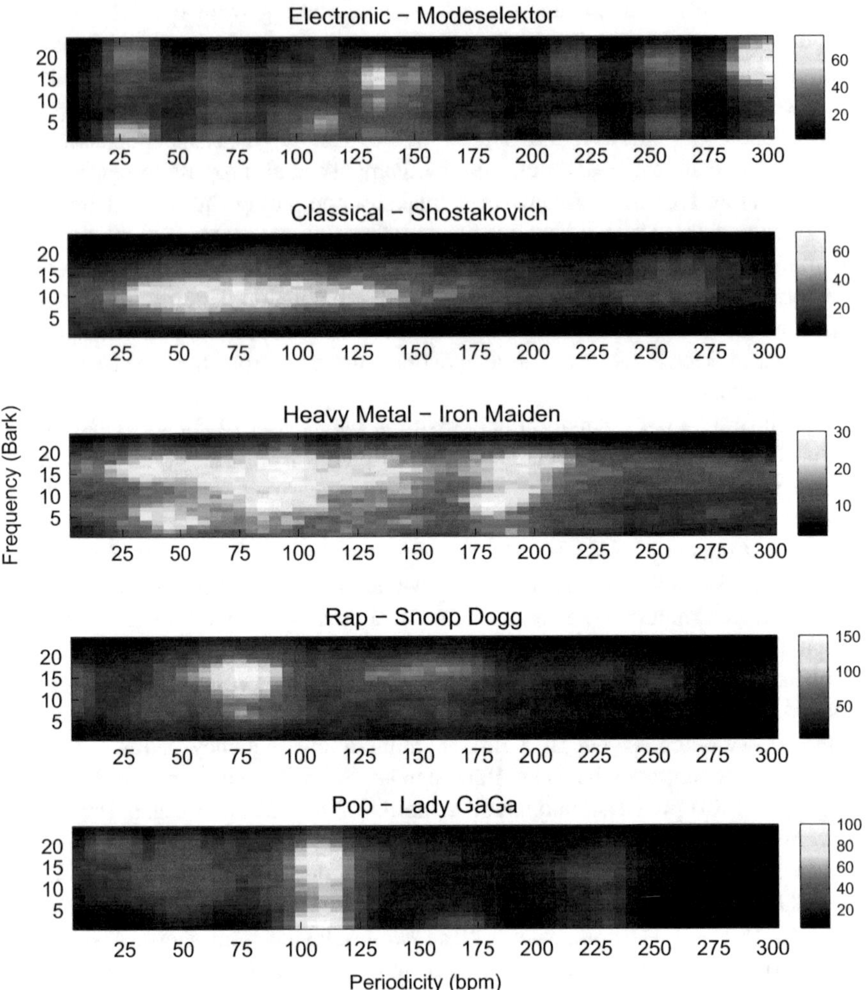

Fig. 3.12 Fluctuation patterns for selected pieces from the toy music data set

3.3.2 *Logarithmic Fluctuation Pattern*

Variants of the fluctuation pattern have been explored over the years, for instance, using other frequency scales, different psychoacoustic models, or onset detection techniques, among others by Pohle et al. [366]. One example is the *logarithmic fluctuation pattern*, which is part of the *block-level feature framework*, proposed by Seyerlehner et al. [439, 440]. This block-level framework includes several individual feature extractors, some of which will be explained in the following. We selected this framework because of its high performance in audio music similarity tasks (cf. Sect. 1.5), although it has also been successfully applied in classification tasks, for instance, in genre or composer classification [441], and in more atypical tasks such as localizing the origin of music [406].

The logarithmic fluctuation pattern and the block-level framework in general use the Cent scale instead of Bark critical bands to represent frequency. The Cent scale is a logarithmic scale used to measure pitch intervals. Since the same notes in adjacent octaves are exponentially spaced in the Hertz scale, the Cent scale uses a logarithmic transformation function to linearly space them. By doing so, the notes C2, C3, C4, and so on are equidistant. A distance of one octave covers 1200 Cents. Seyerlehner et al. convert Hertz into Cent using Eq. (3.17), where c is the Cent value and f is the frequency in Hertz. Unlike the Bark and Mel scales, the Cent scale reflects musically meaningful concepts, i.e., notes.

$$c = 1200 \cdot \log_2 \frac{f}{440 \cdot \left(\sqrt[1200]{2}\right)^{-5700}} \tag{3.17}$$

While the original fluctuation pattern proposed by Pampalk et al. employs linear scaling of periodicities, the logarithmic version uses a logarithmic scale to represent periodicities. This logarithmic scaling allows the representation of periodicities in a more tempo invariant way. To this end, a log filter bank is applied to the periodicity bins in each block, which results in a representation where periodicity in Hertz is doubled every 6 bins [366].

Figure 3.13 shows visualizations of the logarithmic fluctuation pattern features for selected pieces of our toy music data set, one piece for each genre. In the block-level feature framework, 98 frequency bands and 37 periodicity bins are used. The frequency bands are spaced 100 Cents apart, i.e., each band covers a semitone. What immediately sticks out is the much clearer separation between periodicity bins, compared to the original fluctuation patterns shown in Fig. 3.12. In fact, the energy activations blur to a much smaller extent to neighboring periodicity bins. What can still be seen is the high concentration of energy in only very few bins for the Electronic and Pop songs; cf. Fig. 3.13a, e. For the Classical, Metal, and Rap pieces, in contrast, we recognize several regions of high intensity.

3.3.3 Spectral Pattern

The *spectral pattern (SP)* is one of the simplest features described in the block-level feature framework. It is computed from the spectrograms of all blocks in the piece of interest and relates to timbre. A block is defined as ten consecutive frames and a hop size of five blocks is used, which shifts the window of investigation by half of the window size from block to block. Frequencies are measured again on the Cent scale, binned into 98 bands. To obtain a time-invariant representation, each frequency band is sorted within the block under consideration. This is performed for all blocks in

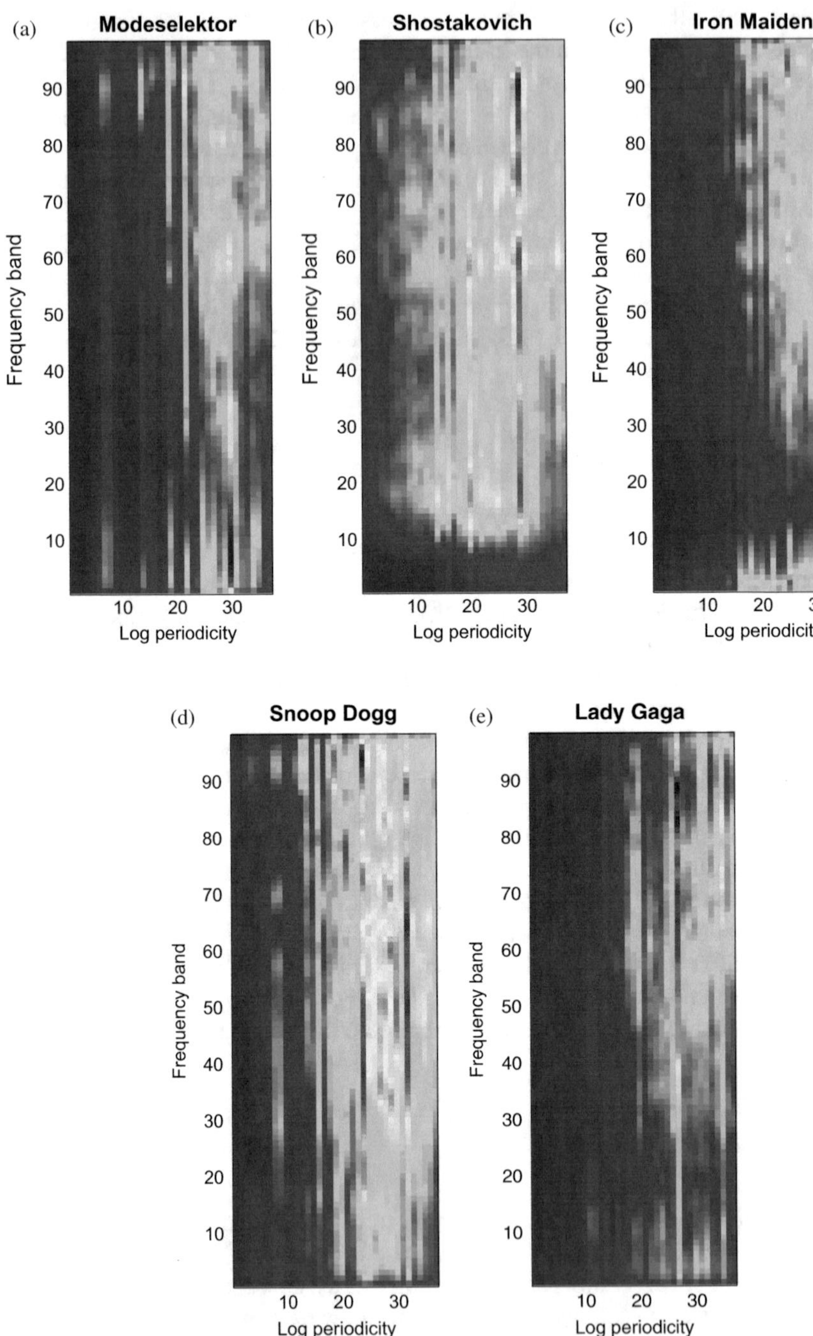

Fig. 3.13 Logarithmic fluctuation patterns for selected pieces from the toy music data set. (**a**) Electronic. (**b**) Classical. (**c**) Metal. (**d**) Rap. (**e**) Pop

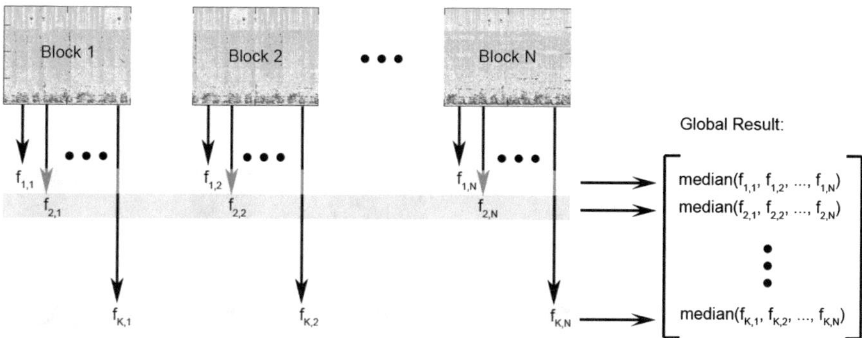

Fig. 3.14 Overview of how to compute a global feature representation from individual blocks via summarization function

the piece. Doing so yields for each block a 98×10 matrix. The spectral pattern of a piece is then calculated as the 0.9-percentile over each dimension in the set of 98×10 matrices for all blocks. The 0.9-percentile is thus used as summarization function, as illustrated in Fig. 3.14, where the median is shown as summarization function, which is the function used for the original fluctuation patterns; cf. Sect. 3.3.1.

Figure 3.15 depicts visualizations of the spectral pattern features for selected pieces of our toy music data set, one piece for each genre. Please note that energy values are sorted within every frequency band. We can see that the Metal piece by Iron Maiden shows high activations over almost all frequency bands and frames, i.e., high energy levels which are invariant of time. This is a typical pattern for music featuring distorted guitars. It can also be seen that the Classical piece by Shostakovich shows a quite different characteristic from the others, as the high activations concentrate in a limited number of bands, between 45 and 55, approximately corresponding to the frequency range [700, 1600] Hz. Also very low frequencies below 40 Hz are missing in the piano piece, which is no surprise. The Electronic, Pop, and Rap pieces show a rather similar overall spectral structure, a reason for which might be a similar production style. However, subtle differences can be made out. For instance, comparing the low frequency bands up to about 25 (approximately 130 Hz) of Lady Gaga to the ones of Snoop Dogg shows that the former has constantly high activations (red and orange shades over almost the entire frame range), whereas high activations around the frequency bands close to 20 can only be found in half of the frames for the latter.

Variants of the spectral pattern are occasionally used to describe particular aspects of the spectrum. For instance, Seyerlehner et al. [440] define *delta spectral pattern* and *variance delta spectral pattern*. The delta spectral pattern emphasizes note onsets, which can be detected by considering changes in the spectrogram over time. A block consists of 25 frames in this case, and a hop size of five frames is used. The pattern is obtained by first computing the difference between the original Cent spectrum and the spectrum which is delayed by three frames. Then again each frequency band of a block is sorted. The differences in the spectra to be compared

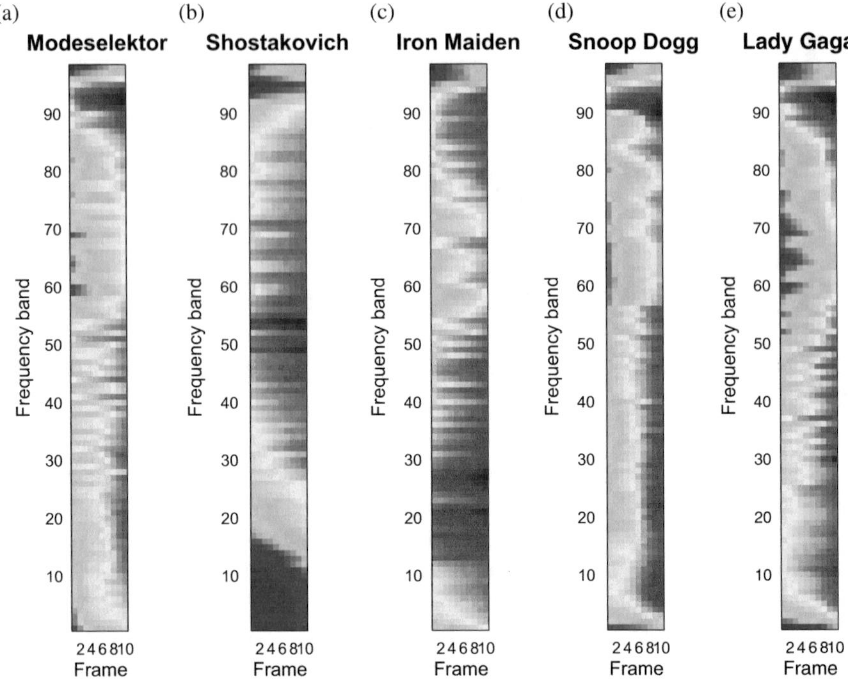

Fig. 3.15 Spectral patterns for selected pieces from the toy music data set. (**a**) Electronic. (**b**) Classical. (**c**) Metal. (**d**) Rap. (**e**) Pop

can be positive or negative. Positive values indicate note onsets, which we are interested in. As the values within each band are sorted, the leftmost columns of the 98×25 matrix are all negative. To save space and computing power, only the rightmost 14 frames with largest values are taken into account. The global representation for a piece, the 0.9-percentile is computed for each dimension in the 98×16 feature matrices of the individual blocks. Figure 3.16 shows typical patterns for music pieces from different genres.

The feature extractor for variance delta spectral pattern applies almost the same approach as the extractor for the delta spectral pattern. The only difference is that variance is used as summarization function, instead of the 0.9-percentile. This choice is supposed to capture variations in onset strength over time.

3.3.4 Correlation Pattern

The *correlation pattern (CP)* is another feature defined in the block-level framework. It was designed to capture harmonic relations of frequency bands in the presence of sustained tones. To this end, pairwise correlations between frequency

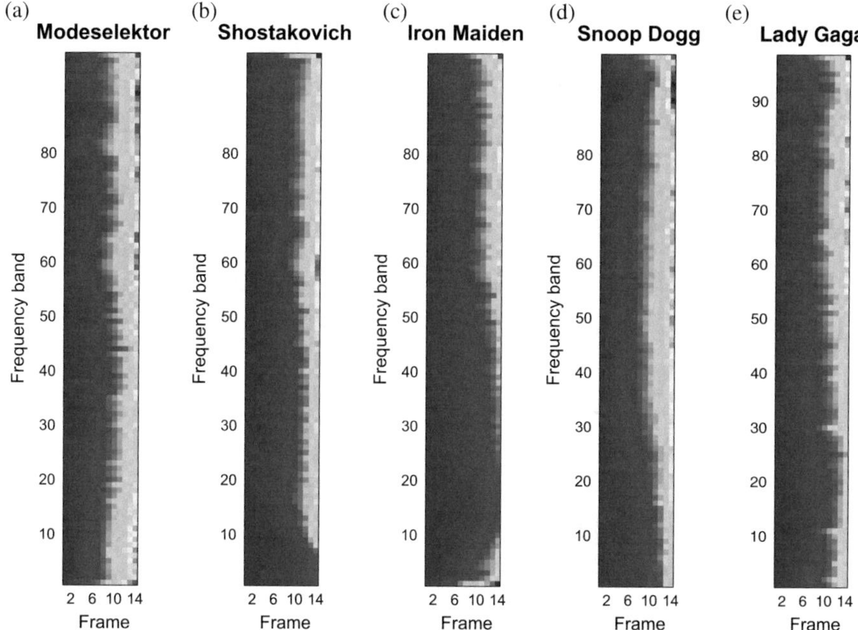

Fig. 3.16 Delta spectral patterns for selected pieces from the toy music data set. (**a**) Electronic. (**b**) Classical. (**c**) Metal. (**d**) Rap. (**e**) Pop

bands are computed. More precisely, the number of frequency bands is first reduced from 98 to 52 by reducing the frequency resolution. This number was found by Seyerlehner to represent a good trade-off between computing time and quality of results [437, 439]. Using a block size of 256 frames and a hop size of 128 frames, Pearson's correlation coefficient between the 256-dimensional feature vectors given by the block's spectrogram is calculated. This linear correlation coefficient is shown in Eq. (3.18), where f_1 and f_2 correspond to the 256-dimensional feature vectors of the two frequency bands to compare, and $cov(f_1,f_2)$ denotes the covariance between the two.

$$r_{f_1,f_2} = \frac{cov(f_1,f_2)}{\sqrt{cov(f_1,f_2) \cdot cov(f_1,f_2)}} \tag{3.18}$$

The correlation is computed for all pairs of frequency bands, which results in a 52×52 matrix for each block. All matrices of a piece's blocks are then aggregated using the median as summarization function.

Figure 3.17 shows visualizations of the correlation pattern features for selected pieces of our toy music data set, one piece for each genre. We can make out big differences in the patterns of different pieces. The correlation pattern of the Classical piece reveals clear harmonies, illustrated by the few orange and yellow

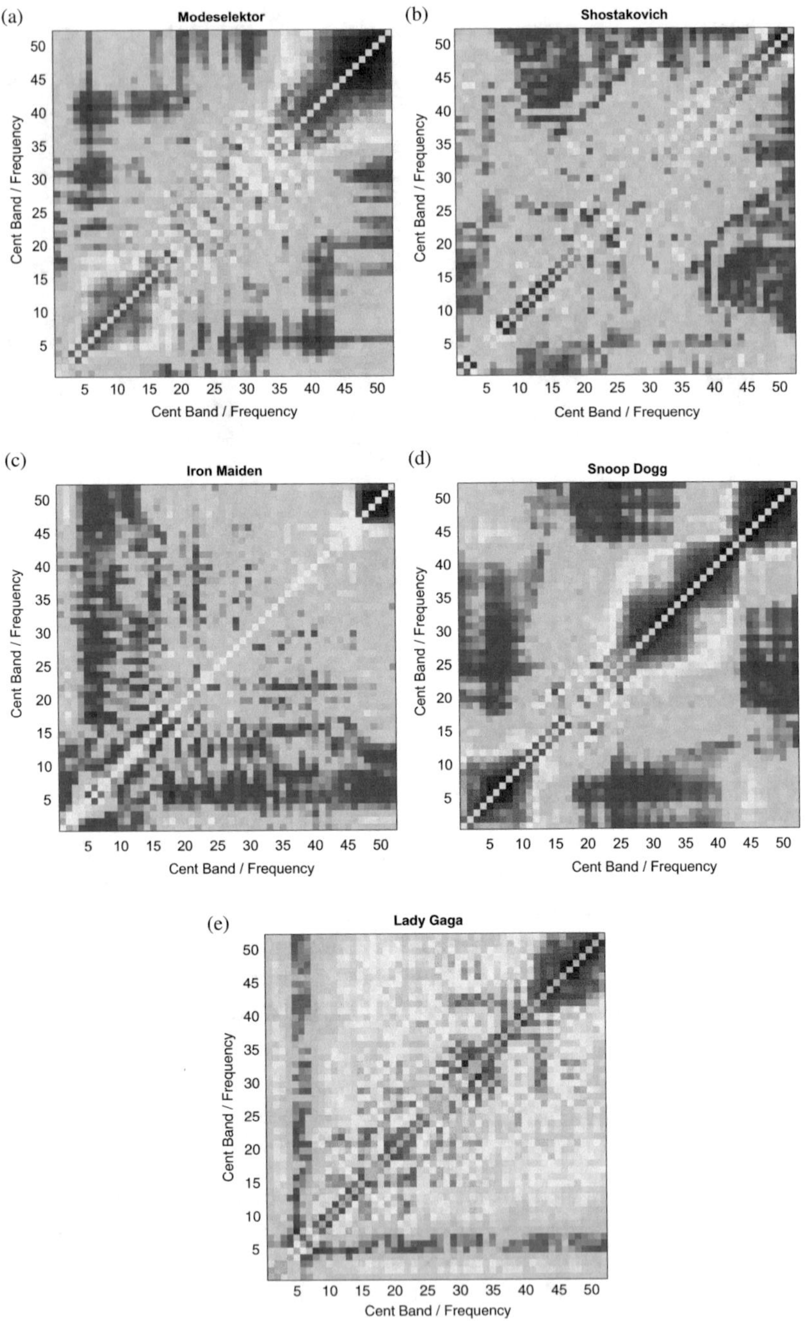

Fig. 3.17 Correlation patterns for selected pieces from the toy music data set. (**a**) Electronic. (**b**) Classical. (**c**) Metal. (**d**) Rap. (**e**) Pop

squares outside of the diagonal. In addition to reflecting harmonic structure, the correlation pattern is capable of identifying rhythmic relations over frequencies. If, for instance, hi-hat and bass drum are always hit simultaneously, this translates to a high correlation between the corresponding low and high frequency bands, which can nicely be seen in the Metal piece. For the Electronic and even more the Rap piece, larger square areas of high correlations in nearby frequency bands reflect high interrelations between similar notes.

3.3.5 Similarity in the Block-Level Feature Framework

Now that we have introduced most features in the block-level framework, an obvious question is how to combine and use them to estimate similarity between music pieces. As already illustrated in Fig. 3.14, we can create a global representation of each block-level feature by applying a summarization function to the patterns of all blocks in the piece under consideration. Flattening the resulting matrix to a vector for all patterns described in the framework, we eventually obtain a high-dimensional feature vector, whose dimensionality varies with the pattern. For instance, a logarithmic fluctuation pattern is of dimensionality 3626 ($= 98 \times 37$). We can then compute a similarity estimate for each pattern and pair of pieces in the collection, for instance, via L_1-norm distances. Doing so results in N similarity matrices, where N is the number of patterns considered. Figure 3.18 illustrates how these N matrices are combined to eventually obtain a single similarity matrix. As the distances in the various feature patterns have different scales, they first need to be normalized. To this end, Seyerlehner et al. [439] propose to perform distance space normalization: Each distance in a distance matrix is normalized via Gaussian normalization over its corresponding row and column elements. That is, given a $p \times p$ distance matrix D for p music pieces, mean and standard deviation are computed over the elements $\{d_{m,i=1\ldots p}, d_{i=1\ldots p,n}\}$, denoted $\mu_{m,n}$ and $\sigma_{m,n}$, respectively. The normalized distance $|d_{m,n}|$ is in turn computed as $|d_{m,n}| = \frac{d_{m,n}-\mu_{m,n}}{\sigma_{m,n}}$, i.e., by subtracting the mean and dividing by the standard deviation. This distance space normalization strategy operates locally and has been shown to improve classification accuracy in music similarity tasks [366]. Distance space normalization for all patterns in the framework yields N normalized distance matrices $D_{1\ldots N}$, which can now be summed to yield an overall distance matrix; cf. Fig. 3.18. Each of the N components can also be given a weight $w_{1\ldots N}$, which makes the overall distance matrix a linear combination of the individual patterns' distance matrices. As a final step, the resulting matrix is once more normalized using the described distance space normalization.

To investigate the performance of the block-level feature similarities, Fig. 3.19 shows a visualization of the similarity matrix computed on the toy music data set. A noticeable separation between genres can already be made out visually, in particular compared to Figs. 3.5 and 3.9. Looking at the intra-/inter-genre similarity ratios supports this impression. Only the separation between genres Electronic and Pop

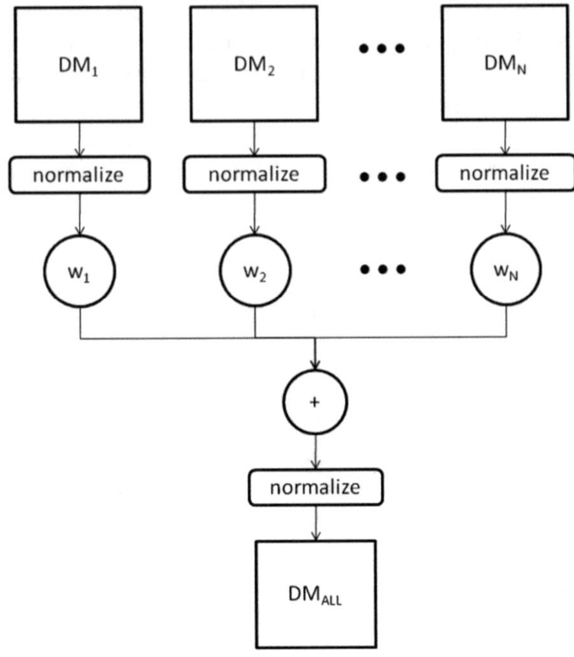

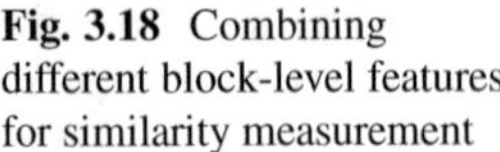
Fig. 3.18 Combining different block-level features for similarity measurement

seems to perform slightly inferior. On the other hand, Classical and Rap music can easily be separated from other genres. The mean intra-/inter-genre similarity ratio is 1.708, while applying the median as aggregation function yields a value of 1.521.

3.4 Hubness and Distance Space Normalization

First described by Aucouturier [11] and further investigated by Pohle et al. [364], using the described audio similarity measures, some particular pieces, so-called hubs, are frequently similar (i.e., have a small distance) to many other pieces in the collection without actually sounding similar. On the other side, some pieces are never similar to others.

While originally these effects were blamed on the features used for audio representation and the similarity measures used, this is in fact a problem that occurs when calculating distances in high-dimensional spaces. More precisely, two phenomena emerge as dimensionality increases, as pointed out by Nanopolous et al. [328]: First, pairwise distances between all data points tend to become very similar, known as *distance concentration*. Second, as a consequence of this, hubs, i.e., data points which have a small distance (or high similarity) to disproportionally many other data points, appear. Both effects are considered to be a new aspect of the *curse of dimensionality* [372, 428, 474]. A number of publications have investigated the influence of hubs on several machine learning tasks, such as classifi-

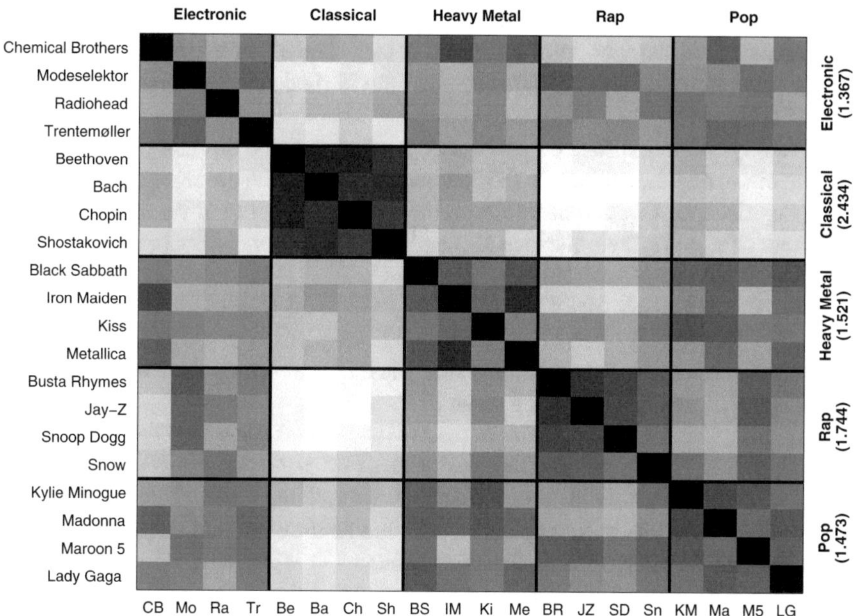

Fig. 3.19 Visualization of similarity matrix computed via Manhattan distance on BLF

cation [372], clustering [477], content-based item retrieval [129], and recommender systems [227, 328], and have shown that (1) they have a negative impact and (2) the mentioned tasks benefit from methods that decrease their influence.

One approach to deal with this is a simple rank-based correction called *proximity verification* (PV) that replaces the distances in the distance matrix D with a rank-based measure [364]. The entries of each row of the distance matrix D are sorted in ascending order, and each original entry of the row is replaced with its rank. The resulting distance matrix (denoted D_1 here) is transformed into the final matrix by adding the transpose (resulting in a symmetric matrix): $D_{PV} := D_1 + D_1'$. This matrix has a better distribution of distances than the original distance matrix, reduces the "hubness," and seems to be better suited as input for subsequent steps such as clustering. As a consequence of this modification, all subsequent steps can only utilize the ranking information of audio similarity, i.e., audio similarity information can only be used to the extent of whether a piece is most similar, second most similar, third most similar, etc., to a piece under consideration and not to which numerical extent the two pieces are considered to be similar since the resulting values are not grounded in any distance space.

Another approach to distance space normalization that builds upon the idea of proximity verification is *mutual proximity* (MP) [428]. MP has been shown to improve nearest neighbor retrieval in a number of domains, including audio retrieval where it originates from. Its general idea is to reinterpret the original distance space so that two objects sharing similar nearest neighbors are more closely tied to each other, while two objects with dissimilar neighborhoods are repelled from each other.

This is done by reinterpreting the distance of two objects as a *mutual proximity* (MP) in terms of their distribution of distances. It was shown that by using MP, hubs are effectively removed from the data set while the intrinsic dimensionality of the data stays the same.

To apply MP to a distance matrix, it is assumed that the distances $D_{x,i=1..N}$ from an object x to all other objects in the data set follow a certain probability distribution; thus, any distance $D_{x,y}$ can be reinterpreted as the probability of y being the nearest neighbor of x, given their distance $D_{x,y}$ and the probability distribution $P(X)$:

$$P(X > D_{x,y}) = 1 - P(X \leq D_{x,y}) = 1 - \mathscr{F}_x(D_{x,y}). \tag{3.19}$$

MP is then defined as the probability that y is the nearest neighbor of x given $P(X)$ and x is the nearest neighbor of y given $P(Y)$:

$$MP(D_{x,y}) = P(X > D_{x,y} \cap Y > D_{y,x}). \tag{3.20}$$

Due to the probabilistic foundation of MP, the resulting values range from 0 to 1 and can again be interpreted as similarity values.

Figure 3.20 shows the similarity matrix from the single Gaussian MFCC model with KL-divergence (cf. Fig. 3.9) after applying MP. The effects become clearly visible as artists generally get more similar to artists from the same genre. Using MP, Classical as a genre is now even perfectly separated from all other genres,

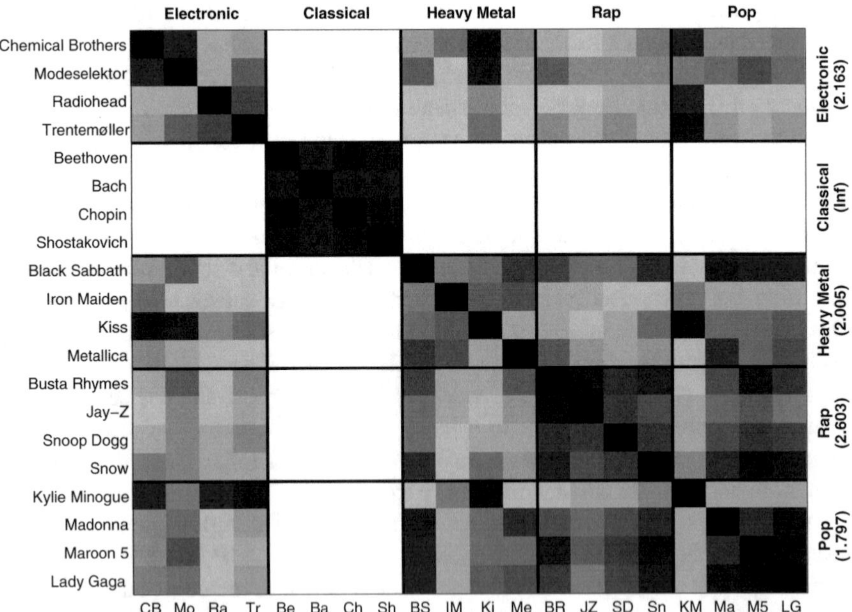

Fig. 3.20 Visualization of similarity matrix computed on symmetric KL-divergences using a single Gaussian model (cf. Fig. 3.9), post-processed by mutual proximity

leading to an infinite intra-/inter-genre similarity ratio (which also results in the mean similarity ratio being infinite). The median over all genres amounts to 2.162 (as compared to 1.464 without MP normalization). Emphasizing common neighborhoods has also a strong influence on Rap where artists are now more similar to each other and leads to the formation of a cross-genre cluster bridging Electronic artists, Kiss (Heavy Metal), and Kylie Minogue (Pop) which can be interpreted as a disco- and club-music-related effect.

Since MP is a general approach to normalize distances stemming from high-dimensional feature spaces, it can be applied to all methods discussed in this book, including those built on textual features (Sect. 6.1) as well as collaborative filtering (Sect. 8.4), and is known to have a positive effect on them. However, in the remainder of this book, we will not apply MP to the resulting distances, as we want to study the effects of the existing approaches without post-processing. Nonetheless, for all presented distance measures, the reader should keep in mind that investigation of hubs and, if applicable, their removal are important steps that can improve performance of retrieval systems substantially.

3.5 Summary

In this chapter, we first introduced basics in psychoacoustics, in particular, discussing important facts on the relationship between physical and perceptual aspects of audio signals. We presented different measures and models for sound intensity (decibel), perception of loudness (phon and sone), and frequency (Hertz, Bark, and Mel scales).

After this short introduction to psychoacoustics, we elaborated on mid-level features and categorized them into frame-level and block-level features. While the former are computed on individual frames (typically a few tens of milliseconds in length), the latter aggregate frames into blocks or segments (usually a few seconds in length) and are therefore capable to capture short-term temporal aspects of the signal to a certain extent.

We presented one of the most popular frame-level descriptor, Mel frequency cepstral coefficients (MFCCs), which is often used to model timbre. We elaborated on various ways to statistically model the MFCC vectors of an entire piece of music, from taking the means and (co)variances and interpreting them as feature vector to vector quantization to more sophisticated methods such as Gaussian mixture models (GMMs), which describe the piece by a certain number of Gaussians given by their parameters and mixture weights. Subsequently, we showed how similarity, one main theme of this book, can be computed from these models. In case of feature vector representations, Manhattan or Euclidean distance (or any other L_p-norm distance) can be calculated in a straightforward manner. If MFCCs are modeled by GMMs, in contrast, methods that compare probability distributions are required. As an example, we discussed Earth Mover's Distance (EMD) and Monte Carlo sampling. Finally, we presented a simple but frequently used probabilistic model to summarize MFCC features: the single Gaussian model in conjunction with the Kullback–Leibler divergence.

We proceeded with a discussion of the block-level feature (BLF) extraction framework, in which features are computed on sequences of frames of fixed length. The block-level features of a piece are then aggregated by statistical summarization (e.g., computing variance or percentiles over all dimensions), which yields a feature vector representation that describes the entire piece under consideration. Given these feature vectors of two music pieces, again L_p-norm distances can be used to compute similarities. As examples, we presented fluctuation pattern (FP), spectral pattern (SP), and correlation pattern (CP). While FP describes intensities over frequency bands and periodicities (beats per minute), hence relate to rhythm, SP derives timbre information from the spectrogram, in which frequency bands are sorted within each block. Harmonic aspects are to some extent captured by the CP feature, which encodes the correlations between all pairs of frequency bands.

Table 3.1 summarizes the overall intra-/inter-genre similarity ratios for selected approaches presented in this chapter. As expected, the simple statistical summarization approach performs worst. In contrast, the MFCC-based approach using a single Gaussian for aggregating MFCC vectors and the block-level features yield much better intra-/inter-genre similarity ratios. The MFCC approach works particularly well to tell Classical music apart from all other genres, while the block-level features approach has a more balanced distribution of intra-/inter-genre ratios over genres. This is reflected by the mean value being higher for the former and the median value being higher for the latter. In practice, it is common to combine different feature categories and approaches, not only audio features described here but also music context features, such as collaborative tags or term vectors extracted from web pages. Such contextual features are explained in Part II.

As of writing this book, the presented methods, in particular, the single Gaussian model and the BLF, represent the state of the art in similarity-based music retrieval, as demonstrated in the MIREX *Audio Music Similarity and Retrieval* tasks from 2009 to 2014. Despite their high performance in this task, it needs to be kept in mind that, while psychoacoustically motivated, they only cover specific aspects of acoustic properties of music. Furthermore, as discussed in Sect. 1.5.2, algorithms evaluated under the specific circumstances of the MIREX evaluation task have reached an upper bound. Therefore it is an open question whether these methods are really "as good as it gets" for timbre-based retrieval or if "innovation" in this area foremost depends on improved evaluation strategies.

Table 3.1 Average intra-/inter-genre similarity ratios for different approaches presented in this chapter

Approach	Figure	Mean	Median
Statistical summarization-L_2	3.5	1.386	1.390
MFCC-single Gaussian	3.9	1.761	1.464
BLF-L_1	3.19	1.708	1.521
MFCC-SG (MP-norm.)	3.20	–	2.162

Considered functions to compute a summarization are arithmetic mean and median

At the end of this chapter, we addressed the problem of "hubness" that affects all tasks in which similarities are computed in high-dimensional feature spaces. In the music retrieval and recommendation domain, it can be characterized by the fact that certain music pieces occur as nearest neighbor for many other pieces, without actually sounding similar. To circumvent this problem, we discussed distance space normalization, for instance, via mutual proximity, in which case two items are only considered similar if they share similar nearest neighbors. We have seen that this simple post-processing measure is capable of substantially improving similarity measures. Hence, with different feature representations to be explored and developed, the effects of hubness should always be investigated and mitigated, where necessary, to ensure that the potential performance of algorithms is not throttled by symptoms of the curse of dimensionality.

3.6 Further Reading

Extending our brief introduction to psychoacoustics, i.e., the processing of sound by the human auditory system, Fastl and Zwicker [124] provide a detailed study of the topic. Generally speaking, the authors discuss how sound stimuli translate to hearing sensation. More precisely, they present mathematical models to describe masking effects, loudness sensation, and sensory pleasantness, among others. In addition to the detailed explanation of information processing in the human auditory system, also higher-level concepts such as pitch, rhythm, and fluctuation strength are discussed. The book is written in a quite technical style and can certainly be regarded as a landmark for the topic of auditory information processing and models thereof.

For readers interested in more details on the statistical modeling of features, in particular via probability distributions, we highly recommend the textbooks by Hastie et al. [166] or Bishop [38]. The authors take a machine learning perspective, hence present the respective approaches in the context of classification, clustering, and related tasks.

Of the books already discussed as further reading in the previous chapter, it is worth mentioning that both Lerch [262] and Müller [326] also address the problem of music similarity computation, in different contexts, Müller for audio identification and matching and Lerch for classification tasks (e.g., genre and mood). For more details on frame-level and block-level features and their use in music similarity computation, we refer the reader to Seyerlehner's PhD thesis [437]. He discusses the respective techniques in the context of a content-based music recommender system. Focusing on MFCCs, Logan [281] discusses in some detail their benefit for modeling music. She starts by presenting their origin in speech recognition and further investigates adapting their standard computation process. She eventually assesses MFCCs for music/speech discrimination, concluding that the use of the Mel scale is at least not harmful to the task.

The topic of music retrieval is of course a very wide one, and plenty of other types of features and retrieval models tailored to their representations are conceivable and have been proposed. For instance, representations of songs using GMM supervectors [70] and, as a recent extension, I-vectors [117] modeled over MFCC distributions have been successfully used for similarity measurement. Alternative approaches, for instance, make use of other features related to the spectral shape to describe timbre [321]; rhythmic aspects, such as beat histograms, beat structure, tempo, or onset rate (see, e.g., [122, 242]); tonal features, such as pitch class profiles (PCP) (also known as chroma vectors) [137, 148, 433], key [197], scale, distribution of chords, or inharmonicity; acoustic events, such as the occurrence of instrumental or vocal content; or a combination thereof [44, 63, 265, 286, 289]. Even this short list of but a few references shows that a lot of effort has been put in designing features for representation of various aspects of music.

Finally, we want to point the reader to a recent trend that bypasses the need for "feature crafting," i.e., manually investigating and optimizing representations in order to achieve high performance in a given task, such as those presented in Chaps. 2 and 3. Instead, (hierarchical) feature representations should be learned automatically, in an unsupervised or semi-supervised manner, from a large set of unlabeled training data. This trend has become particularly popular through the application of *deep learning* architectures [32, 98] and shown highly effective in tasks such as image classification [244] or speech recognition [177]. Also in music processing, deep learning of audio representations from large music repositories has shown very promising results in music feature learning [164], similarity estimation [426], and classification [102] and will undoubtedly play a central role in future music retrieval research.

Chapter 4
Semantic Labeling of Music

So far, we have dealt with the basics of signal processing and low-level features in Chap. 2 and methods that allow the retrieval of similar-sounding pieces based on mid-level acoustic characteristics such as timbre and rhythmic activations in Chap. 3. In this chapter, we focus on capturing musical properties that bear semantic meaning to the average listener. These semantic descriptors of music can be considered high-level features. Following up on the discussion of the term "semantic" in Sect. 1.3.2, in this context, the notion of "semantic" should be pragmatically considered to refer to "words that make sense to humans when describing music" and that are therefore of help when searching or browsing for music or when matching a given context.

Automatically learning or inferring such semantic descriptors is typically performed via supervised machine learning. We therefore discuss classification approaches to estimate semantic concepts and tags such as genre, style, and mood. While the importance of genre classification is lately seeing a slight decline in MIR, we nevertheless start by discussing some approaches, since the task has received considerable attention (Sect. 4.1). Subsequently, we review in Sect. 4.2 auto-tagging approaches, which differ from genre or style or other classification tasks in that they do not predict a single well-defined class label, but learn relationships between (typically audio) features and collaboratively created or expert-based annotations, which cover a wide range of aspects, from instruments to meaning of lyrics to evoked emotions. The resulting models can then be used to predict a set of semantic labels (or tags) for unseen pieces. A topic that has been receiving particularly high attention in the last few years is music emotion recognition (MER). Due to this fact, we devote a considerable part of the chapter to a discussion of different models of human emotion and techniques to computationally learn and recognize emotional responses to music (Sect. 4.3).

P. Knees, M. Schedl, *Music Similarity and Retrieval*, The Information Retrieval Series 36, DOI 10.1007/978-3-662-49722-7_4

4.1 Genre Classification

We have discussed the concept of genre already in the context of algorithm evaluation, where genre labels are often used as ground truth and as a proxy for music similarity (see Sect. 1.5.1). Musical genre is a multifaceted concept that, over the years, has caused a lot of disagreement among musicologists, music distributors, and, not least, MIR researchers. Fabbri defines a musical genre as "a set of musical events (real or possible) whose course is governed by a definite set of socially accepted rules" [123]. These socially accepted rules that define genres can be categorized into five types, which McKay and Fujinaga [310] summarize as[1]:

- Formal and technical: Content-based practices.
- Semiotic: Abstract concepts that are communicated (e.g., emotions or political messages).
- Behavior: How composers, performers, and audiences appear and behave.
- Social and ideological: The links between genres and demographics such as age, race, sex, and political viewpoints.
- Economical and juridical: The laws and economic systems supporting a genre, such as record contracts or performance locales (e.g., cafés or auditoriums).

Furthermore, "the notion of set, both for a genre and for its defining apparatus, means that we can speak of sub-sets like 'sub-genres', and of all the operations foreseen by the theory of sets: in particular a certain 'musical event' may be situated in the intersection of two or more genres, and therefore belong to each of these at the same time" [123].

MIR research has often tried to overcome the "ill-defined" concept of genre [338] as definitions "have a strong tendency to end up in circular, ungrounded projections of fantasies" [14], and the "Aristotelian way" of making music "belonging to genres" denies the fact that genre cannot be defined without considering the users' individual creation and consumption contexts of music [472]. While none of these points can be contested, genres have nonetheless been important concepts in production, circulation, and reception of music in the last decades [114], and their relevance is evidenced by studies that show the existence (at least to some extent and level of granularity) of common ground between individuals, e.g., [276, 442], their importance in users' music similarity assessment [334], and their recognizability within fractions of seconds (values reported range from 250 ms [145] to 400 ms [245]). This makes genre an important semantic music descriptor and, consequently, music genre classification a central task in MIR research [310, 399]. In fact, in the last 20 years, almost 500 publications have dealt with the automatic recognition of musical genre [472].

However, the general shortcomings of genre definitions have a strong impact on the modus operandi of genre classification methods, as *intensional* genre definitions,

[1] Of these five types, only the first actually relates to qualities of sound and content; cf. Sect. 1.3.

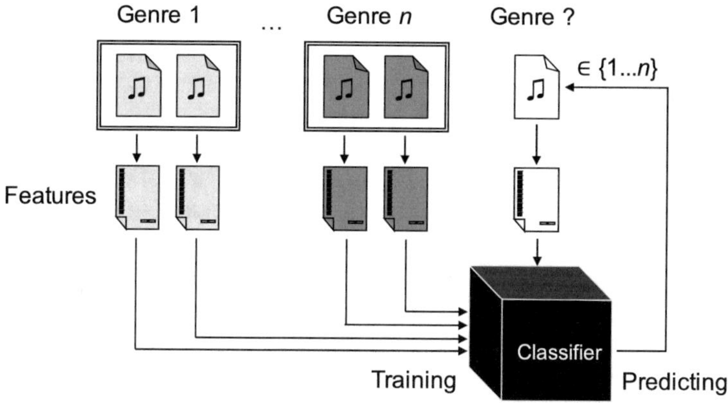

Fig. 4.1 Schematic illustration of a music genre classifier

i.e., "what makes a genre" (see above), are subordinate to *extensional* definitions, i.e., specifying all examples that belong to the genre. As a consequence, genre classification is (more or less exclusively) carried out as a black box machine learning task with pre-defined target classes defined via examples.

Figure 4.1 shows a schematic overview of such a machine-learning-based music genre classification approach. First, genres are defined by examples, i.e., assigning music pieces to one genre (class). Following the definition by Fabbri [123], one music piece can be assigned to more than one genre; however, in the case of multiple genre labels per piece, it is advised to change the overall classification architecture to contain a binary classifier for each individual genre label and learn to predict whether this specific genre label applies to the piece or not (cf. Sect. 4.2). Second, for all the given examples, audio (or other) features are extracted. Third, these features, together with the assigned class labels, are used to train a machine learning classifier, which in turn is then used to predict a genre label for new, unseen music pieces. Genre classification approaches proposed in the literature follow this overall structure but differ in all three steps, i.e., definition of genre classes, extraction of features, and chosen machine learning technique. As this is an extensively covered area, in the following, we only review exemplary approaches and focus on these three aspects.

With regard to used **features**, it can be observed that information on timbre has shown to be the most informative for distinction of genres and styles (we consider the latter to be fine-grained and nuanced subgenres in the following; see [322] for a discussion on the ambiguous use and the relation of the terms genre and style). More specifically, in virtually all state-of-the-art approaches to genre classification (as well as tag prediction and music similarity, see Sect. 4.2 and Chap. 3, respectively), timbre properties represented via MFCCs or a variation thereof are at the core. This finding or implicit assumption is rather disturbing considering the existing literature on style analysis using symbolic representations of music, where elements such as the preference for certain rhythm patterns, chord progressions, and melodic patterns have been noted as style discriminators [368]. This further indicates that current audio-based approaches, despite their effectiveness, are very poorly informed in terms of musical understanding [470] and that automatically derived intensional definitions of genre and style are still a big challenge.

Beside MFCCs, often-used features include a pool of low-level features such as energy, zero-crossing rate, or spectral flux (e.g., [43, 52, 309, 487]; see Chap. 2), as well as musically more informed features modeling rhythmic and tonal aspects. These comprise features such as beat histograms [52, 269, 309, 487], rhythm histograms [270], percussive and bass-line patterns [1, 479], pitch statistics [43, 487], pitch-class profiles [43, 269, 309], or harmony features derived from chord transcription [7]. Feature extraction has also shown to benefit from performing the analysis steps on separated streams obtained by dividing a polytimbral signal using the spatial information present in stereo recordings [45].

Alternative approaches to genre detection make use of cultural features, i.e., meta-data derived from web sources (cf. Part II). For genre classification, meta-data-based approaches, in principle, follow the same overall structure. Instead of extracting features from the audio signal, descriptors of a music piece or a music artist are extracted from related sources external to the actual piece of music. Typical sources for meta-data comprise editorial meta-data [17, 309], manually annotated features [370], artist web pages [219], product reviews [191], and collaborative tags [263, 506]. Meta-data-based features are often combined with acoustic features in order to overcome limitations of the individual domains, e.g., [17, 24, 507]. Optionally, and regardless of the feature origin, the dimensionality of the (combined) features is reduced using data projection methods such as principal component analysis (PCA) [187] or statistics such as mean and variance.

In terms of **genre taxonomies** used, existing approaches are highly heterogeneous due to available collections and intentions of the work, making comparisons of reported results difficult. For instance, the most used data set GTZAN, introduced in the archetypal work in the field of genre recognition by Tzanetakis and Cook [487], contains ten musical genres, namely, classical, country, disco, hip-hop, jazz, rock, blues, reggae, pop, and metal. Despite its wide acceptance and reuse in subsequent studies, it exhibits inconsistencies, repetitions, and mislabelings [469]. Taxonomies in other work comprise similar, broad genres, for instance, alternative, blues, electronic, folk/country, funk/soul/RnB, jazz, pop, rap/hip-hop, and rock [185]; country, folk, jazz, blues, RnB/soul, heavy, alternative/indie, punk, rap/hip-hop, electronic, reggae, classical, rock 'n' roll, and pop [219]; or specific ballroom styles such as cha-cha-cha, jive, quickstep, rumba, samba, tango, Viennese waltz, or waltz [157]. Burred and Lerch propose a hierarchical approach to genre classification that divides music into classical and nonclassical music, with classical music further divided into chamber and orchestral music (each with further subgenres) and nonclassical music divided into rock (with further subgenres), electronic/pop (with further subgenres), and jazz/blues [52].

Hörschläger et al. perform electronic dance music (EDM) style classification as a step towards tempo estimation [186] on a collection comprising 23 different and partly highly specific EDM styles, such as glitch hop, progressive house, dubstep, or psytrance. As audio descriptors, they use PCA-compressed BLF features (cf. Sect. 3.3) and—as suggested when using BLF, for reasons of efficiency [437]—a *random forest* as style classifier. A random forest [46] consists of multiple *decision trees* and predicts the class that is obtained by voting on the predictions made by the individual trees (*ensemble classifier*).[2] Random forest classifiers incorporate the principles of *bagging* (multiple drawing of samples with replacement) and *random feature selection* (finding the most discriminating features from a randomly selected subset of features) for learning the decision trees. While automatic genre classification systems reportedly yield classification accuracies around 70 % on very small taxonomies with, say, six broad genres (e.g., classical, dance, pop, rap, rock and "other" [276]), classification becomes more difficult with an increase of genres (which are often also less well defined and less clearly distinguishable) and more specific and narrow styles, as can also be seen in subgenre taxonomy classification experiments within heavy metal music [478].

In terms of used **machine learning methods**, other than the mentioned random forest classifier, k-nearest neighbor (k-NN) and support vector machine (SVM) classifiers are often used to predict musical genre, e.g., in [219, 478] and [266, 268, 522], respectively. The *k-NN classifier* [86] is a so-called lazy learner since training consists only of storing the given training examples alongside the class labels. At the time of label prediction, the k closest stored training examples to the music piece to be classified are found according to a chosen distance measure.[3] From the k nearest neighbors, the most frequent class label is used to predict the genre of the piece to be classified.

Support vector machines [89, 493] learn to separate classes by finding hyperplanes that have maximum margins between representative instances (support vectors) of the classes. In order to find such hyperplanes, if the classes are not already linearly separable in the feature space, feature vectors are projected into a higher dimensional space, in which classes are linearly separable. Since for higher dimensional operations, only calculating the inner product of two data points is required, the projection can be implicit, which is computationally cheaper than an explicit transformation into higher-dimensional coordinates ("kernel trick"). In order to employ support vector machines, a similarity function (kernel) over pairs of data points in the feature space needs to be specified. Although originally applied for artist classification [293], the KL-kernel [324] (based on Kullback–Leibler divergence) over single Gaussian MFCC models (cf. Sect. 3.2.5) has shown to be a good choice for the task of genre classification as well.

The range of machine learning techniques that can be applied to genre classification tasks is broad. In addition to the previously discussed random forests, k-NN, and SVM classifiers, methods reported range from AdaBoost [478] (cf. Sect. 4.2.2) to linear discriminative analysis [266] to Markov model classifiers [299]. Deep learning methods have also shown to improve performance in genre classification

[2]For a detailed description of a widely used decision tree classifier (*C4.5*) see [371].

[3]This is basically a retrieval task as described in the previous chapters.

when using large collections of music represented by a variety of features to pretrain a convolutional deep belief network (unsupervised learning step) and initializing a convolutional multilayer perceptron with the learned parameters before further optimizing parameters on the convolutional multilayer perceptron using the labeled genre information (supervised learning step) [102].

4.2 Auto-tagging

Auto-tagging is related to genre classification in that it relies heavily on machine learning techniques in order to predict semantic labels for music pieces. These semantic labels, however, are not limited to genre and style terminology, and typically a music piece is relevant to more than one semantic label. Hence, genre classification can in fact be seen as a special case of auto-tagging where only label from a set of genre and style descriptors is predicted for each music piece.

Tagging music with arbitrary descriptors has become popular with the advent of collaborative tagging platforms, such as *Last.fm* in the music domain. Tags have since become a popular way to search for music. Retrieving music by tags works well for music items that are well known and, for that reason, tagged by many users. For the vast majority of music items, however, user-generated tags are scarcely available. For instance, Lamere [248] showed that the average number of tags per track on *Last.fm* is only 0.26. Even though this study dates back to 2008, lesser known artists and tracks still suffer from a smaller amount of attached tags. This in turn results in a lower retrievability of such items. Auto-tagging, i.e., automatically inferring semantic labels from the audio signal of a music piece (or related data; cf. Sect. 6.2.3), is a proper way to overcome this shortcoming. Before presenting auto-tagging methods, despite all commonalities, we want to deepen the discussion on differences between genre classification and tag prediction.

4.2.1 Differences to Classification

Auto-tagging differs in several regards from the aforementioned tasks of classification by genre and style (as well as mood, as discussed in Sect. 4.3). The most outstanding difference is that the labels predicted by an auto-tagger are not drawn from only one category, i.e., such as terms describing musical style or mood. Instead they are typically given either by a folksonomy[4] or by a pool of expert annotations and range from indications of instruments to content of the lyrics to

[4]A folksonomy is the result of a collaborative tagging process, which yields an unstructured set of labels—in contrast to a taxonomy, which is hierarchically structured. Each tagger can attach an arbitrary number of tags to an arbitrary number of music items.

personal opinions about the performer (cf. Sect. 5.3.2). This high variety of tags in turn requests that auto-taggers do not only predict a single label for a piece but a set of labels, likely from different categories, e.g., genre, language of the lyrics, and musical era. From a technical point of view, predicting multiple tags for a song can be achieved in two ways: (1) training a binary classifier for each tag in the folksonomy, which predicts whether or not the label applies and (2) using a probabilistic classifier that assigns each label a likelihood that it applies to the song under consideration; cf. [440]. In the latter case, all tags with likelihoods above a predefined threshold are predicted.

Another difference between music auto-tagging and classification relates to the human tagging process itself, which results in the training data for the classifier. Regarding creating human annotations, we can distinguish between weak labeling and strong labeling; cf. [482]. In strongly labeled data sets, which are used in music classification, the fact that a label is not attached to an item indicates that this label certainly does not apply to the item. In contrast, when annotations are the result of a collaborative tagging process, the absence of a tag may not necessarily mean that the tag does not apply to the respective item. In fact, taggers could have simply not thought of the tag, which is bigger a problem the lower the number of taggers for the item under consideration.

When tags are given by a folksonomy, the weakly labeled nature of the music items needs to be borne in mind. A strongly labeled tagging process is possible, on the other hand, if human experts are given a list of predefined tags to choose from when annotating music collections. An example of weakly labeled data sets is a set that draws from *Last.fm* tags; strongly labeled sets are, in contrast, created by commercial services such as Pandora but are also freely available for research, for instance, the CAL-500 set [480]. When feeding an auto-tagger with a weakly labeled data set, performance measures that involve precision are problematic as we do not know for sure whether a tag predicted by the auto tagger does not apply or has just been not thought of by the taggers.

4.2.2 Auto-Tagging Techniques

An illustration of the auto-tagging process and its evaluation is shown in Fig. 4.2, proposed by Sordo [458]. In general, auto-tagging involves training a classifier that learns relations between music pieces and tags from a training set, i.e., a set of songs with predefined labels, for instance, identified by music experts. To this end, the classifier creates a model that can then be employed to assign tags to unseen music pieces, given in the test set. Since both the training set and the test set come with predefined labels, but the labels of the test set are not shown to the classifier, it is possible to compare the predictions of the auto-tagger with the actual tags of the pieces in the test set and compute performance measures of the auto-tagging algorithm.

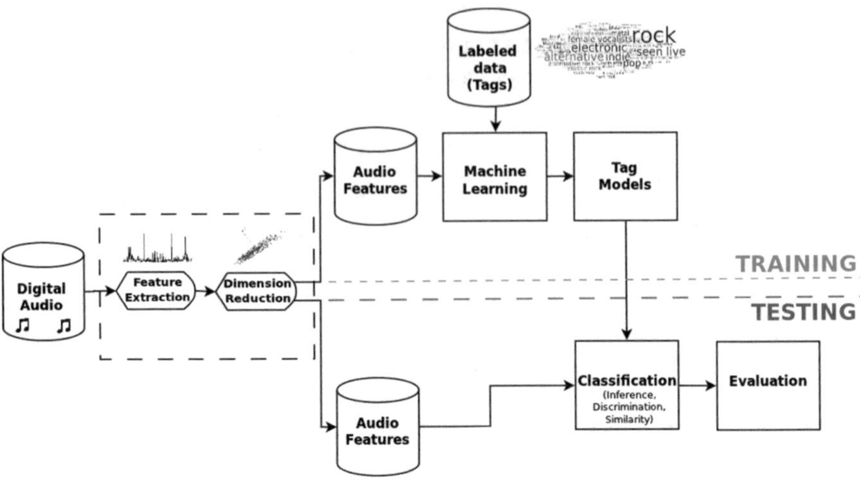

Fig. 4.2 Schematic illustration of a music auto-tagger, according to Sordo [458] (reprinted with permission)

To be more specific, the first step in the process of music auto-tagging is to extract features from the audio, which can be followed by a dimensionality reduction or feature selection step, in order to increase computational performance. As shown in [458], dimensionality reduction of down to 5 % of the original number of dimensions does not yield significantly worse classification results in terms of accuracy, on typical music collections. Features commonly used in auto-taggers include rhythm and timbre descriptors [296]. Based on the extracted features, tag models are learned as indicated above. For this purpose, a wide range of machine learning algorithms is available; cf. Sect. 4.1. After the training process is finished, the classifier can be used to predict tags for previously unseen music items. In the following part, we exemplify some common and recent approaches to music auto-tagging.

An early approach to auto-tagging is presented by Whitman and Ellis [511]. The tags to be learned from audio features are automatically extracted from corresponding record reviews.[5] As audio features, block-level-like features based on MFCCs are calculated. For learning the relation of audio features to extracted tags, regularized least-squares classification (which works in a fashion similar to SVM) is applied.

In the Autotagger system [35, 115], a combination of MFCCs, autocorrelation coefficients, and spectrogram coefficients is aggregated over 5-second segments by calculating means and standard deviations. The issue that the used *Last.fm* tags are very unequally distributed is dealt with by predicting whether "a particular artist has 'none', 'some' or 'a lot' of a particular tag relative to other tags" [115] instead of

[5]Therefore, this method produces "automatic record reviews" for arbitrary music pieces. Predicting descriptive tags to generate automatic reviews is also a goal of the work of Turnbull et al. [480].

simply predicting whether a tag applies or not. For learning the tag classes for each 5-second segment, a multi-class extension of AdaBoost is utilized. AdaBoost [133] is a meta-learner that uses a set of simple classifiers ("weak learners," e.g., binary thresholds on one of the features, as used in [115]) and constructs a strong classifier from these in an iterative way, i.e., by adapting weights of individual weak learners based on their performance on the training set. The output for each class is a real-valued number that is a weighted combination of the weak learners' outputs. For predicting tags for songs, the output scores over all 5-second segments of the song are averaged, and the difference between the predicted values of the classes "a lot" and "none" is used to predict the tag. The value of the class "some" is discarded.

Hoffman et al. use vector quantized MFCC features (cf. Sect. 3.2.3) from the CAL-500 set [480] and apply the codeword Bernoulli average model (CBA), a probabilistic generative model that models the conditional probability of a tag appearing in a music piece conditioned on the empirical distribution of the piece's codebook words [182]. Fitting of the model is performed using expectation–maximization (cf. Sect. 3.2.4). Despite being a simple model, the performance of CBA is on par with Autotagger when using the same features but yields significantly worse precision than Autotagger trained on additional data and features.

An even simpler classifier frequently used in auto-tagging algorithms is k-NN or variants thereof (cf. Sect. 4.1). For instance, Kim et al. [206] employ a k-NN classifier for artist auto-tagging. The authors investigate different artist similarity measures, in particular, similarities derived from artist co-occurrences in *Last.fm* playlists, from *Last.fm* tags, from web pages about the artists, and from music content features. Using an artist collection tagged by music experts, Kim et al. found that the similarity measure based on *Last.fm* co-occurrences yields highest recall and precision. Sordo [458] proposes an approach named weighted vote k-nearest neighbor classifier. Given an unknown song s to be tagged and a training set of labeled songs, the k closest neighbors N of s according to their feature vector representation are first identified. The frequencies of each tag assigned to any of the neighbors N are subsequently summed up, and the most frequent tags of N are predicted for s.

Unlike Kim et al. and Sordo, who auto-tag artists and songs, respectively, Mandel et al. [296] propose an approach that learns tag language models on the level of song segments, using conditional restricted Boltzmann machines [455]. They investigate three vocabularies: user annotations gathered via *Amazon*'s *Mechanical Turk*, tags acquired via the tagging game *MajorMiner* [295], and tags extracted from *Last.fm*. In addition to tags on the level of song segments, the authors further incorporate into their model annotations on the track level and the user level.

Combining different audio features described within their block-level framework (cf. Sect. 3.3) Seyerlehner et al. investigate the use of support vector machines [440] and random forests [441] for auto-tagging songs. Both classifiers are parametrized to yield probabilistic output, which in turn results in tag affinity vectors or binary feature vectors for each song. The audio features considered are spectral pattern, delta spectral pattern, variance delta spectral pattern, logarithmic fluctuation pattern, correlation pattern, and spectral contrast pattern.

Recent methods for music auto-tagging often include two-stage algorithms. In a first step, they derive contextual information from music content features or meta-data, for instance, weights of descriptive terms mined from web pages. This derived additional information is then fed into a classifier to learn tags; cf. [87, 317]. Sometimes both original and derived features are jointly input to the classifier.

4.3 Mood Detection and Emotion Recognition

Detecting emotion in music has become a popular research topic within the field of MIR. This task is typically referred to as music emotion recognition (MER) [526]. Emotions certainly represent high-level concepts according to our categorization. Although the terms "mood" and "emotion" have different meanings in psychology, they are commonly used as synonyms in MIR research. In contrast, in psychology, "emotion" refers to a short-time reaction to a particular stimulus, whereas "mood" refers to a longer-lasting state without relation to one specific stimulus (an overall inclination).

Before digging deeper into the techniques used for MER, we first have to define what is meant by "music emotion," as this is not obvious. Is it the emotion the composer or songwriter had in mind when creating the piece? Is it the emotion the performer intends to convey? Is it the emotion that is perceived by the listener during active music consumption? Or is it the emotion induced in and felt by the listener? As it may be argued that the third interpretation is least influenced by contextual factors and in turn the least subjective one, it is the one commonly applied in MIR and MER [527]. Please note, however, that all these definitions of emotion come with a much higher degree of subjectivity than other semantic music descriptors (for instance, instrumentation or music style).

Since MER is a multidisciplinary topic, connecting computer science and psychology, we will first review the most popular categorization models of human emotion, which are also used in MER. Subsequently, we will present the most common MER techniques.

4.3.1 Models to Describe Human Emotion

Describing and understanding human emotion is a fundamental research topic in psychology. From a general point of view, it is common to distinguish between categorical models and dimensional models of emotion. The *categorical models* describe the emotional state of a person via distinct emotional words, assuming that there are a limited number of canonical and universal emotions, such as happiness, sadness, anger, or peacefulness. An advantage of categorical models is that the task of MER can be regarded as a classification task and thus rely on the same classification techniques as for music classification into other categories (e.g., genre or style). On the other hand, putting each human emotion into one of a limited set of categories may severely confine the variety of emotional sensations. Furthermore, words used to describe emotions can be ambiguous, and subtle differences of the words can be hard to make out, not only for nonnative speakers. The *dimensional models*, in contrast, presume that every human emotion can be placed in a continuous space spanned among several dimensions, thereby assigning a certain value to each emotional dimension. Both kinds of models suffer from the general problem that human emotion is highly subjective. For this reason, the same music piece can be perceived very differently by different users, in terms of emotion. This as well as the aforementioned challenges render the task of MER particularly difficult, compared to other music classification or regression tasks. Frequently used categorical and dimensional models are reviewed in the following section.

4.3.1.1 Categorical Models

These are models that are based on the assumption that there are a limited number of basic emotions, into which each music piece can be categorized. There is, however, no agreement among researchers about what is considered a basic emotion [526]. Nevertheless, the use of four basic emotions that are directly related to the dimensional valence–arousal model described below [204] is supported by psychological theories. These emotions are *happy* (high valence, low arousal), *relaxed* (high valence, low arousal), *angry* (low valence, high arousal), and *sad* (low valence, low arousal). They are frequently used in the context of MER, for instance, in [252, 457].

To overcome the problem that particular choices of basic emotions are highly disputed, another approach is to cluster affective terms into categories, without attaching one particular basic emotion label to each cluster. One of the best known models of this kind is Hevner's [174]. The author investigated affective terms and clustered emotions specific to music into eight categories. The result can be seen in Fig. 4.3. Each cluster is described by a set of emotion terms that give the reader a good impression about the corresponding kinds of emotional sensations, even in the absence of a dedicated name for each cluster. Based on editorial information and collaborative tags rather than on psychological theory, Hu and Downie [188] propose a categorization of music emotions into five clusters, which they derive from mood labels given by *allmusic* and data from *Epinions*[6] and *Last.fm*.[7] Table 4.1 gives examples of the most important emotions of each cluster. These categories are also used in the MIREX Audio Music Mood Classification task.[8] Although this five-cluster model is used in several publications, it is also criticized for not being supported by psychological studies, e.g., [457].

---

[6] http://www.epinions.com/.

[7] http://www.last.fm/.

[8] http://www.music-ir.org/mirex/wiki/2010:Audio_Music_Mood_Classification.

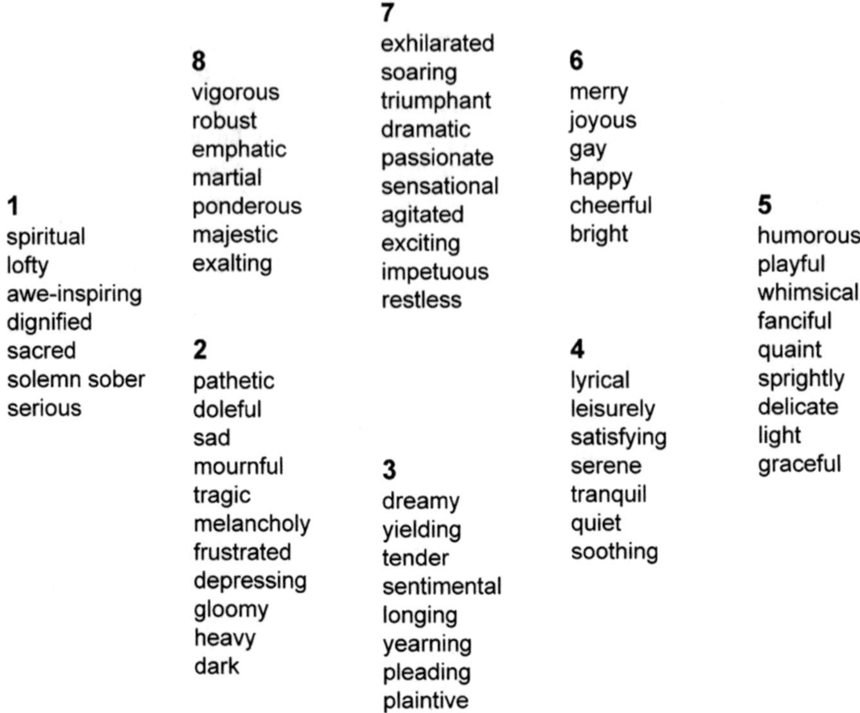

Fig. 4.3 Hevner's eight categories of emotion terms [174], adapted from [526]

Table 4.1 The five-cluster model of music emotion, according to Hu and Downie [188]

Cluster	Emotions
1	Passionate, rousing, confident, boisterous, rowdy
2	Rollicking, cheerful, fun, sweet, amiable/good natured
3	Literate, poignant, wistful, bittersweet, autumnal, brooding
4	Humorous, silly, campy, quirky, whimsical, witty, wry
5	Aggressive, fiery, tense/anxious, intense, volatile, visceral

Laurier et al. [253] compare the model of four basic emotions, Hevner's model, and Hu and Downie's model. Investigating different ways to cluster collaboratively generated emotion tags from *Last.fm*, the authors find that using four clusters yields the most stable results. Interestingly, the emotion terms assigned to these clusters correlate very well to the four basic emotions from psychological theory. The authors further investigate whether the expert models (Hevner's and Hu and Downie's) and the community model are consistent. To this end, they compute similarities between the gathered *Last.fm* emotion tags and use these similarities to estimate the quality of the clusters defined by experts, in terms of *intra-cluster similarity* and *inter-cluster dissimilarity* of the emotion labels. The former measures how consistent the clusters are, by computing the mean cosine similarity between emotion terms in the same cluster. The latter quantifies how well the individual clusters are separable, calculating the mean cosine distance between the cluster centers. The four basic emotions are thus reflected in the community data from *Last.fm*. Particularly high consistency values are found in Hevner's clusters 1 and 6, but very low consistency values in cluster 8; cf. Fig. 4.3. In Hu and Downie's model, intra-cluster similarity (consistency) is quite high, except for cluster 2; cf. Table 4.1. The average consistency of Hevner's clusters is 0.55 and of Hu and Downie's clusters 0.73. Inter-cluster dissimilarity (separability) is on average 0.70 and 0.56 for Hevner's and for Hu and Downie's model, respectively. The lowest separability in Hu and Downie's model can be observed between clusters 1 and 5 (0.11) whereas the highest between clusters 2 and 5 (0.87). In Hevner's model, clusters 7 and 8 as well as 1 and 2 are best separable (>0.95), whereas separability of clusters 1 and 4 is worst (0.09). To summarize, in Hevner's model, clusters are less consistent but better separated than in Hu and Downie's model.

4.3.1.2 Dimensional Models

A shortcoming of categorical models is that the basic emotions according to psychological theories are too few to capture the various sensations of human emotion. On the other hand, increasing the number of emotion terms can easily lead to overwhelming the user and can further introduce ambiguities among the labels. A remedy for these issues are models that assume a continuous emotion space, spanned among several dimensions of human emotion sensation. The most prominent of this kind of models is the *valence–arousal model* proposed by Russell [387]. The dimension of *valence* refers to the degree of pleasantness of the emotion (positive versus negative) and *arousal* refers to the intensity of the emotion (high versus low). Some dimensional models further incorporate a third dimension, typically called *dominance*, which refers to the distinction of whether an emotion is dominant or submissive [313]. To give an example, both anger and fear are unpleasant emotions (low valence). However, anger is a dominant emotion and fear is a submissive one. The former can also be seen as active while the latter as passive. Other examples are disgust versus humiliation and excitement versus joy.

Russell's two-dimensional circumplex model [387] represents valence and arousal as orthogonal dimensions; cf. Fig. 4.4. Categorical emotion labels can then be mapped to certain points or regions within the continuous emotion space. The author further proposes to organize a number of eight affect terms around an approximate circle in the valence–arousal space, where opposing emotions are inversely correlated in terms of valence and arousal, for instance, depressed versus excited or aroused versus sleepy. Please note that emotion terms are only approximately placed in Fig. 4.4.

There are also hybrids between categorical and dimensional models, in which the dimensions are specific emotional categories, each of which is given a degree

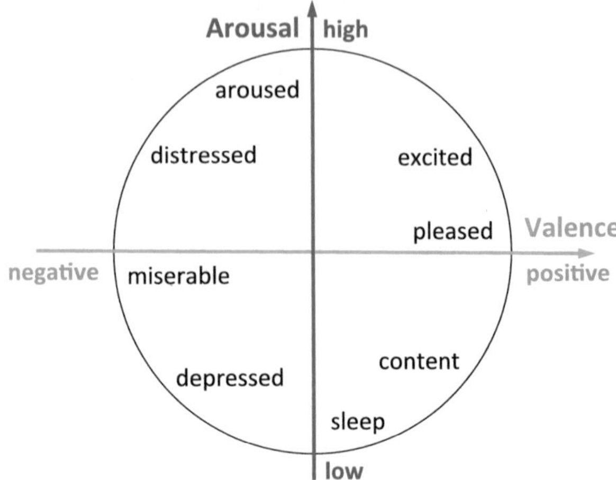

Fig. 4.4 Eight emotional concepts placed in the valence–arousal space, according to Russell [387]

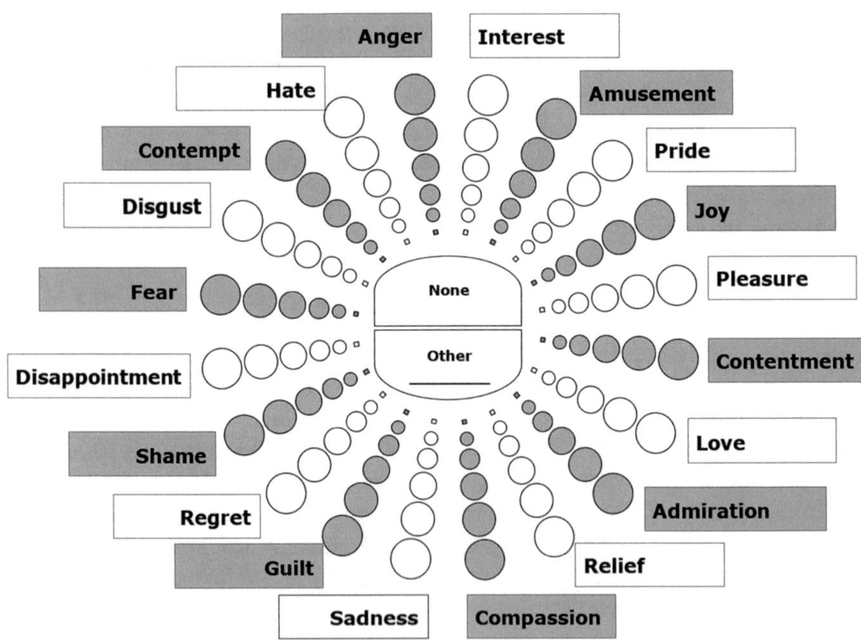

Fig. 4.5 Geneva emotion wheel, according to Scherer [425] (reprinted with permission)

of intensity. The degree is chosen from a limited number of values. An example for such a hybrid model is the "Geneva emotion wheel"[9]; cf. Fig. 4.5.

[9]http://www.affective-sciences.org/gew.

4.3.2 *Emotion Recognition Techniques*

Reviewing the literature on MER reveals that it is still a young and experimental albeit highly popular topic. This is underlined by the fact that researchers focus on investigating a variety of features that already proved to work for other tasks. A variety of (non-standardized) music collections is used, and the number of emotion groups highly differs between publications; cf. [526, Table II]. These factors make it hard to compare the different approaches proposed.

As for the features exploited in MER, most works use either audio or lyrics features or a combination of both. Audio features include aspects related to energy, tempo, rhythm, spectrum, harmony, or melody. Spectral features, such as MFCCs, spectral centroid, and spectral flux (cf. Chaps. 2 and 3), were found to perform best among the content-based features [190, 457]. Features extracted from lyrics include generic n-grams (sequences of n consecutive words), affective terms, and stylistic features (e.g., number of unique words, number of words per minute, or ratio of repeated lines; cf. Sect. 6.5). Identifying affective terms in lyrics is commonly performed using predefined word lists from various sources, such as psycholinguistic dictionaries or *WordNet*.[10] To obtain numeric feature vectors that represent the lyrics of a song, either simple word counts or TF·IDF weights (cf. Chap. 6) are subsequently computed for the n-grams or affective terms.

Corresponding to categorical and dimensional models of human emotion, research on the topic of MER can be categorized into categorical and dimensional MER techniques. While the former employ classification methods, the latter use regression techniques to predict emotion in music.

4.3.2.1 Categorical Emotion Recognition Using Classification

Categorical emotion recognition relies on classifiers that assign each input song to one or more emotion classes. Frequent choices are k-NN [190, 252], SVMs with different kernels [189, 252, 457], and random forest classifiers [252].

Song et al. [457] investigate different groups of audio content features (dynamics, rhythmic, spectral, and harmonic) and combinations thereof. Each individual feature, e.g., chromagram centroid for harmony or spectral flatness, is computed as mean and standard deviation over all frames in the song under consideration. The authors created a collection of about 3000 songs, which were selected as a subset of tracks tagged with the four basic emotions *happy*, *relaxed*, *angry*, and *sad* on *Last.fm*. The distribution of songs per emotion category is approximately uniform. Audio snippets of 30 s length have been gathered from *7digital* to extract audio features from. The authors then perform tenfold cross-validation using SVMs as classifiers. They investigate two different kernels for the SVM: a polynomial

[10]http://wordnet.princeton.edu.

kernel and a radial basis function (RBF). When only one category of features is used, spectral features outperform the others, achieving 52 % accuracy with the polynomial kernel that turned out to perform better than the RBF kernel in all experiments. Harmonic features reach 48 % accuracy, while rhythm and dynamics features underperform (both about 37 % accuracy). Song et al. eventually show that fusing rhythm, spectral, and harmony features outperforms any of the individual features, yielding an accuracy of 54 %.

Hu and Downie [189] look into both spectral and lyrics features, although they do not combine the two. Concerning the latter type of features, the authors extract n-grams, identify affective words, and model stylistic features such as special punctuation or number of unique words. They use a collection of about 5300 songs categorized into 18 emotion classes, which they derive from *Last.fm* tags. Performing tenfold cross-validation and using a SVM with a linear kernel, Hu and Downie achieve an average accuracy, over all 18 emotion classes, of around 60 % for text-based lyrics features (n-grams and affective words). Audio features yield on average 58 %, while stylistic lyrics features underperform with 53 % accuracy. Analyzing the performance for the individual emotion classes, the authors find that text features perform particularly well when there is a strong semantic connection between the lyrics and the respective emotion category, in particular for categories *romantic*, *angry*, *cheerful*, *aggressive*, *anxious*, *hopeful*, and *exciting*.

Laurier at al. [252] perform experiments that differ from those previously discussed in that the authors (1) combine lyrics and audio features, (2) ask experts to validate *Last.fm* tags they build their collection upon, and (3) perform binary classification experiments, i.e., they train one binary classifier for each emotion class; songs can hence be assigned none, one, or more emotion classes. Similar to [457] and to human emotion research, the authors adopt the four classes of basic emotions. They retrieve songs from *Last.fm*, tagged by the respective emotion labels. Annotators then listen to snippets of the songs and are asked to confirm or falsify the labels. In around 70 % of the cases, *Last.fm* labels are confirmed. The author's data set eventually comprises of 1000 songs, 125 for each class and its complement, e.g., *happy* vs. *not happy*. In terms of features, Laurier et al. extract timbral, rhythmic, tonal, and temporal audio descriptors. Lyrics are represented as TF·IDF vectors, in a latent semantic space (cf. Sect. 6.1.1), and using a language model of 100 terms for each category. As classifier, Laurier et al. use k-NN and a SVM with a polynomial kernel. They perform tenfold cross-validation and already reach an average accuracy of almost 90 % using only audio features. Note however that balanced class distribution and binary classification mean that the baseline is 50 %. Lyrics features alone reach up to 81 % accuracy. To combine both audio and lyrics features, the authors concatenate the respective feature vectors. By doing so, a significant further boost in accuracy is found (up to 93 %). The two data sources thus complement each other.

4.3.2.2 Dimensional Emotion Recognition Using Regression Models

Dimensional emotion recognition requests regression models that output a value in a continuous range, rather than a fixed class. If the dimensional model under consideration is the valence–arousal model with a range of $[-1, 1]$, for instance, a regression algorithm could predict a value of 0.15 on the valence and 0.7 on the arousal scale for a given song, instead of the label *excited*. Regression methods used in MER include logistic regression [252], multiple linear regression [116, 427], and support vector regression [195]. Approaches to dimensional MER typically model each dimension separately, i.e., one regression model is learned for each dimension, such as valence, arousal, tension, or dominance.

Being one of the most widely used techniques, we review *linear regression*, in particular, least-square estimation. The input to a regression model in dimensional MER consists of a set of feature vectors $\mathbf{X}$, which is an $n \times d$ matrix, where n is the number of music pieces and d is the dimensionality of the feature vectors. Each music piece is further assigned a continuous variable y_i related to the emotion dimension to be modeled. Consequently, the set of all of these dimensional ground truth emotion values for data set $\mathbf{X}$ is denoted as $\mathbf{y}$, where the scalar value y_i corresponds to the feature vector $\mathbf{x_i}$. The rationale of regression in general is then to learn a function $f(\mathbf{X})$ that fits the ground truth variable $\mathbf{y}$ as well as possible. To this end, an error term ϵ_i is used to model the deviation of $f(\mathbf{x_i})$ from the ground truth y_i for all instances i in the training data. Linear regression, as the name suggests, assumes that the relationship between the variables constituting the feature vector and the dependent variable y_i is linear. The regression problem can thus be formulated as shown in Eq. (4.1), where $x_{i,j}$ denotes the jth dimension of feature vector x_i and β_j are called regression coefficients, which need to be learned from the data $\mathbf{X}$ and the ground truth $\mathbf{y}$. Equation (4.2) depicts the problem in matrix denomination.

$$y_i = \beta_1 \cdot x_{i,1} + \ldots + \beta_d \cdot x_{i,d} + \epsilon_i \tag{4.1}$$

$$\mathbf{y} = \mathbf{X}\beta + \epsilon \tag{4.2}$$

Once the regression coefficients β have been learned from the training data, i.e., the regression model has been fitted to $\mathbf{X}$ and $\mathbf{y}$, they can be used to predict the dimensional emotion value y_k for an unseen feature vector $\mathbf{x_k}$, using again Eq. (4.1) (typically setting $\epsilon_k = 0$). There are various techniques to learn the regression coefficients. One of the simplest and most common techniques is least-square estimation, which provides a closed-form solution to the regression problem. Minimizing the sum of the squared errors, it estimates the β coefficients as shown in Eq. (4.3) and in Eq. (4.3) again in matrix denomination.

$$\hat{\boldsymbol{\beta}} = \left(\sum_{i=1}^{n} \mathbf{x_i} \cdot \mathbf{x_i}^{\mathsf{T}} \right)^{-1} \cdot \sum_{i=1}^{n} \mathbf{x_i} \cdot y_i \tag{4.3}$$

$$\hat{\boldsymbol{\beta}} = (\mathbf{X}^{\mathsf{T}}\mathbf{X})^{-1}\,\mathbf{X}^{\mathsf{T}}\mathbf{y} \tag{4.4}$$

A detailed discussion of more sophisticated techniques can be found, for instance, in the book by Hastie et al. on statistical learning [166].

Evaluation of regression models is most frequently performed by computing the R^2 statistics that measure the amount of variance in the data that is explained by the learned regression model, as a fraction of the total amount of variance in the input data. These statistics are also known as coefficient of determination and is defined in Eq. (4.5). You may notice that the R^2 statistics correspond to the squared Pearson's correlation coefficient; cf. Eq. (3.18).

$$R^2(\mathbf{y},f(\mathbf{X})) = \frac{cov(\mathbf{y},f(\mathbf{X}))^2}{\sigma_y \cdot \sigma_{f(\mathbf{X})}} \tag{4.5}$$

R^2 values always fall into the range [0, 1], where a value of 0 indicates that none of the variance in the input data is explained by the regression model, whereas a value of 1 means that the learned regression model perfectly explains the data. For what concerns the task of MER, predicting valence is more difficult than predicting arousal. This is reflected by the fact that state-of-the-art methods achieve R^2 values of around 0.70 for arousal recognition but only 0.25 for valence prediction [195]. This may be due to valence being a more subjective dimension than arousal.

As for features used in dimensional music emotion recognition, they largely correspond to those used in categorical MER, i.e., they commonly relate to timbre, energy, rhythm, spectrum, harmony, or tonality. In particular, studies showed that timbre, energy, and tempo seem important factors to model arousal, while harmony and mode (major or minor) more strongly relate to valence [204, 526].

Dimensional approaches to MER request other ground truth data sets than categorical methods. For instance, Yang and Chen collected valence and arousal values for 1240 Chinese Pop songs via crowdsourcing, involving 666 annotators [525]. Each song received on average 4.3 subjective annotations. Kim et al. present a collaborative game called *MoodSwings* to obtain continuous valence and arousal annotations for a collection of about 1000 popular songs [208]. The songs were drawn from the frequently used *uspop2002* data set [33]. However, similarly to other MIR tasks, there is still no standardized, agreed on data set for neither dimensional nor categorical MER.

4.4 Summary

In this chapter, we focused on three types of semantic labels representing high-level concepts in music perception, namely, genre and style, tags, and mood.

Genre and style classification is a traditional task, originally manually performed by librarians and musicologists. Automatic systems follow this tradition by trying to assign music to the best suited category within a given taxonomy. However, with the features available, a musically informed process, i.e., a process that knows "what makes a genre," as performed by librarians and musicologists, is currently not possible, resulting in a "glass ceiling" for existing methods. In order to yield more meaningful, contextually appropriate classifications, the central challenge remains to develop genre recognition systems that build upon intensional rather than extensional definitions of genres. For *auto-tagging*, intensional definitions of concepts are even more difficult to achieve; however, since labels typically stem from a folksonomy, it is not even clear whether this is possible. As it stands, auto-tagging is carried out as a traditional machine learning task and has to deal with many of the problems that come from a biased ground truth generated in a collaborative effort. For *mood and emotion detection*, the quest for ground truth is even more difficult. While genre, style, and arbitrary tags, covering instrumentation, country information, language, etc., are at least to a certain extent "objective" as humans adhere more to conventions and a common understanding, for mood, labels are far more subjective.

Going back to the introduction of this chapter, all three types of descriptors can play an important role in retrieval processes, e.g., when browsing or when searching for music for a specific context. The systems presented learn to predict "words that make sense to humans when describing music" for music pieces or artists and therefore, despite all justified criticism, add a "semantic" layer on top of audio feature extraction processes.

4.5 Further Reading

All approaches discussed in this chapter strongly draw from machine learning research, in particular from unsupervised techniques like *classification* and *regression*. We refer readers who want to deepen their knowledge of these tasks to one of the reference books on machine learning, e.g., Mitchell [318], Hastie et al. [166], or Bishop [38]. A broader and comprehensive introduction to artificial intelligence, including but not restricted to machine learning, is given by Russell and Norvig [388]. Reference implementations of most of the discussed classifiers can be found in the WEKA package [162].

The topic of music emotion recognition (MER) is connected to psychological work about human *mood* and *emotion*, a detailed discussion of which is beyond the scope of this book. Among the most important reference works on the topic, not already mentioned above, is Ekman's treatment of emotions expressed in faces [118], in which he identifies and discusses six basic emotions present in all cultures: anger, disgust, fear, happiness, sadness, and surprise. He later extends the list by amusement, contempt, contentment, embarrassment, excitement, guilt, pride in achievement, relief, satisfaction, sensory pleasure, and shame, not all of which can be detected in faces, though [119]. Plutchik presents the *wheel of emotions*, in which he defines eight bipolar basic emotions, joy vs. sadness, anger vs. fear, trust vs. disgust, and surprise vs. anticipation, as well as more complex ones derived from the basic emotions [362]. Scherer provides in [425] a detailed discussion of what emotions are and how they can be categorized. He further proposes the *Geneva emotion wheel*, introduced above in Sect. 4.3.1.2. Taking a more computer science-oriented perspective, *affective computing* deals with the recognition, modeling, and interpretation of human emotions by computers. Reference works in this field include Picard's book [358] and Calvo et al.'s edited book [58].

In addition to emotion, genre, style, and descriptive tags, there obviously exists other semantic concepts to describe music, for instance, rhythm, tonality, harmony, or instrumentation. While a detailed treatment of such topics is beyond the scope of this book, we refer the interested reader to Gouyon and Dixon's survey article on automatic *rhythm* detection [155] and to Gómez' Ph.D. thesis [147] for computational approaches to recognize *tonality* aspects, such as pitch detection and key estimation. We further recommend de Haas' Ph.D. thesis [97] for a detailed treatment of *harmony* and computational methods to chord detection and transcription from audio as well as harmonic similarity. As for *instrumentation*, the automatic recognition of instruments is addressed comprehensively, among others, in Fuhrmann's Ph.D. thesis [135].

Part II
Music Context-Based MIR

Part II provides a detailed description and comparison of contextual data sources on music (e.g., web pages, blogs, microblogs, user tags, and lyrics), where both music context and user context are covered. We start with a discussion of related methods to obtain music context data (web mining, games with a purpose, etc.). The main data formats used by music web services are introduced. So are the various types of data, such as annotations, lyrics, web texts, or collaborative tags (Chap. 5). Subsequently, we address the task of indexing context information about musical entities gathered from the web, followed by a review of methods for music similarity estimation and retrieval from music context sources (Chap. 6).

Chapter 5
Contextual Music Meta-data: Comparison and Sources

In the methods discussed in Part I, music descriptors are extracted directly from digital audio using signal processing techniques. The presented content-based approaches provide powerful methods for retrieving and discovering music with similar acoustic properties. However, as already mentioned in Chap. 1, for the perception of a music piece, the acoustic content is just one factor among many. In particular, the *cultural context* of music should not be neglected as it contains important aspects not directly included in the audio signal.

For instance, the image and current popularity of performing artists might have a much stronger influence on the personal preference of music consumption than the sound—after all, shaping and manipulating such factors is the goal of music marketing strategies and the business behind. Other examples of aspects that need to be considered and that can influence the attitude towards a music piece are the language of a song, the message or contents of the lyrics, the popularity, the city or country of origin, the era in which it was composed or performed, or the members of a band (particularly if an already popular artist joins or splits from a group), to name but a few.[1] Much of this surrounding information can be found on the Internet, which can be considered the central source of today's common knowledge.

For advanced music retrieval applications, the goal is to automatically exploit such music-relevant *contextual meta-data* present on the Internet and derive characteristic descriptions and features.[2] In practice, modeling of contextual factors is

[1]Note that the language and lyrics of a song can be categorized as both music content and music context. Although being present in the audio signal, they cannot be automatically detected or transcribed to a textual representation (at least with current technology). On the other hand, lyrics can be retrieved from the web which also allows for easy language detection. For that reason, we discuss both of them in this chapter.

[2]This kind of data is also referred to as "cultural features," "community meta-data," or "(music) context-based features." For clarification, in this and the next chapter the term *context* specifically refers to the music context and not to the user's or the usage context, expressed through parameters such as location, time, or activity (cf. Chap. 7).

P. Knees, M. Schedl, *Music Similarity and Retrieval*, The Information Retrieval Series 36, DOI 10.1007/978-3-662-49722-7_5

important as similarity measures relying solely on content analysis have shown to be limited in terms of classification accuracy [15] and user acceptance [450]. Despite their potential for intelligent applications, it can be frequently observed that content-based methods will yield unexpected, in the best case serendipitous, results due to their strictly "objective" and descriptive nature with regard to sound characteristics (cf. the semantic gap; Sect. 1.3).

In this and the next chapter, we look into approaches that make use of meta-data and music-context data. In this chapter, we focus on contextual aspects of music primarily accessible through web technology and discuss different sources of context-based data for individual music pieces, albums, and artists. Chapter 6 focuses on the modeling and processing of these sources to facilitate the estimation of similarity and the indexing of musical entities for music retrieval.

5.1 Web-Based Music Information Retrieval

Methods that exploit some of the cultural context of music make use of data that is present on the Internet and can thus be accessed easily in an automatic fashion. Therefore, these methods and, generally, this area of research are sometimes also referred to as *web-based MIR*.

5.1.1 The Web as Source for Music Features

The type of information about music that can be found on the Internet is diverse and ranges from tags to unstructured bits of expressed opinions (e.g., forum posts or comments in social media) to more detailed reviews to rich encyclopedic articles and fan-maintained archives and compendia (containing, e.g., biographies, discography release histories including artwork of all versions as well as the type of packaging and included paraphernalia, liner notes, links to statements made by artists in various accompanying media, etc.) to databases of musical meta-data. From the platforms and services that provide structured information on music artists, albums, songs, and other entities, *Last.fm*, *MusicBrainz*,[3] *Discogs*,[4] and *Echonest*[5] stand out for providing public access to their records via custom APIs.

The types of features that can be extracted from these sources are strongly related to traditional text IR, e.g., TF·IDF-weighted vector space model features or co-occurrence analysis. The extraction of features from web sources follows the same

[3]http://www.musicbrainz.org.

[4]http://www.discogs.com.

[5]http://the.echonest.com.

overall scheme as with content-based approaches:

1. **Accessing the (raw) data source**: consists of methods from web data mining to automatically identify, request, and download related resources
2. **Segmenting the data stream**: preprocessing the downloaded sources
3. **Extracting features**: applying methods, typically from text IR, to the preprocessed documents
4. **Entity modeling**: applying aggregation methods to and/or build models from the extracted features
5. **Computing similarity**: comparison of the built entity models

In this chapter, we address the first two steps in detail, namely, accessing and preprocessing different data sources, whereas Chap. 6 deals with the remaining steps.

Again, we can categorize extracted features into low-, mid-, and high-level features. Low-level features are strongly tied to the source they are extracted from. From textual sources, such as web pages, this is typically statistics of extracted segments, such as n-gram frequencies, or part-of-speech tags. Mid-level features are a step closer to more meaningful models of user data and comprise semantic tags, concept or topic clusters, and sentiments. These representations aim at capturing the high-level concepts of musical context that make up one of the facets of human music perception such as the message, i.e., the meaning, of lyrics or the mood intended to be conveyed. Examples within the hierarchy of music-context features are given in Fig. 5.1.

It needs to be mentioned that even low-level features extracted from contextual sources such as record reviews can be perceived to be more "semantic" than high-

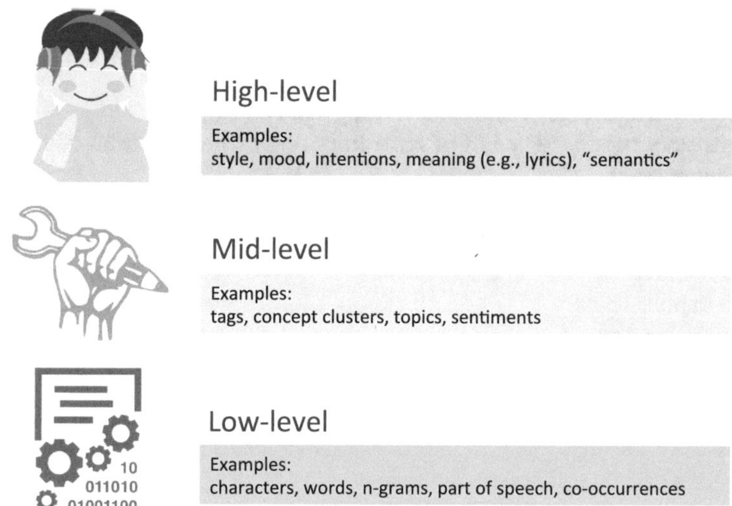

Fig. 5.1 The different levels of feature abstraction for music context

level features extracted from the audio signal due to originating from user-generated data and therefore an inherently high-level perspective. However, even if music-context-based low-level features appear more comprehensible than content-based features, particularly to consumers without musical education, this does not change the fact that the creation of a true high-level representation remains very difficult and has not been solved yet. Despite the similarities of content- and context-based features in the extraction process and the feature hierarchy, there are several differences which are discussed next.

5.1.2 Comparison with Content-Based Methods

Before examining sources for context-based features in detail, we want to discuss general implications of incorporating context-based similarity measures, especially in comparison to content-based measures; cf. [482]. In contrast to content-based features, to obtain context-based features it is not necessary to have access to the actual music file. Hence, applications like music information systems can be built without any acoustic representation of the music under consideration by having a list of artists [417].

On the other hand, without meta-information like *artist* or *title*, most context-based approaches are inapplicable. Also, improperly labeled pieces and ambiguous identifiers pose a problem. When dealing with music data, frequently occurring examples of such problems in the identifying meta-data as well as in the external cultural sources are:

- Artists with identical names (and even identical song titles); cf. [141]
- Different versions of songs (live versions, remixes, etc.)
- Inconsistent meta-data structures, such as in listings of multiple performing artists or orchestras with soloists and conductors
- Inconsistent spellings of names, including nicknames
- Inconsistent naming schemes of recordings (e.g., when indicating the structure of a symphony)
- Inconsistent language use (e.g., Quasi una fantasia vs. Moonlight Sonata vs. Mondschein Sonate)
- Abbreviations
- Typographical errors
- Special characters
- Inconsistent character sets

However, it needs to be accepted that user-generated data will always exhibit such artifacts and that dealing with non-standardized input is one of the central challenges when exploring contextual data.

The more severe aspect is that in fact all contextual methods depend on the existence of data sources to provide this meta-data as well as on the existence of available meta-data within these sources. Music not represented in the used

Table 5.1 A comparison of music-content- and music-context-based features

Features	Music content	Music context
Prerequisites	Music files	Users
Meta-data required	No	Yes
Cold-start problem (sparsity)	No	Yes
Popularity bias	No	Yes
Community bias	No	Yes
Noise	Audio artifacts	User artifacts
Features	Objective	Subjective
	Stable	Time-dependent
	Direct	Indirect
	Numeric	"Semantic"

sources is virtually nonexistent to these approaches. Even when present, a lack of a sufficient amount of associated data renders context-based approaches ineffective. Such *data sparsity* is particularly a problem for up-and-coming music and low-populated data sources ("cold start") as well as for music in the so-called long tail [5]. This is tied to the more general *popularity bias* that can be observed with principally all user-generated data sources. The effect of the popularity bias is that disproportionately more data is available for popular artists than for lesser known ones, which often distorts derived similarity measures. Furthermore, methods that aim at taking advantage of user-generated data are prone to include only participants of existing communities in a broad sense (from very specific services, like a certain platform, to the web community as a whole), while it is a common effect that users of certain communities tend to have similar music tastes. In general, this phenomenon is known as *community* or *population bias*.

To sum up, the crucial point is that deriving cultural features requires access to a large amount of unambiguous and non-noisy user-generated data. Assuming this condition can be met, community data provides a rich source of information on the musical context. Table 5.1 gives a brief comparison of content- and context-based feature properties.

5.1.3 Applications Using Web Data

The application scenarios in which music-context-based features and content-based features can be used overlap to a large extent. For instance, exploiting context-based information permits, among others, *automatic tagging* of artists and music pieces [115, 459], *enriched user interfaces* for browsing music collections [222, 345], *automatic music recommendation* [249, 531], *automatic playlist generation* [13, 363], as well as building *music search engines* [65, 223, 484]. Context-based features stem from different sources than content-based features and represent different aspects of music. Ultimately, to complement and outperform unimodal approaches, these two categories should be combined (cf. Sect. 6.4). Examples of beneficial combinations are the *acceleration of playlist creation* [221]; *improvement of classification accuracy* according to certain meta-data categories like genre, instrument, mood, or listening situation [17]; or *improvement of music retrieval* [534]. Although the proposed methods and their intended applications are highly heterogeneous, again, they have in common that the notion of music similarity is key.

In the remainder of this chapter, we describe principles of web mining that frequently occur in context-based MIR and discuss and analyze different web-based sources that can be used to extract contextual parameters.

5.2 Data Formats for Web-Based MIR

To access the contextual information of music present on the web, one has to deal with methods from web technology and web data mining. In the following section, we briefly discuss and describe the most important web technologies that occur when accessing web resources.

The foundation of all data communication on the World Wide Web is the **hypertext transfer protocol** (HTTP). HTTP is designed as a request-response protocol in which a server delivers network resources identified by **uniform resource locators** (URLs) starting with `http` to a client. Clients are web browsers, web crawlers, or any software that accesses such web content. The resources provided by servers typically consist of (semi-)structured data formatted in HTML, XML, or JSON.

HTML (short for **hypertext markup language**) is the standard language for providing web content. The central elements of HTML are tags that are enclosed in angle brackets and contain meta-information for formatting and structuring the web page as well as hyperlink information to other web resources. Besides text, images, scripts (foremost JavaScript), and other objects can be embedded. The set of valid HTML tags is defined in the W3C HTML specification.[6] Removing all HTML tags results in a plain text representation that is often used for text feature extraction and further processing. An example of a web page in HTML format can be seen in Fig. 5.2.

While HTML deals with the presentation of content to humans, XML (short for **extensible markup language**) only deals with the description of content with the goal of being both human and machine-readable. As with HTML, XML markup tags are enclosed in angle brackets. The content of the resource is to be found outside of markup tags or within dedicated `CDATA` tags. Markup tags can be freely defined and nested to allow for a "semantic" description of the contained data. Optionally, an XML document can contain a description of its grammar, making it a *valid*

[6]http://www.w3.org/TR/1999/REC-html401-19991224/.

```
<!DOCTYPE html>
<html lang="en">
  <head>
    <title>Kraftwerk</title>
  </head>
  <body>
    <p><b>Kraftwerk</b> are a German
    <a href="http://en.wikipedia.org/wiki/Electronic_music">
    electronic music</a> band.</p>
  </body>
</html>
```

Fig. 5.2 A sample HTML page

```
<?xml version="1.0" encoding="utf-8"?>
<lfm status="ok">
  <artist>
    <name>Kraftwerk</name>
      <mbid>5700dcd4-c139-4f31-aa3e-6382b9af9032</mbid>
      <url>http://www.last.fm/music/Kraftwerk</url>
    <stats>
      <listeners>812954</listeners>
      <playcount>19639647</playcount>
    </stats>
    <bio>
      <summary>
        <![CDATA[Kraftwerk is an electronic band.]]>
      </summary>
    </bio>
  </artist>
</lfm>
```

Fig. 5.3 A sample XML response obtained via the *Last.fm* API (edited and shortened for readability)

document according to the W3C XML 1.0 specification (5th ed.).[7] An example of an XML document is given in Fig. 5.3.

JSON (short for **JavaScript object notation**) is a data transfer format for attribute-value pairs organized in hierarchical objects.[8,9] Like XML, JSON is used for data structuring and object serialization. JSON objects are delimited with curly brackets. Commas are used to separate each contained attribute-value pair. Within each pair, the attribute name, which should be a unique string within each object, is separated from the value by a colon. In comparison to XML, JSON encoding entails

[7] http://www.w3.org/TR/2008/REC-xml-20081126/.

[8] http://tools.ietf.org/html/rfc7159.

[9] http://www.ecma-international.org/publications/files/ECMA-ST/ECMA-404.pdf.

```
{"created_at":"Mon Feb 10 05:26:16 +0000 2014",
 "id":416239763108017526,
 "text":"On Air Playing Kraftwerk - The Man-Machine
         #nowplaying #music #hitmusic",
 "user":{"id":532260500,
         "followers_count":293,
         "lang":"es",
        },
 "entities":{"hashtags":
                [{"text":"nowplaying","indices":[43,54]},
                 {"text":"music","indices":[55,61]},
                 {"text":"hitmusic","indices":[62,71]}],
             }
}
```


Fig. 5.4 A sample JSON-formatted "tweet object" returned by the *Twitter* API (IDs randomized; edited and shortened for readability)

less data overhead. An example of a JSON formatted server response is given in Fig. 5.4.

5.3 Tags and Annotations

One method for describing and indexing musical entities is to directly involve humans in labeling or tagging. This meta-data can either be used as a ground truth to learn these labels based on automatically derived features, i.e., auto-tagging (Sect. 4.2), or as features of the tagged entity itself.

As in other domains, a tag can be virtually anything, but it usually consists of a short description of one aspect typical to the item. For music, this may refer to genre or style, instrumentation, mood, or the performer but also to personal categories that mark whether, e.g., an artist has been seen live or is one of the favorite artists of a user. The set of associated tags comprises a compact and highly relevant description of an entity using a music-targeted vocabulary. Additionally, tags are usually assigned a weight, either explicitly (expert-assigned) or implicitly (e.g., corresponding to the number of users who assigned the tag). This makes tags particularly easy to be integrated into term-weighting frameworks such as the vector space model. We discuss three different approaches to obtain tags in the following: expert annotations, collaborative tagging, and games with a purpose.

5.3.1 Expert Annotations

The most intuitive approach to obtain meta-data for music is simply by asking humans. Since most high-level categories cannot be sufficiently modeled algorithmically, it seems straightforward to rely on expert judgments for semantic labeling. Two examples of such efforts are *allmusic*[10] and the *Music Genome Project*.[11] While *allmusic* is more focused on editorial meta-data, the *Music Genome Project* aims at representing songs in a feature space consisting of manually defined musically meaningful attributes (called "genes") that are manually evaluated by musical experts by analyzing each song with a consistent set of criteria, "designed to have a generalized meaning across all genres (in western music) and map to specific and deterministic musical qualities" [370]. While the actual dimensions and full list of attributes remain undisclosed, it is at least confirmed that the descriptors comprise elements of the timbre (vocals, instrumentation, sonority), rhythm, melody, harmony, genre, and subgenre, among others; cf. [370].

This type of feature extraction results in the most qualitative description possible. Moreover, it leads to a strongly labeled data set on the track level, which is the most valuable but also most rare type of data set. Over a period of 15 years, the *Music Genome Project* was able to collect such descriptions for over a million songs [370]. However, manual annotation is a very labor-intensive task, and evaluation of a single song by a musician/expert takes approximately 20–30 min. It is obvious that the vast number of (commercial) music pieces available now and in the future is highly unlikely ever to be fully annotated by musically trained experts. Thus, this approach does not entirely qualify for automatic indexing. Another aspect concerns the objectivity of expert judgments. Although description categories aim at capturing properties that are objectively either present or not, certain attributes (or even the used genre taxonomy) inevitably are inherently subjective. To compensate for an expert's biases, multiple opinions may be included—which further increases annotation costs.

5.3.2 Collaborative Tagging

Another strategy is to rely on the "wisdom of the crowd" rather than on individual experts, i.e., to "crowdsource" the task of annotation (also known as *social tagging*). Such a collaborative approach is one of the characteristics of the so-called Web 2.0 where websites encourage (and for the sake of their own existence even require) their users to participate in the generation of content. On Web 2.0 platforms, items such as photos, films, news stories, blog posts, or music can be tagged by every

[10]http://www.allmusic.com/, formerly *All Music Guide (AMG)*.

[11]http://www.pandora.com/mgp.shtml.

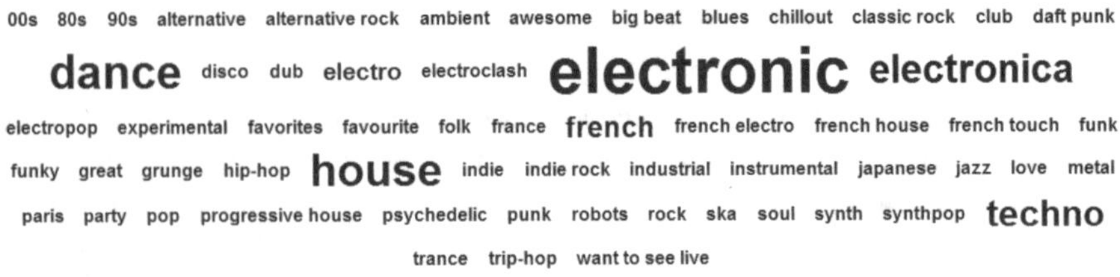

00s 80s 90s alternative alternative rock ambient awesome big beat blues chillout classic rock club daft punk
dance disco dub electro electroclash electronic electronica
electropop experimental favorites favourite folk france french french electro french house french touch funk
funky great grunge hip-hop house indie indie rock industrial instrumental japanese jazz love metal
paris party pop progressive house psychedelic punk robots rock ska soul synth synthpop techno
trance trip-hop want to see live

Fig. 5.5 *Last.fm* tag cloud for the band Daft Punk. The size of the tag corresponds to the number of users that have assigned the tag

member of the user community. The vocabulary used is typically open, i.e., users can assign arbitrary labels ("folksonomy"; cf. Sect. 4.2). The more people who label an item with a tag, the more the tag is assumed to be relevant to the item. For music, the most prominent platform that makes use of this approach is *Last.fm*. Since *Last.fm* also allows user to access the collected tags via an API,[12] it is a valuable, however weakly labeled, source for context-related information. Figure 5.5 shows an example of collaborative tags given to the French electronic band Daft Punk on *Last.fm*.

One advantage of the collaborative approach is that entities on all levels can be tagged. However, the more specific an entity gets, the more sparse the annotations. In a study from 2008, Lamere shows that on average, an artist is tagged 25 times, while a track is only tagged 0.26 times, i.e., 100 times less often [248]. Furthermore, Lamere demonstrates that the frequency distribution of tags on *Last.fm* is highly skewed and follows a classic long-tail shape. In fact, over 90 % of the artists in the study are not tagged at all. Thus, while the influence of expert biases is marginalized due to their collaborative nature, tag-based entity descriptions suffer heavily from popularity biases. Within tagging platforms, community biases (or tagger biases [248]) are inherent.

In addition, openness to the public comes with the risk of being the target of malicious manipulation attempts such as spamming and hacking. These attacks, where irrelevant tags are assigned to entities, are performed in order to *push* an entity, i.e., to make an item more attractive or more likely to be displayed by assigning popular tags, and to *nuke* an entity, i.e., to make an entity less attractive, or for personal reasons such as mere entertainment or to express a particular, typically negative, sentiment; cf. [248, 319].

Other challenges posed by social tags are typos, spelling variations, synonyms, polysemy, and noise. While inconsistencies, on one hand, contribute to the data sparsity of annotations, they can also help to improve the retrievability when modeled. To remove noisy tags for artists, the hierarchical nature of musical entities (i.e., *artist–album–track*) can be exploited by matching an artist's tags with the tags assigned to the artist's tracks [142]. To deal with synonymy and polysemy, *latent*

[12] http://www.last.fm/api.

topic modeling approaches have shown to be very effective [263]. We deal with term filtering and topic modeling in Sect. 6.1.

While providing valuable meta-data for music, relying on sources like *Last.fm* has the disadvantage that data might not be made available anymore at any point in time or that the service as a whole might cease to exist, rendering approaches that build on it useless. However, since, for instance, collaborative tagging as a principle is widely distributed and applied, one can assume that other services with similar functionality would emerge, as there is a demand as evidenced by the large user community.

5.3.3 Games with a Purpose

A strategy to efficiently collect tags and annotate media is to take advantage of people's ludic nature and to capitalize on their resources for playing games. Originally, this idea has been proposed by von Ahn and Dabbish to label images on the web. In the so-called ESP game,[13] two players on the web are randomly paired and have to guess what (free-form) label will be applied to an image by the other player ("output agreement") [501]. The faster an agreement is reached, the more points the players get. These incentives, i.e., the joy of gaming and yielding successful results, lead to quick inputs of highly relevant and semantic labels to the system. When an item is already labeled with a tag, it can be used as a "taboo word" to enforce additional and more diverse tags. Since a productive activity that solves problems difficult for computers is carried out while playing, games like the ESP game are referred to as *games with a purpose*.

The same idea can be applied to harvest tags for music. The biggest advantage of this method over both the expert and the collaborative tagging approaches is the high throughput in qualitative tag generation (5–6 tags per user-minute vs. 1 tag per user-month on *Last.fm* [248]). Examples of games with the purpose of tagging music are *MajorMiner* [294], *Listen Game* [481], *MoodSwings* [208], and *TagATune* [256]. In *MajorMiner*, output is agreed on and points are earned for a single player when the player enters a tag for a 10-second clip that matches a previously entered tag. The *Listen Game* follows the multiplayer design of the ESP game for tagging 30-second clips but only allows a player to select the best and the worst tag from a given list. New tags are created in so-called freestyle rounds that allow free-form input. In *MoodSwings*, a two-dimensional *valence-arousal* grid serves as user interface for mood annotation—a traditionally difficult task in MIR; cf. Sect. 4.3. Output-agreement and points are achieved based on the distance of the players' mouse-click positions on the grid.

After also initially following the ESP game design in *TagATune*, Law and von Ahn encountered some problems of output-agreement-based games when used for

[13] *ESP*: acronym of extra-sensory perception.

music tagging [255]. First, players seem to have more problems agreeing on tags for acoustic material than for images. While images usually contain recognizable objects or can be assigned some concept, the tags collected for music were very abstract (referring to "sound temperature," mood, blurry genres, or evoked images). We can safely attribute this phenomenon to the semantic gap being larger for music than for images; cf. Sect. 1.3. Second, this increased difficulty resulted in overall less fun for players and a higher rate of opt outs, i.e., less successful taggings, or the application of very general tags that describe only broad categories of music such as high-level genres.

To deal with these problems, Law and von Ahn proposed another mechanism for *TagATune*, called "input-agreement," which they describe as follows: "In this mechanism, two players are shown either the same object or different objects and each is asked to type a description of their given object. Unlike output-agreement games, where all communication is forbidden, all of the players' descriptions are revealed to each other. Based on these descriptions, the players must decide whether they have been given the same object." [255]

This change in strategy results in better verification strategies, prevention of cheating, a more rewarding experience for players, and, more complex tags. A screenshot of the game interface that incorporates the input-agreement mechanism can be found in Fig. 5.6. Using this interface, a collection of 30-second clips from the online record label *Magnatune* has been annotated. The resulting weakly labeled

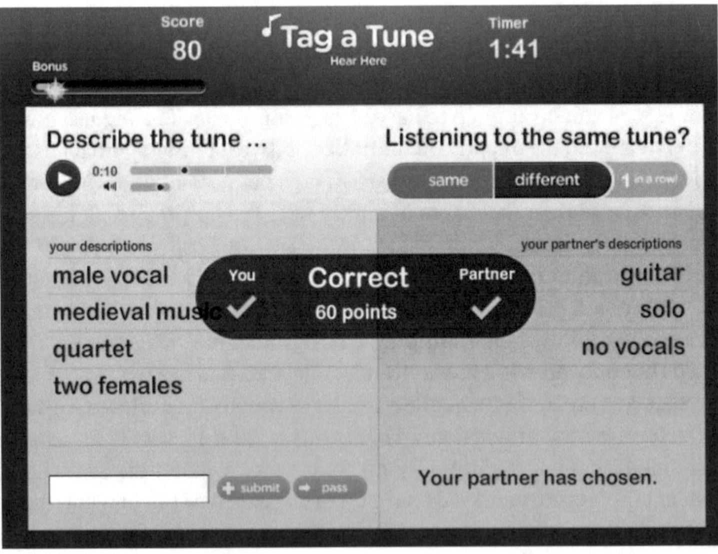

Fig. 5.6 Screenshot of the *TagATune* game interface based on input-agreement [255] (reprinted with permission)

data set on the track level has been released as *MagnaTagATune* data set.[14] Since the corresponding audio content is made available, it presents a very valuable source for research, particularly for auto-tagging systems.

5.4 Web Texts

The sources previously discussed all consist of structured information that is explicitly assigned to musical entities. For the sources discussed in this section, this is not always the case. First, the data we are dealing with is not necessarily explicitly assigned to a music piece or an artist but rather needs to be connected, e.g., by identifying mentioned entities and thus artificially creating a context. Second, the data sources are unstructured in that they consist of natural language texts from which the information needs to be extracted first.

The general idea of using related texts for music description is that the difficult problem of semantic music feature calculation is transferred from the audio domain to the text domain. Within the text domain, the full armory of text-based information retrieval and computational linguistics techniques is available. Here, we will not detail all of these but rather outline the main steps (for more details, we refer to further sources in Sect. 5.7). For preprocessing texts from the web, the following steps are commonly performed:

- **Removal of markup**, e.g., HTML tags, is performed in order to obtain a plain text representation; cf. Sect. 5.2.
- **Case folding** converts all characters to lowercase.
- **Tokenization** splits up the text into a sequence of single term chunks where any white space, line break, or punctuation mark serves as word delimiter. The resulting sequence is then stored in a data structure that can be accessed efficiently at the word level, such as an *inverted index*; cf. [68, 277].
- **Bag-of-words** models the text based only on the presence of terms and fully discards their ordering in the text, e.g., as a preprocessing step for applying the vector space model (Sect. 6.1.1).
- ***n*-gram extraction** extracts contiguous sequences of n terms in order to preserve the order of terms to some extent and further to model the statistical properties of the underlying language. Typically, uni-, bi-, or tri-grams ($n = 1, 2, 3$, respectively) are extracted.
- **Part-of-speech tagging** (POS tagging; also grammatical tagging) assigns each word in a text the part of speech that it corresponds to [48, 71]. Depending on the used nomenclature, one can distinguish between 50 and 150 distinct parts of speech for the English language. Examples of common part-of-speech categories are *article*, *preposition*, *adjective*, *plural proper noun*, and *past-tense verb*.

[14] http://mi.soi.city.ac.uk/blog/codeapps/the-magnatagatune-dataset.

- **Stopword removal** filters frequently occurring words that are needed for constructing grammatically correct language but are otherwise not related to the content of the text. This includes articles, prepositions, conjunctions, some verbs, and some pronouns, i.e., in English, words such as *a*, *and*, *are*, *for*, *of*, *in*, *is*, *the*, *to*, or *with*. Such a step can be extended to a more general blacklisting approach to filter all types of unwanted terms.
- **Vocabulary extraction**, as a whitelisting approach and thus as alternative to stopword removal or other exclusion strategies, only considers terms that are included in a predefined list of desired expressions ("vocabulary").
- **Stemming** removes all prefixes and suffixes from a word and thus reduces it to its stem. Since syntactic variations are removed, different forms of a word can be mapped to the same stem and facilitate retrieval of relevant documents. On the other hand, important discriminating information may be lost.

In practice, not all of these steps need to be applied. Rather, it depends on the application scenario which steps to perform in which order.

In the following section, we discuss four types of web text sources that usually benefit from being processed using the above-described techniques and how to obtain these sources. These four types are music-related web pages, artist biographies, product reviews, and RSS feeds.

5.4.1 Web Pages Related to Music

Possibly the most extensive source of cultural data—also on music—are the zillions of available web pages. When dealing with web data, some unique characteristics that distinguish the web from traditional databases and texts have to be taken into account (cf. [277, pp. 4–5]):

- The web contains a huge amount of most diverse information that covers virtually any topic.
- The web is highly heterogeneous in terms of contained data types (i.e., all types of multimedia data), structuring of data (e.g., tables vs. unstructured text), and information (e.g., different formulations of redundant information).
- Information is typically linked via hyperlinks.
- Web data is noisy, first, due to heterogeneous content that often includes navigation links or advertisements and, second, because web content is arbitrary, i.e., information is often low in quality, contains errors, or is (intentionally) misleading. Especially given that the web is an enormous economic factor, some issues that have to be taken into consideration are spam and other techniques which aim at exploiting the nature of information ranking algorithms to acquire more attention—and consequently more revenue.
- The web is used commercially and for (public) services.

- Information on the web is dynamic, i.e., the web is subject to constant change. However, outdated information may persist on the web and impose difficulties to web mining approaches if not contextualized chronologically.
- The web is a social medium, i.e., people communicate via forums, blogs, or comments and participate in activities (often referred to as Web 2.0; cf. Sect. 5.3.2).

To access the massive amount of available web data and to find those pages that contain music-related information, there are different strategies of which we discuss the two most straightforward, namely, a web crawler focused on music-specific pages and retrieval of web pages via common search engines.

5.4.1.1 Music-Focused Web Crawler

Indexing and retrieving data from the web is usually accomplished automatically by a *web crawler*. Starting from a seed URL, a crawler follows the contained hyperlinks recursively to find other web resources. Crawlers that collect as many pages as possible with the aim of indexing the complete web (as used by web search engines) are called *universal crawlers*, as opposed to *focused crawlers* that try to download and index only web pages of a certain type [69]. We are interested in the latter, i.e., a crawler focused on web pages that are related to music. The principal architecture of a focused web crawler is described in the following (cf. [277, pp. 274ff, 286ff]).

A central component of any crawler is the so-called frontier. The frontier is a list of unvisited URLs that is initialized with one or many seed URLs and constantly updated, while new URLs are extracted from the crawled web pages. In case the frontier is implemented as a priority queue that reorders contained URLs such that more promising pages are visited first, the crawler is said to be *preferential* or a *best-first* crawler. To estimate whether a page is promising, different strategies can be applied, for instance, calculation of *PageRank* [341]. For a focused crawler, the goal is to prefer pages that are topic related, which is difficult to assess before having access to the actual content. To this end, the textual context of the link (as well as the full text of the page where the link was extracted from) can be classified by a pretrained topic classifier (e.g., [351]).

For *fetching* the web data, the "crawler acts as a web client [i.e.,] it sends an HTTP request to the server hosting the page and reads the response" [277, p. 277]. Typically, fetching of data is performed in parallel by multiple threads which each refer to a single server to keep connections alive and minimize overhead. Issues like slow or non-responding servers, large files, broken links, or (cyclic) redirections have to be handled. After fetching the data (and extracting the contained hyperlinks that are inserted into the frontier), several *preprocessing steps* (cf. Sect. 5.4) are carried out before a document is stored and/or indexed. Upon having access to the content of the page, again, a focused crawler can reassess whether the page is relevant or not and if outgoing links from this page need to be further considered.

After collecting data, the mere application of traditional, purely content-based text IR methods to web documents is not sufficient. First, the number of available documents on the web is too large, i.e., it is difficult to select a small subset of the most relevant web pages from millions of pages that all contain the desired keywords. Second, content-based relevance ranking methods are prone to spamming due to the nature of the used algorithms. Thus, for retrieving the indexed content, the link structure, i.e., relations between the web pages, gives important additional information to determine the "authority" and thus the relevance of a page. Two examples of ranking algorithms that incorporate information on hyperlink structure are *hypertext induced topic search (HITS)* [215] and *PageRank* [341], which is the foundation of the ranking scheme of the *Google* search engine.[15]

To retrieve pages that are relevant for a music entity from the crawled corpus of music-related web pages, one can query the resulting index with the entity's description, e.g., the name of the artist or composer, the title of the song, etc. Using the text-content- and structure-based authority measures mentioned above will result in a ranking of music-related documents according to relevance with regard to the query. The assumption is that this ranking also reflects the relevance of the pages with regard to the entity. More detailed information on indexing, querying, and retrieving text-based documents can be found in Chap. 6. More details on the challenges of retrieving related web pages from an index are given in the next section, where we deal with approaches that make use of existing commercial web indexes, namely, general web search engines, and content filtering approaches.

5.4.1.2 Page Retrieval Using a Web Search Engine

Creating a music-focused web crawler and an accompanying index prior to accessing web page information is a resource-intensive task. As an alternative, web-based MIR approaches take a shortcut, so to speak, and resort to commercial web search engines, such as *Google*, *Yahoo!*,[16] or *Bing*,[17] as underlying data source.[18] In contrast to a music-focused crawler, the restriction to musically relevant content is not made while crawling but rather needs to be made after retrieving the content from the index or, preferably, already at query time.

In order to target the search to relevant pages at query time, additional keywords need to be specified when searching the index. To this end, different schemes have been proposed, typically comprising of the artist's name extended by the query terms

[15] http://www.google.com.

[16] http://www.yahoo.com.

[17] http://www.bing.com.

[18] The estimated numbers of indexed web pages containing the term *music* for these three services as of September 2014 are 3.5 billion for *Yahoo!*, 725 million for *Google*, and 455 million for *Bing*. The quality of the used search engine index affects the quality of the subsequent steps [216].

Table 5.2 *Yahoo!* result page count estimates in thousands for different degrees of specificity in music-targeted queries (as of Sept. 2014); query scheme abbreviations: `A`...*"artist name"* (as phrase), `T`...*"track name"* (as phrase), `m`...music

Artist	Track	Query scheme		
		`A`	`A+m`	`A+T+m`
Chemical Brothers	Burst generator	1350	1230	5
Radiohead	Idioteque	7760	6900	87
Frédéric Chopin	Nocturne for piano ...	2530	1770	5
Dmitri Shostakovich	Prelude and fugue ...	464	421	5
Black Sabbath	Paranoid	6470	5540	664
Kiss	I was made for loving you	137,000	72,400	201
Snoop Dogg	Drop it like it's hot	17,100	14,200	163
Snow	Informer	175,000	53,300	969
Madonna	Hung up	44,800	23,500	497
Lady Gaga	Bad romance	41,100	27,600	2070

music [216], *music review* [27, 512], or *music genre style* [219].[19] These additional keywords are particularly important for artists whose names also have a meaning outside the musical context, such as 50 Cent, Hole, or Air. Approaches relying on web pages are usually targeting the artist level since the more pronounced data sparsity and popularity biases become, the more specific the target entity (and thus the query) gets.

Table 5.2 compares different degrees of specificity in music-targeted queries by means of raw page count estimates (in thousands) from the commercial search engine *Yahoo!*.[20] Several effects can be observed from the page counts of the queries *"artist name"* (`A`), *"artist name" music* (`A+m`), and *"artist name" "track name" music* (`A+T+m`). First, we can observe the mentioned problem of common-speech named artists. The highest page counts are found for Kiss and Snow, followed by Madonna. Adding the keyword *music* reduces the number of pages significantly for these artists, whereas it remains comparable for artists with a unique name, e.g., Chemical Brothers, Radiohead, or Snoop Dogg. Still, adding the keyword *music* does not necessarily yield desired web pages only. For instance, a query like *"Snow" music* is not solely yielding pages about the artist Snow, particularly as the word snow already has multiple meanings outside the music context. In another example, the query *"Kiss" music* results in many musically relevant web pages that are not related to the band, for instance, to several radio stations named *Kiss.FM*.

Adding the name of a specific track by the artist should result in highly specific pages; however, the number of available pages is comparably low, particularly for less popular artists and tracks. Furthermore, for less popular tracks, the information

[19]A comparison of different query schemes can be found in [225].

[20]Note that the raw page count estimates as well as the relations between artist and track page counts can be used for the task of popularity estimation, which we address in Sect. 9.4.

contained is not as useful as on artist-specific pages as track-specific pages are merely entries by digital record or ringtone shops, as well as lyrics pages (see Sect. 5.5) which per se are not suited for contextual feature extraction. Thus, both general and highly specific queries result in data containing particular types of noise, making it generally difficult to find an optimal query scheme that finds the balance between these two extremes. Furthermore, the quality of retrieved pages depends more on the popularity of the artist (and track) than on the detailed selection of query keywords.

Despite the effects of varying popularity, we can also see the effects of a language bias when investigating the numbers for composers Chopin and Shostakovich. For Chopin, although the name is already rather specific for music, adding the keyword *music* results in a relatively large drop of pages (about 30 %). Indeed, the keyword *music* implicitly restricts the result set to the English part of the web. On one hand, this leads to a loss in information; on the other hand, it is a requirement in order to have comparable sources (and features, in subsequent steps). For Shostakovich, the number of retrieved pages is initially rather low and does not change as drastically with the inclusion of an English keyword since this particular spelling of the name already refers to the English transliteration of the Russian name.

5.4.1.3 Web Content Filtering

Upon retrieval of a set of candidate web pages from an index, a major challenge is to determine the relevance of the retrieved web content for the following steps and to filter irrelevant information and spam, such as commercials and hub pages.

To this end, Baumann and Hummel [27] propose to discard retrieved web pages with a size of more than 40 kilobytes after removing HTML markup. Furthermore, to exclude advertisements, they ignore text in table cells if it does not comprise at least one sentence and more than 60 characters. Finally, they perform keyword spotting in the URL, the title, and the first text part of each page, where each occurrence of the initial query constraints (i.e., in their setup, the artist's name and the terms *music* and *review*) contributes to a page score. Pages with a low score are filtered out, leading to improved performance in a text-based similarity task.

The following are the three complementary strategies for removing irrelevant content that have shown to be effective for context-based indexing of music collections [216, 226, pp. 80ff]:

1. **Alignment-Based Noise Removal:** To exclude irrelevant page content, such as navigation structures and site-specific text and annotations, this strategy consists in finding text-segment overlaps in randomly sampled pages originating from the same domain or website and to ignore these text portions on all pages from this domain.
2. **Too Many Artists Filter:** The goal is to reject pages that provide ambiguous content by dealing with several types of music, as well as hub and directory pages. Some of these pages can be identified easily since they contain references

to many artists. Starting from a list of potential artists, pages that contain more than a maximum threshold of artists are filtered. Experiments have shown that removing all pages containing more than 15 distinct artists yields best results.
3. **Classifier-Based Page Filter:** Similar to the decision process of a preferential crawler, a classifier can be trained to detect irrelevant pages. In order to train the classifier, sets of relevant and irrelevant pages need to be provided. As feature representation for web pages, characteristic values such as the length of the HTML content, the length of the parsed content, the number of different terms occurring on the page, the number of contained artist names, as well as ratios between these numbers can be used together with statistics of word occurrences in the title and the URL.

Depending on the task performed, different filtering strategies or combinations thereof are required. However, it needs to be noted that heavy filtering of retrieved pages, i.e., a strict configuration of filters or a combination of filters aiming at different criteria, can lead to a lack of remaining data and therefore harm the approach [216].

5.4.1.4 Content Aggregation

Traditional IR tasks deal with modeling entities at the document level. In contrast, retrieving and selecting web pages about a musical entity results in a set of related documents—or rather a ranking that should reflect the relevance of the documents with regard to the entity. Typically, ranking information is exploited as far as to select the N top-ranked documents, with N ranging from 10 to 100; cf. [225]. The following are three common strategies to deal with this scenario:

1. **Single Document Selection:** From the retrieved set, the most promising document, e.g., the top-ranked result, is chosen as textual representation of the entity. This strategy bears the danger of neglecting the diversity of results and biased entity modeling.
2. **Pseudo Document Generation:** The selected documents or parts thereof are joined into one large pseudo document (or virtual document). This strategy ignores the relevance information given by the ranking and might aggregate unrelated data. A shortcut version of this approach consists in directly processing the query result page returned from the page index which typically contains text snippets of the relevant parts of the corresponding pages [225].
3. **Set- and Rank-Based Modeling:** Each selected document is modeled independently and associated with the entity it was retrieved for. This strategy requires additional steps and methods in subsequent steps, such as indexing and assessing similarity between entities.

While exploiting general web pages allows the coverage of the most diverse range of information and sources, due to the discussed issues (the presence of noisy data, the preference of the approach for the artist level, the need to deal with multiple results), there exist a number of approaches that exploit specific web resources to avoid some of these challenges, which we briefly discuss next.

5.4.2 Biographies, Product Reviews, and Audio Blogs

We consider three contextual text sources that are suited for description of entities on a specific level, namely, artist biographies, album reviews, and audio blogs delivered via RSS feeds.

Artist Biographies. Biographies are textual sources that are strongly tied to one specific artist. In contrast to general biographies, the biographical texts used in music information retrieval are characterized by a dense prosaic description of factual knowledge about an artist. Typical elements are stylistic descriptions, geographical origin, important dates, the history of band members, noteworthy collaborations and influences, discographical information, optionally together with indicated critical reception and sales figures, as well as particular distinguishing aspects such as all-time records and public controversies. These types of texts can be found, for instance, on *Wikipedia*.[21,22] In comparison to web texts, biographies contain no or only a small amount of unrelated data and allow the modeling of an entity by one descriptive document. The quality and quantity of information contained is strongly tied to the popularity of the artist. In research, biographies have served as a source for artist similarity estimation by extracting and linking entities and calculating distances in semantic graphs or feature spaces [336] or in combination with other sources [305], for extraction of geographical information for radio station playlist visualization [158], as well as for band member extraction and discography extraction and as a reference set to identify the central artist or band in a text about music [217].

Product Reviews. Customer reviews are most suited for description on the album level, as an album is still the preferred format for physical music retailing and considered a conceptual unit that apparently provides enough facets and content to stimulate critique and articulation of opinions. Review data can be found online from most major music stores, such as *Amazon*,[23] as well as from non-retail websites explicitly devoted to customer reviews, such as *Epinions*. Mining album reviews entail dealing with a diverse set of highly subjective texts and comments, permitting the analysis of the sentiments of reviews, as well as the measurement of the controversiality of an album. Hu et al. [191] utilize album reviews gathered from *Epinions* and demonstrate the applicability of this source for the tasks of genre classification and rating prediction. Vembu and Baumann [496] aggregate album

[21] http://www.wikipedia.org.

[22] Exemplary excerpts of such texts for the toy music data set can be found in Appendix A.

[23] http://www.amazon.com.

review texts from *Amazon* to model and cluster artists with the goal of structuring music collections for browsing.

Audio Blogs. To access descriptive texts for individual tracks, Celma et al. [65] propose to crawl so-called audio blogs, which can be accessed via RSS feeds.[24] In these blogs, the authors explain and describe music pieces and make them available for download (whether legally or illegally depends on the blog). Hence, the available textual information that refers to the music, together with the meta-data of the files, can be used for textual indexing of the pieces, text-based calculation of music similarity, and tagging (see also [452]).

5.4.3 Microblogs

A—in comparison to the before-mentioned sources—rather atypical textual source of music information are microblogs. Microblogs are short status update messages posted within a social network. The use of such microblogging services, *Twitter*[25] in particular, has considerably increased during the past few years. In their status updates, many users share their music listening activities, making them also a data source for contextual music information [415]. One of the particularities of microblog data is the limitation to short texts—tweets, messages on *Twitter*, are limited to 140 characters, for instance.

One consequence is that messages are drastically abbreviated, for instance, by using acronyms and emoticons. This imposes further challenges to the preprocessing, such as the need to resolve links and other content expansion steps. On the other hand, compared to web pages, text processing can be performed in little time due to the limited amount of text. In addition, the inclusion of indexing terms (called hashtags) plays an important role in both message contextualization and content generation; cf. Fig. 5.4. For instance, tweets covering music listening activity are often marked with the hashtag `#nowplaying`.

It needs to be kept in mind that, as microblogging platforms are closed infrastructure, access to tweets is not fully available. This introduces an uncontrollable bias caused by the service provider. Furthermore, microbloggers might not represent the average person, which potentially introduces a community bias. Using microblogs as a text source, good results can be achieved when modeling entities on the artist level; however, song level experiments have also been conducted [532]. Since textual content in tweets is limited (typically a large share of this information is consumed by the artist and track identifiers), also the mere occurrence of identifiers within a user's status feed can serve as an indicator of user preference. This makes

[24]An XML-formatted web feed containing full or summarized text, meta-data, and links to resources. RSS stands for Rich Site Summary and Really Simple Syndication.

[25]http://www.twitter.com.

microblogs also a valuable source for co-occurrence analysis in interaction traces, as discussed in Sect. 7.3.

5.5 Lyrics

The lyrics of a song represent an important aspect of the semantics of music since they usually reveal information about the "meaning" of a song, its composer, or the performer, such as the cultural background (via different languages or use of slang words), political orientation, or style (use of a specific vocabulary in certain music styles). Also the fact that a piece is solely instrumental, i.e., has no lyrics associated, presents valuable descriptive information. While automatic extraction of the lyrics directly from the audio signal is a very challenging and still unsolved task, as mentioned in the beginning of this chapter, lyrics for a good portion of available commercial songs can be found online. Strictly speaking, they are no contextual data source as they present information contained in the signal. Since access and processing requires methods from web data mining and text IR, we treat them like contextual information for retrieval and discuss them here.

There are a number of online lyrics portals that allow the sharing and accessing of the lyrics of songs. However, even though the portals have comprehensive databases, none is complete. Another fact is that very frequently the lyrics for a song differ among the portals, e.g., due to simple typos, different words, or different annotation styles. The reason for this is that in many cases, lyrics are just transcriptions by fans, who submit their versions. The following section describes some observations made when analyzing lyrics.

5.5.1 Analysis of Lyrics on the Web

The following, not necessarily complete, listing gives an overview of various annotation characteristics in lyrics.

- **Different spellings of words:** Besides unintended typos, words, especially in more "recent" songs, can have different morphologic appearances. For example, the slang term *cause* is often found to be written as *'cause*, *'coz*, *cuz*, or even correctly as *because*. Although the semantic content is the same, these variations cannot be handled with simple string comparison. Similar observations can be made for numbers (e.g., *40's* vs. *forties*) and censored words (*f**k*).
- **Differences in the semantic content:** These result from the simple fact that not all people contributing to lyrics pages understand the same words. Many

perspicuous examples of misheard lyrics can be found on the web.[26] For this reason, most of the lyrics portals offer the possibility to rate the quality of lyrics or to submit corrections.

- **Different versions of songs:** Querying the web with the name of the track may result in a variety of lyrics which are all "correct," but highly inconsistent, e.g., due to cleaned versions or changes over time. Also translations of lyrics can be found frequently.
- **Annotation of background voices, spoken text, and sounds:** For songs where phrases are repeated, for example, by background singers, in some lyrics versions these repetitions are written out. Sometimes even comments (*get him!*, *yeah I know*, etc.) or sounds from the background (**scratches**, *gun fires*, etc.) are explicitly noted.
- **Annotation of performing artist:** Especially, but not exclusively, in duets, the artist to sing the next part is noted (e.g., *[Snoop - singing]*).
- **Annotation of chorus or verses:** Some authors prefer to give meta-information about the structure of the song and explicitly annotate the type of the subsequent section, e.g., *chorus*, *verse 1*, *hook*, *pre-chorus*, etc.
- **References and abbreviations of repetitions:** To keep the lyrics compact, to minimize redundancies, and to avoid unnecessary typing effort, lyrics are rarely found to be written completely. Rather, references to earlier sections (*repeat chorus*) and instructions for multiple repetitions (*x4*) are used. In general, this poses no problem for human readers but can be very difficult for a machine since, for example, the chorus may be referenced without proper annotation of the corresponding section. Also repetitions may be difficult to detect, since they can occur in variable form (*x2*, *(4x)*, *repeat two times*, or even *chorus x1.5*). A lot of prior knowledge is necessary to correctly expand the shortened content.
- **Inconsistent structuring:** Identifying sections or finding coherent parts within the lyrics could be facilitated if the lyrics were similarly structured. Differences can be observed in line breaking, paragraph structuring, or even in the general structure of the lyrics, since some authors prefer to write whole verses into one line.

Besides these differences in content and structure, also trivial deviations like case inconsistencies or bracketing of words can be observed.

5.5.2 Retrieval and Correction

As can be seen from the above listing, lyrics retrieved from the web are highly inconsistent and potentially contain errors. As a consequence, a user may be forced sometimes to examine different sources and to investigate the "correct"

[26]For example, http://www.amiright.com/misheard/.

version, e.g., by meta-searching lyrics portals (and the rest of the web) by simply using a standard search engine like *Google*. This process of comparing different versions and constructing a version that is as correct as possible can also be automatized [220]. The main idea behind this method is to identify and extract the lyrics by finding large segments of text that are common to lyrics web pages. The consistent parts among these pages are considered to be the lyrics. The inconsistent rest is usually irrelevant page-specific content like advertisements, site navigation elements, or copyright information. This automatic approach can be summarized in three main steps: gathering and selecting the lyrics from the web, aligning the lyrics pages on the word level to find the corresponding parts, and producing an output string based on the alignment.

An initial set of candidates is gathered through a web search engine using the query *"artist name" "track name" lyrics*. When a returned page does not contain the query terms in its title, the links contained on the page are followed to find a page that fulfils this requirement. If none is found, this page is discarded. This process is applied until ten candidates are found, which are all converted to plain text. The lyrics present on the different candidate pages typically are consistent in the beginning and differ after the second verse due to some pages only referring to the first chorus (e.g., *repeat chorus*), while others spell out the chorus each time. Therefore, in a preprocessing step, (multiple) references to choruses are expanded with the first paragraph labeled as chorus. Also lines that are marked for repetition are expanded. Additionally, in this step, already some unrelated text fragments can be removed from the pages using fingerprinting [241].

In most cases, the lyrics are surrounded by advertisements, information about the page, links to other lyrics, links to ringtones, or notices concerning copyright. On average, about 43 % of the content on pages is irrelevant. In order to extract the actual lyrics, multiple sequence alignment, a technique used in bioinformatics to align DNA and protein sequences, is performed. For the task at hand, it can be used to find nearly optimal matching word sequences over all lyrics pages and thus allows to discover the consistent parts in the pages, i.e., the lyrics. To align two texts, the Needleman–Wunsch algorithm [330] finds the globally optimal alignment of two word sequences for a given scoring scheme based on dynamic programming. For aligning more than more than two sequences, different approaches have been proposed (cf. [220, 241]), including hierarchical alignment [85].

Given the alignment of multiple lyrics pages, a final string containing the lyrics can be produced. To this end, every column of aligned words is examined, and the most frequent word above a threshold is output, filtering nonmatching parts over the pages. Figure 5.7 depicts a sample section from one such alignment. It can be seen that different pages provide different lyrics. Despite these variations, the voting over columns in the bottom row results in the correct version.

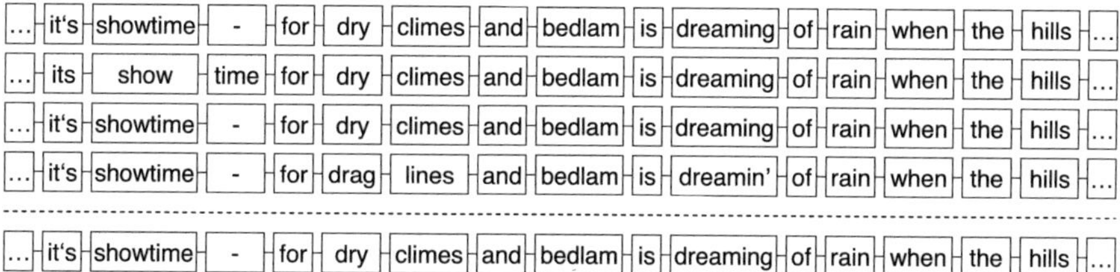
... it‘s showtime - for dry climes and bedlam is dreaming of rain when the hills ...
... its show time for dry climes and bedlam is dreaming of rain when the hills ...
... it‘s showtime - for dry climes and bedlam is dreaming of rain when the hills ...
... it‘s showtime - for drag lines and bedlam is dreamin’ of rain when the hills ...
... it‘s showtime - for dry climes and bedlam is dreaming of rain when the hills ...

Fig. 5.7 Example of a multiple text alignment on song lyrics (*Los Angeles is burning* by *Bad Religion*); cf. [220]. The *four rows on the top* are word sequences extracted from the web; the *row at the bottom* is the result obtained by majority voting

5.6 Summary

In this chapter, we discussed data sources for cultural meta-data of music, divided into tags, texts, and—as a specialty when dealing with music—lyrics. Obtaining this meta-data almost always involves web technologies. In addition to common data formats used on the web, we therefore also discussed issues that arise from dealing with web data and its specifics. The most frequent observation in all data sources and the most persisting issue for approaches that make use of them are inconsistencies in identifiers. This affects the meta-data identifying the music under consideration as well. However, it needs to be accepted that user-generated data will always exhibit such artifacts and that dealing with non-standardized input is one of the central challenges when exploring contextual data.

To deal with noise in available data, we discussed and presented methods to increase reliability for all types of sources. For collaborative tags, comparing tags on different levels for plausibility (e.g., comparing the tags given to a track and its performer) or by including a control step that requires some inter-person agreement already in the tagging process, such as a game with a purpose, has shown to increase the quality. For web texts, noise can be detected by comparing multiple documents or identifying texts that provide ambiguous content by dealing with several types of music, as well as hub and directory pages. For lyrics, we presented a technique that aligns multiple instances found on the web to find a most plausible version by majority voting.

Based on the textual representations obtained as described in this chapter, indexing, feature extraction, and similarity calculation of music can be carried out, as we detail in the next chapter. However, it needs to be kept in mind that meta-data approaches rely on the availability of related data and, even more, on the existence of sources that provide data. While individual sources might stop providing information or cease to exist at all (as has happened to collaborative music tagging platforms before), the general principles explained in this chapter will—most likely—also apply to similar services and data sources in the future.

5.7 Further Reading

More detail on dealing with textual data sources and the web is to be found in dedicated web IR books, e.g., [68, 277, 298]. Chakrabarti [68] focuses on web crawling techniques in general and on building focused crawlers in particular. He further addresses the topics of social networks, graph search, relevance ranking, and learning to rank, next to text indexing and similarity methods. Liu [277] addresses the topic of web mining first from a data mining perspective in order to elaborate on web mining and link analysis. Further topics addressed are structured data extraction, information integration, opinion mining, and web usage mining. Emphasis is put on spam detection and removal methods for the different mining areas. The second edition of the book additionally deals with related topics such as social network analysis and sentiment analysis [277]. Another comprehensive overview is provided by Markov and Larose [298] who detail the technical foundations of diverse crawling and mining techniques. In addition to the topics covered by the other books mentioned, they also investigate exploratory data analysis methods in the context of web mining.

Chapter 6
Contextual Music Similarity, Indexing, and Retrieval

In this chapter, we focus on the modeling and processing of contextual data from various sources. Since all of the obtained information is expressed in some form of text, the methods presented are strongly tied to text-based IR methodology. In Sect. 6.1, we review standard approaches to text-based similarity, in particular the vector space model, TF·IDF feature weighting, and latent space indexing methods. Section 6.2 discusses the use of contextual sources for indexing and building text-based music search engines. In Sect. 6.3, we address similarity calculation by analyzing co-occurrences of entities in the investigated data sources. Section 6.4 highlights ways of combining contextual with content-based information. Finally, in Sect. 6.5, we discuss aspects of stylistic analysis—concretely, by analyzing lyrics—to show the greater potential of contextual data beyond similarity and indexing.

6.1 Text-Based Features and Similarity Measures

From any of the sources discussed in Chap. 5, a plain text representation can be obtained. Depending on the type, it might be necessary to perform additional preprocessing steps and/or aggregate multiple texts into one document (cf. Sect. 5.4).

In the following section, we assume documents to be tokenized at the word level. Potentially, there might be different term sets associated with the same document, either by extracting n-grams and POS information (resulting, e.g., in individual term sets for unigrams, bigrams, noun phrases, artist names, and adjectives [512]) or by modeling different sections of documents (e.g., different terms sets for the title and the body section of an HTML document). In this case, similarities between the corresponding sets need to be calculated individually and combined into an overall similarity score; cf. [150].

P. Knees, M. Schedl, *Music Similarity and Retrieval*, The Information Retrieval Series 36, DOI 10.1007/978-3-662-49722-7_6

For reasons of simplicity, we assume the presence of a single set of terms T, typically unigrams [219], to model a document. Furthermore, when assuming that each musical entity is represented by a single text document, calculation of similarity between musical entities is achieved by calculating the similarity between their associated text documents [225]. Since most frequently the following techniques are applied to model music artists, we also consider the case of artist similarity. However, the same methods can be applied to other entities without modification.

6.1.1 Vector Space Model

The idea of the *vector space model* (VSM) is to represent each document (therefore each artist) as a vector in a high-dimensional feature space. In this space, each dimension corresponds to a term from the global term set (cf. the "bag-of-words" model, Sect. 5.4). To calculate the feature value for this term for a specific artist, several approaches have been proposed for document retrieval that are built on descriptive statistics and follow three underlying monotonicity assumptions (cf. [95]; quotes from [539]):

1. **Term Frequency Assumption:** "Multiple appearances of a term in a document are no less important than single appearances," or—in our scenario—the more often a term occurs within an artist's document, the more important it is as a descriptor for the artist.
2. **Inverse Document Frequency Assumption:** "Rare terms are no less important than frequent terms." Moreover, terms that occur only with few artists are important descriptors as they are very specific and have high discriminative power.
3. **Normalization Assumption:** "For the same quantity of term matching, long documents are no more important than short documents." The overall length of a document (i.e., the number of contained terms) should have no impact on the relevance with regard to a query or similarity calculation.

The first two assumptions are reflected in the well-established TF·IDF measure that takes into account the number of times a distinct term $t \in T$ occurs within an artist's document $a_i \in A$ [term frequency $tf(t, a_i)$], the number of documents in which t occurs [document frequency $df(t)$], and the overall number of artists $n = |A|$. The general idea of TF·IDF is to consider terms more important for a document when they occur often within the document but rarely in other documents. Technically speaking, terms t with high $tf(t, a_i)$ and low $df(t)$ or, correspondingly, a high inverse document frequency $idf(t)$ are assigned higher weights $w(t, a_i)$. The general scheme of all TF·IDF approaches is

$$w_{tfidf}(t, a_i) = tf(t, a_i) \cdot idf(t) \tag{6.1}$$

with the impact of the document frequency often reduced using a logarithmic dampening within $idf(t)$, i.e.,

$$idf(t) = \log \frac{n}{df(t)}. \tag{6.2}$$

As an example of the effects of this formulation consider the following. When modeling an artist like Bob Marley, we want to put emphasis on the term *reggae*, as this is a typical term (high *tf*) that has high discriminative power as it does not often appear with other types of artists (low *df*). The same would hold for the terms *country* and *prison* when modeling Johnny Cash. On the other hand, a term like *music* will occur in texts associated with all types of artists (high *df*) and thus has very low discriminative power. As a result, the weight w given to the term *music* should be low for all artists to have a low impact when comparing artists. In accordance with other text-based IR tasks, it can be seen that proper names, i.e., artist names, as well as terms from album and track titles typically have very high discriminative power [225].[1]

In the context of artist similarity, several variants of TF·IDF have been proposed. One of them is the *ltc* variant as shown in Eq. (6.3) (cf. [95, 219, 392]).

$$w_{ltc}(t, a_i) = \begin{cases} [1 + \log_2 tf(t, a_i)] \log_2 \frac{n}{df(t)} & \text{if } tf(t, a_i) > 0 \\ 0 & \text{otherwise.} \end{cases} \tag{6.3}$$

Whitman and Lawrence [512] propose another variant of weighting in which rarely occurring terms, i.e., terms with a low *df*, are also weighted down to emphasize terms in the middle *idf* range. The idea is that rarely occurring terms actually often just consist of words with typos or are otherwise noise that should be removed. Equation (6.4) shows this alternative version using Gaussian smoothing where μ and σ need to be empirically chosen.

$$w_{gauss}(t, a_i) = tf(t, a_i) \frac{e^{-(\log df(t) - \mu)^2}}{2\sigma^2} \tag{6.4}$$

[1] In order to enforce highly discriminative terms, optionally, feature selection, e.g., using the χ^2-test [524], can be performed. Such a step requires additional knowledge about the contained items such as class information like genre [219]. However, for artist similarity computation, information about the genre is a priori unknown. Hence, we will not further detail this possibility.

In order to account for the third monotonicity assumption, i.e., the normalization assumption, the ranges of the obtained feature vectors need to be equalized before comparison. Again, there are several methods to accomplish this. A simple method consists of normalizing each term value between the local minimum and maximum values for each artist vector [512], i.e.,

$$w_{rangenorm}(t, a_i) = \frac{w(t, a_i) - \min_{u \in T}(w(u, a_i))}{\max_{u \in T}(w(u, a_i)) - \min_{u \in T}(w(u, a_i))}. \tag{6.5}$$

A more common strategy is to normalize the feature vector to unit length [Eq. (6.6)].

$$w_{unitnorm}(t, a_i) = \frac{w(t, a_i)}{\sqrt{\sum_{u \in T} w(u, a_i)^2}}. \tag{6.6}$$

Table 6.1 shows the ten highest scoring terms for artists from the toy data set after TF·IDF weighting [Eq. (6.3)] and unit length normalization. The term descriptors have been obtained by joining the top 20 pages from *Google* for the query *"artist name" music* into a pseudo document, extracting unigrams, and removing stopwords as well as terms appearing in less than three or more than 50 % of the documents. The total number of distinct terms used is about 7000.

As expected, among the most important terms, we see many stemming from the artist's name (*shostakovich*, *snow*), band members or real names (*simmons*, *stanley*, and *frehley* for Kiss, *brien* for Snow), or related artists (*mahler*, *jeezy*, *timbaland*), as well as from record names (*destroyer* for Kiss, *blueprint*, *takeover*, *throne*, *shoulder*, and *knock* for Jay-Z, *inches* for Snow). Other terms are descriptive for the type of music or context of the musician, such as *concerto*, *symphonies*, *moscow*, or *jewish* for Shostakovich; *makeup* and *spitting* for Kiss; *vignette*, *verse*, and *gangster* for Jay-Z; and *jamaican* for Snow. Interestingly, terms related to books play an important role as well. Testimony is the title of Shostakovich's memoirs. Leaf and Gill are both authors of biographies on Kiss. The book-related features contained

Table 6.1 Highest scoring artist terms after TF·IDF calculation and unit length normalization

Dmitri Shostakovich		Kiss		Jay-Z		Snow	
shostakovich	0.135	simmons	0.120	blueprint	0.111	lynne	0.116
testimony	0.081	makeup	0.094	takeover	0.078	snow	0.109
concerto	0.078	frehley	0.080	throne	0.074	array	0.107
symphonies	0.077	kiss	0.076	shoulder	0.074	inches	0.094
philharmonic	0.075	gill	0.075	knock	0.073	rae	0.084
mahler	0.072	stanley	0.071	jeezy	0.072	jamaican	0.082
moscow	0.071	destroyer	0.070	vignette	0.072	brien	0.082
jewish	0.071	subscriber	0.068	timbaland	0.072	embed	0.080
operas	0.069	leaf	0.067	verse	0.072	canoe	0.077
conservatory	0.068	spitting	0.066	gangster	0.069	fish	0.079

in the term set for Snow are completely disconnected from the artist. Lynne Rae Perkins (all the terms are ranked high) is the author of a book titled Snow Music, which perfectly matches the query initially sent to *Google*. Generally, it can be seen that the features for Snow are of low quality. Despite further confusions with other music (Go Fish's album titled Snow), the remaining terms are not descriptive. This effect is caused by the artist's generic name and the inconsistent set of retrieved web pages (cf. Sect. 5.4).

6.1.1.1 Similarity Calculation

Calculation of (normalized) feature vectors allows the calculation of the similarity between two artists a_i and a_j. A simple strategy is to compute the overlap between their term profiles, i.e., the sum of weights of all terms that occur in both artists' sets [Eq. (6.7)].[2]

$$sim_{overlap}(a_i, a_j) = \sum_{\{\forall t \in T | w(t,a_i)>0 \wedge w(t,a_j)>0\}} w(t, a_i) + w(t, a_j) \tag{6.7}$$

The most popular measure for similarity estimation in the VSM is the cosine measure. In the cosine measure, the similarity between two term vectors is defined as the cosine of the angle θ between the two vectors [Eq. (6.8)]. If both vectors are unit normalized [cf. Eq. (6.6)], this is equal to calculating the inner product (dot product) of the vectors.

$$sim_{cos}(a_i, a_j) = \cos\theta = \frac{\sum_{t \in T} w(t, a_i) \cdot w(t, a_j)}{\sqrt{\sum_{t \in T} w(t, a_i)^2} \cdot \sqrt{\sum_{t \in T} w(t, a_j)^2}} \tag{6.8}$$

As can be seen, by measuring the angle between two document vectors, explicit normalization of vectors prior to comparison is not necessary as this is already expressed in the similarity calculation. In case of another similarity measure (e.g., Euclidean distance), normalization might still be required to remove the influence of different document lengths. For a comparison of different weighting, similarity, and normalization schemes for artist similarity, see [416].

Figure 6.1 shows the similarity matrix for the toy music data set using the features described before (cf. Table 6.1) and the cosine measure for similarity estimation. The similarity among artists from the same genre is on average higher than for artists from other genres, reflected in the intra-/inter-genre similarity values. The

[2]Note that this simple scoring function can also be used to calculate similarities over multiple feature vectors originating from different term set vectors.

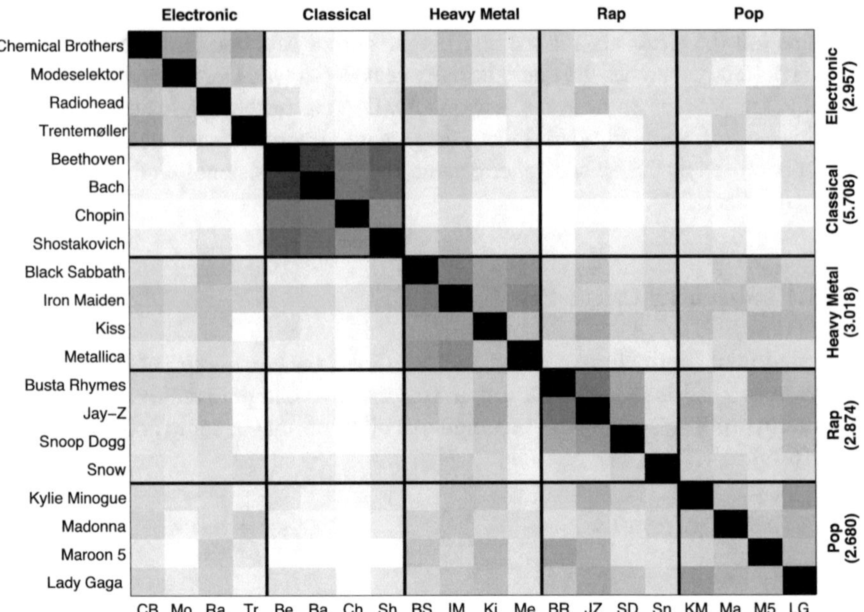

Fig. 6.1 Visualization of similarity matrix computed from the VSM using web pages retrieved via *Google*

average intra-/inter-genre similarity ratios amount to 3.447 for the mean and 2.957 for the median. These values are overall higher than the values achieved with the content-based approaches (cf. Figs. 3.5, 3.9, and 3.19). As with content-based methods, Classical as a genre is discriminated best (ratio value of 5.708), since this genre is clearly different from the others in terms of acoustic qualities and cultural context. Within the genres, Radiohead in Electronic, Chopin in Classical, Kiss in Heavy Metal, and Snow in Rap stick out for showing lower intra-genre similarity. These findings correspond with the general considerations on the toy music data set (Sect. 1.5.4).

Similarity according to Eqs. (6.3) and (6.8) can also be used for clustering of artists in a hierarchical browsing interface as described in [349]. Instead of constructing the feature space from all terms contained in the downloaded web pages, in this approach, a manually assembled vocabulary of about 1400 terms related to music (e.g., genre and style names, instruments, moods, and countries) is used. This domain-specific vocabulary is also displayed in the browsing process and allows for better orientation in a semantic music space.

6.1.2 Latent Semantic Indexing

We have seen that the VSM provides several advantages, such as the uniform representation of entities as vectors, the assignment of weights to terms according to importance, and straightforward score computation based on the cosine measure, making it also applicable in clustering and classification tasks. One of its shortcomings is the assumed mutual independence of the feature dimensions which is not justified when dealing with natural language data which is inherently redundant and ambiguous. Two classic problems arise with natural language that the VSM cannot deal with, namely, *polysemy* and *synonymy*. Polysemy refers to words that have more than one meaning. Examples from the music domain are *bass*, *tune*, and *cover*. Synonymy refers to different words that have the same meaning. An example from the music domain are the words *forte*, *loudly*, and *strong*. Also cases such as abbreviated variations fall into this category, e.g., gangsta rap vs. G-rap. Additionally, there are a number of "quasi-synonyms" or "de facto synonyms" in music that basically have the same meaning in a general discourse, such as black metal and death metal. In the VSM, synonymous words are each represented in a separate dimension and can therefore not be matched. Likewise, polysems are all matched to the same dimension, resulting in distorted feature representations and overestimated similarity values.

In order to deal with these limitations, these latent semantic associations of terms can be modeled by means of co-occurrence analysis, i.e., from a given corpus of documents (e.g., a set of artist representations), clusters of co-occurring terms belonging to a concept can be uncovered. To this end, let us extend the notation of the VSM and consider an $m \times n$ weight matrix W where each row corresponds to a term $t_{j=1\ldots m}$, $m = |T|$ and each column to a document $a_{i=1\ldots n}$, i.e., a described musical entity. The columns of the matrix thus refer to the term vectors already defined for the VSM. The entries of each cell w_{ij} can again be calculated using the TF·IDF measure or simply as a binary matrix with $w_{ij} = 1$ iff t_j appears in document a_i, and $w_{ij} = 0$ otherwise. The clustering is now achieved by approximating W by a matrix W_ℓ of lower rank using truncated *singular value decomposition* (SVD). This not only yields new representations of musical entities based on the uncovered concepts but also reduces the impact of noise in the data by discarding concepts of lower order. This approach is known as *latent semantic indexing* (LSI) or *latent semantic analysis* (LSA) [96].

In case W is a real-valued matrix, as can be assumed when dealing with term weights, the SVD of W is a factorization of the form

$$W = U \Sigma V^T, \tag{6.9}$$

Where:

- U is an $m \times m$ unitary matrix (whose columns are called the left-singular vectors of W).
- Σ is an $m \times n$ rectangular diagonal matrix with nonnegative real numbers σ_{ii} on the diagonal sorted in decreasing order, i.e., $\sigma_{ii} \leq \sigma_{i+1,i+1}, i+1 < \min(m,n)$, known as the singular values of W.
- V is an $n \times n$ unitary matrix (whose columns are called the right-singular vectors of W).

Note that the SVD is related to the eigendecomposition as the left- and right-singular vectors are related to eigenvectors of WW^T and W^TW, respectively, and the singular values to the eigenvalues of both. The interpretation of these matrices when factoring a term-document matrix is as follows:

- U relates to the terms (and is therefore known as the SVD term matrix). Its jth row $\mathbf{u_j}$ represents the SVD vector for term t_j, with each dimension l corresponding to the lth largest singular value.
- V relates to the documents and is therefore known as the SVD document matrix. Analogous to U, its ith row $\mathbf{v_i}$ represents the SVD vector for document a_i, with each dimension l corresponding to the lth largest singular value.

For a low-rank approximation W_ℓ of W, we can now simply truncate these matrices and thus map terms and documents to an ℓ-dimensional space that is defined by the ℓ largest singular values. We obtain $W_\ell = U_\ell \Sigma_\ell V_\ell^T$, where U_ℓ, Σ_ℓ, and V_ℓ are matrices with dimensionality $m \times \ell$, $\ell \times \ell$, and $n \times \ell$, respectively. W_ℓ is therefore still an $m \times n$ matrix, irrespective of the value chosen for ℓ (which is typically set to be between 50 and 350 and significantly below the rank of W). However, unlike W, W_ℓ (as well as U and V) can contain negative cell values. Each dimension of the ℓ-dimensional space can be interpreted as a latent semantic cluster, and the (absolute) values in U_ℓ and V_ℓ can be interpreted as the affinity to this cluster for each term and document, respectively.

We want to exemplify this and the potential effects of LSI on an artist similarity task. To this end, we download the *Last.fm* top tags with a score above 20 for each artist in the toy music data set (resulting in a total number of 167 distinct tags) and use the associated scores as a raw frequency count value. The resulting artist vectors are further unit normalized and the full weight matrix factorized using SVD. Table 6.2 shows the five most important terms per concept cluster ($\ell = 9$). The first five clusters largely correspond to the five genres present in the collection (in order of decreasing associated singular value): pop, rap, heavy metal, electronic, and classical. Going beyond these main genre clusters, we see more specific concepts emerging that accommodate for specific artists and substyles. Cluster 6 covers the "alternative" side of pop and rock music. Clusters 7 and 9 represent specific aspects of so-called classical music, namely, modern compositions by Shostakovich (7) and Baroque music by Bach as well as Romantic music by Chopin (9). Cluster 8 specifically models the artist Snow, who is tagged as *reggae* and with other expected

Table 6.2 Most important tags for the first nine latent concept clusters after applying LSI

Cluster 1		*Cluster 2*		*Cluster 3*	
pop	−0.591	rap	−0.867	heavy metal	−0.660
electronic	−0.540	hip hop	−0.431	hard rock	−0.422
dance	−0.382	east coast rap	−0.114	metal	−0.387
female vocalists	−0.248	west coast	−0.114	rock	−0.295
rock	−0.163	jay-z	−0.079	classic rock	−0.225
Cluster 4		*Cluster 5*		*Cluster 6*	
electronic	−0.506	instrumental	−0.498	alternative	−0.542
pop	−0.466	romantic	−0.362	rock	−0.488
techno	−0.294	composers	−0.349	alternative rock	−0.420
instrumental	−0.241	piano	−0.333	heavy metal	−0.244
minimal	−0.223	electronic	−0.256	pop rock	−0.239
Cluster 7		*Cluster 8*		*Cluster 9*	
russian	−0.590	reggae	−0.788	baroque	−0.720
contemporary classical	−0.533	japanese	−0.378	piano	−0.436
modern classical	−0.407	j-pop	−0.371	romantic	−0.335
composer	−0.354	90s	−0.206	german	−0.260
Piano	−0.188	anime	−0.110	polish	−0.209

related terms but also as *j-pop*, *japanese*, or *anime* due to a mix-up with the Japanese artist SNoW.

Using the same data, we can also calculate the similarities between artists via their latent feature representations v_i truncated to the $\ell = 9$ dimensions discussed before. To this end, we unit normalize all v_i and calculate pairwise Euclidean distances between them. The resulting (inverted and range normalized) similarity matrix can be seen in Fig. 6.2. We see how the compacted feature space almost "binarizes" the similarity assessments, making the nuances in similarity between artists almost disappear. This is not necessarily a general effect of LSI but rather due to the very small value of ℓ chosen for this example on a small collection. The five main genre clusters are well separated and most artists are most similar to the artists from their respective genre, also reflected in comparably high average intra-/inter-genre similarity ratios (mean: 5.942, median: 5.961). However, with the additional four dimensions (clusters 5–9 in Table 6.2), specific separations from the main genres can be observed. First, classical music is not represented as consistent as in all approaches discussed before (content- and context-based) and has the lowest intra-/inter-genre similarity ratio of all genres. There are actually three different subgroups of "classical music" within the nine chosen dimensions. Second, cluster 6 takes the function of grouping all types of "alternative" types of music, even if they are not similar to each other, i.e., Radiohead, Kiss, and Maroon 5, which are in fact the outliers of their respective genre, leading to unwanted high similarity values between them. In contrast to the separation of classical music, this is a case of polysemy that is not handled beneficially by LSI within the chosen compact

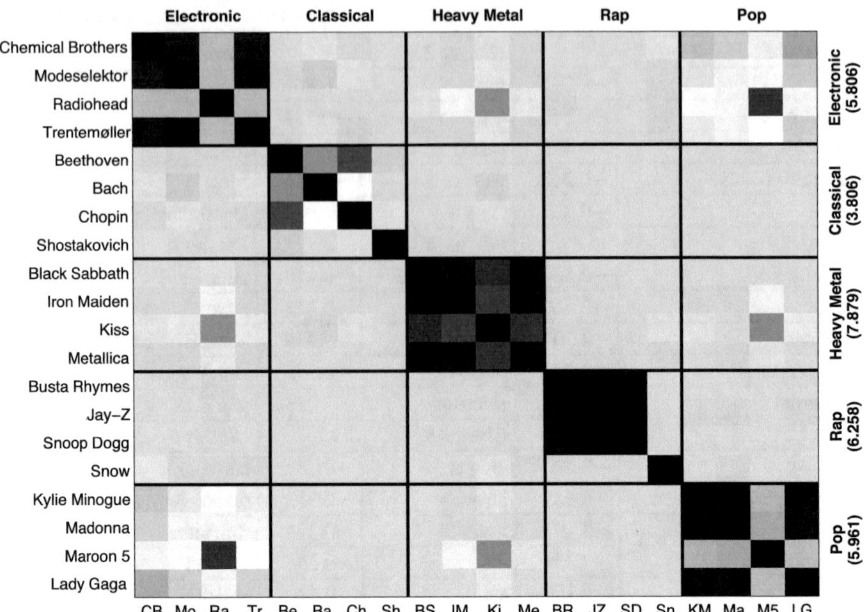

Fig. 6.2 Visualization of the range normalized similarity matrix computed from the SVD document matrix V_ℓ ($\ell = 9$) using *Last.fm* top tags

representation. Adding further dimensions would allow for a more fine-grained distinction of the overall too general (and thus to some extent meaningless) tag *alternative*. Third, the artist Snow is not similar to any other artist in the collection. While this bears some truth, the clear separation observed here is mostly an effect of the artist mix-up in the tagging process that cannot be compensated for with the application of LSI.

We can see that applying LSI results is a compact and more robust feature representation that can model interdependencies of terms considered independent in the VSM. The choice of the parameter ℓ is crucial, however. Truncating too many dimensions by setting ℓ too low results in a loss of important information and limits the capabilities for modeling synonymy and polysemy. Choosing a value that is too high results in an unnecessarily large feature representation and preserves noise contained in the data. In the following section, we review work that has used LSI and related latent factor approaches for processing music context-based data sources.

6.1.3 Applications of Latent Factor Approaches

Similar to the example discussed in the previous section, however, on the track level, Levy and Sandler [263] utilize social tags to construct a semantic space for music pieces. To this end, the sequence of all tags retrieved for a specific track is treated as a normal text description, tokenized, and a standard TF·IDF-based document-term matrix is created. For the TF factor, three different calculation methods are explored, namely, weighting of the TF by the number of users that applied the tag, no further weighting, and restricting features to adjectives only. Optionally, the dimensionality of the vectors is reduced by applying LSI. The similarity between vectors is calculated via the cosine measure [Eq. (6.8)]. For evaluation, for each genre or artist term, each track labeled with that term serves as a query, and the mean average precision [see Eq. (1.4)] over all queries is calculated. It is shown that filtering for adjectives clearly worsens the performance of the approach and that weighting of term frequency by the number of users may improve genre precision (however, it is noted that this may just artificially emphasize the majority's opinion without really improving the features). Using LSI, the mean average precision increases by two percentage points in both genre-based and artist-based evaluation.

LSI can be used in mood categorization or music emotion recognition; cf. Sect. 4.3.2. Laurier et al. [252] strive to classify songs into four mood categories by means of lyrics and content analysis. For lyrics, the TF·IDF measure with cosine distance is chosen with optional dimensionality reduction using LSI (achieving best results when projecting vectors down to 30 dimensions). Audio-based features performed better compared to lyrics features; however, a combination of both yielded best results.

Logan et al. [285] apply *probabilistic latent semantic indexing* (PLSI) [183] to a collection of over 40,000 song lyrics by 399 artists to extract ℓ topics typical to lyrics and determine artist similarity. PLSI, in contrast to LSI, is not based on SVD but assumes a latent class model underlying the observed co-occurrences of words and documents. This latent model is represented as a mixture of conditionally independent multinomial distributions whose parameters are estimated using the EM algorithm; cf. Sect. 3.2.4.[3] This allows the interpretation of each dimension of the reduced feature vectors as the likelihood of the artist's tracks to belong to the corresponding topic. In the next step, ℓ-dimensional artist vectors are created from all lyrics by the artist. The artist vectors are compared by calculating the L_1-norm distance, i.e., Manhattan distance, as shown in Eq. (6.10); cf. Eq. (3.6).

$$dist_{L_1}(a_i, a_j) = \sum_{k=1}^{N} \left| a_{i,k} - a_{j,k} \right| \tag{6.10}$$

This similarity approach is evaluated against human similarity judgments and yields worse results than similarity data obtained via acoustic features (irrespective of the chosen ℓ, the usage of stemming, or the filtering of lyrics-specific stopwords). However, as lyrics-based and audio-based approaches make different errors, a

[3] A generative extension to this approach, namely, *latent Dirichlet allocation* (LDA) is described in Sect. 8.3.

combination of both is suggested also for the task of similarity estimation (cf. above and Sect. 6.4).

In a similar approach, Kleedorfer et al. [214] apply *nonnegative matrix factorization* (NMF) [257] to reveal conceptual clusters within lyrics. NMF is similar to LSI; however, NMF factorizes a matrix W typically only into two matrices U and V such that $W = UV$. A prerequisite for W and a property of U and V is that all of them (must) contain no negative entries. Kleedorfer et al. find that a value of $\ell \geq 60$ is necessary to capture relevant topics and that the use of binary term vectors shows preferable results over TF and TF·IDF weights for NMF clustering.

Pohle et al. [365] apply NMF to a data set of 1979 artist term vectors. Artists are represented by TF·IDF vectors extracted from artist-related web pages using a vocabulary of about 3000 tags retrieved from *Last.fm*. As with LSI, applying NMF to the TF·IDF features results in an implicit clustering of the artists. As NMF yields a weighted affinity of each artist to each cluster, Pohle et al. propose an interface for browsing the artist collection by allowing to set the desired impact of each cluster and displaying the closest artists based on the similarity to this setting (see Sect. 9.2.2). To facilitate this step, expressive labels are manually assigned to the clusters after inspection.

6.2 Text-Based Indexing and Retrieval

The focus so far was on deriving representations for musical entities and calculating similarity values between them. This scenario allows the retrieval of the most similar entities given a seed entity (query by example).

A more general scenario is to use the contextual data and the derived representations for indexing of the entities to automatically build a search engine for music collections that can be queried through natural language beyond a limited set of manually assigned indexing terms as employed by most existing music search systems (e.g., those offered by commercial music resellers; cf. Chap. 1). For example, instead of just retrieving tracks labeled *rock* by some authority, a system should allow for formulating a query like *rock with great riffs* to find songs with energetic guitar phrases [223].

We discuss three approaches to indexing of music collections based on contextual data. The first approach builds upon page aggregation and the VSM representations discussed earlier. The second directly operates on a document index and exploits the connections with the entities to obtain a relevance ranking. The third approach aims at learning a rich set of semantic labels and thus relates to auto-tagging approaches as discussed in Sect. 4.2. For the approaches presented, we assume the case of having a set of (preprocessed) web pages related to each music track in the collection; cf. Sect. 5.4.1. We use the same set of symbols as in the examples before, i.e., a musical entity (here: an individual track) is denoted by a_i.

6.2.1 Pseudo Document Indexing

The basic idea behind this approach to indexing individual music pieces is simply to agglomerate all text information associated to each piece and treat this agglomeration as one document. This straightforward approach builds on the aggregation of the set of related texts D_i for each music piece a_i. To this end, all $d \in D_i$ are concatenated into a single pseudo document ψ_i. The resulting pseudo documents can then be indexed using a standard document indexing approach such as an inverted index to quickly retrieve the subset of documents relevant to a query; cf. Sect. 5.4. Furthermore, all ψ_i are transformed into a VSM representation by calculating term weights.

Given the query q, relevance of the pseudo documents with regard to q can be obtained by applying the cosine similarity measure $sim_{cos}(\psi_i, \mathbf{q})$. A query q can simply be considered a very short document that needs to be processed like any other document (e.g., tokenized and TF·IDF weighted). Using the similarity values, an ordered list of pseudo documents can be obtained. This ranking of ψ_i is directly interpretable as a ranking of a_i, i.e., a ranking of music pieces since there exists a direct and unique mapping between music piece a_i and text document ψ_i. A schematic overview of the pseudo document approach can be seen in Fig. 6.3.

With the discussed approach, it is possible to retrieve music entities for queries that overlap with the contextual data. Thus, the goal is to collect a substantial amount of texts related to the pieces in the music collection to obtain diverse descriptions and a rich indexing vocabulary that allows for a large number of possible queries. In order to extend the space of potential queries to virtually any query, and also to broaden the result sets, a query expansion step involving an online search engine can be integrated [223].

To this end, a given query q is augmented by the constraint *music* and sent to a web search engine. From the x top-ranked web pages (with, e.g., $x = 10$), a pseudo

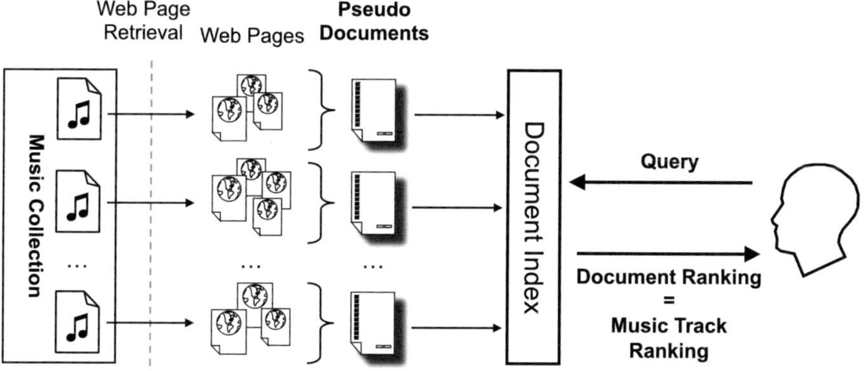

Fig. 6.3 Schematic overview of the pseudo document approach to indexing and retrieval

document ψ_q is constructed. As with a regular query, ψ_q needs to be processed and transferred to the VSM defined by the set of ψ_i. The "expanded," i.e., less sparsely populated query vector is then again compared to all ψ_i in the collection by calculating the cosine similarities to their respective vectors.

As another extension, the VSM of pseudo documents can be transformed to a lower-dimensional space using LSI (see Sect. 6.1.2). In order to cast regular as well as expanded queries into the projected feature space, they first need to be transformed according to Eq. (6.11).

$$q_\ell = \Sigma_\ell^{-1} U_\ell^T q \tag{6.11}$$

Using this representation, similarity measures can again be used to rank candidates. The discussed extensions, query expansion and LSI, could also be used as extensions to the approach presented next.

6.2.2 *Document-Centered Rank-Based Scoring*

This approach to obtain a relevance ranking of music pieces with respect to a given text query operates directly on the indexed set of all individual web pages $I = \bigcup D_i$. Instead of just joining all available information for one music piece without differentiating and constructing pseudo documents that are heterogeneous in structure and content, a simple ranking function is introduced to propagate relevance information from a web document ranking to scores for individual music pieces.

More precisely, relevance of a music piece a_i with respect to a given query q is assessed by querying I with q to obtain a ranking of web pages and applying a technique called *rank-based relevance scoring* (RRS) [224] to the r most relevant web documents. RRS exploits the associations between pages and tracks established in the data acquisition step. A schematic overview of this approach can be seen in Fig. 6.4.

The idea of the RRS scoring function is that if a web page is highly relevant to query q, this might also be an indicator that the music piece(s) for which this web page was relevant in the data acquisition step is also highly relevant. A web page that is not as relevant to q is seen as an indicator that associated music pieces are also not as relevant to q. This relation is expressed in the RRS scheme by exploiting the rank of a web page p within the web page ranking obtained for query q. The relevance score of a music piece is assessed by summing up the negative ranks of all associated web pages occurring in the page ranking as shown in Eq. (6.12).

$$RRS_r(a_i, q) = \sum_{p \in D_i \cap D_{q,r}} -rnk(p, D_{q,r}) \tag{6.12}$$

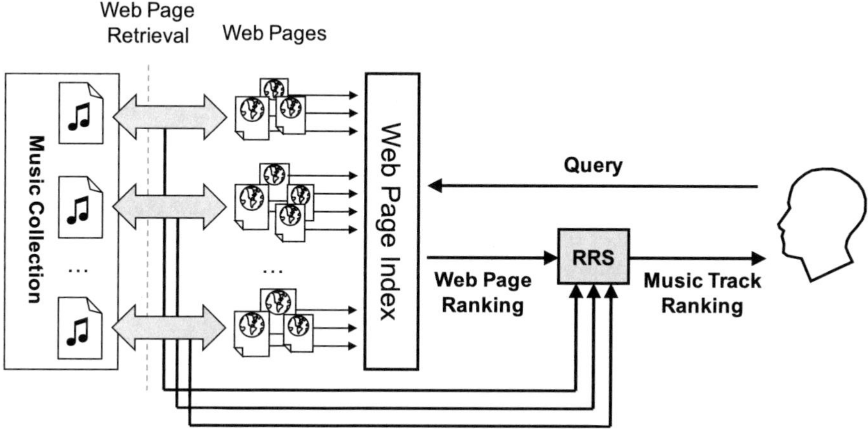

Fig. 6.4 Schematic overview of the web-document-centered approach to indexing and retrieval

Here, r denotes the maximum number of top-ranked documents when querying I, $D_{q,r}$ the ordered set (i.e., the ranking) of the r most relevant web documents in I with respect to query q, and $rnk(p, D_{q,r})$ the rank of document p in $D_{q,r}$. The additional parameter r is introduced to limit the number of top-ranked documents when querying the page index. For large collections, this is necessary to keep response times short. For music retrieval, the final ranking is obtained by sorting the music pieces according to their RRS value in descending order.

Turnbull and Barrington et al. [482, 484] suggest an alternative version of relevance scoring called *weight-based relevance scoring* (WRS) that incorporates the relevance scores of the web page retrieval step rather than the ranks. In their experiments, they show that this can be a beneficial modification.

6.2.3 *Auto-Tag Indexing*

In principle, all contextual data sources discussed in Chap. 5 can be used for indexing of music collections. In particular, this is the case for tags—regardless of whether they are assigned by experts, collected collaboratively, or via a game—which naturally represent labels and indexing terms. Thus, contextual data from these sources can be directly used for indexing and matched against queries.

Another approach to obtain indexing terms that has been discussed earlier in Sect. 4.2 is auto-tagging. As mentioned, auto-tagging employs machine learning techniques to predict a set of semantic labels from a variety of music features. Early approaches use pure audio features for this purpose; however, later work discovered the advantages of integrating contextual data to predict semantic tags.

Turnbull et al. [484] exploit social tags and web-mined tags (extracted from web pages using the WRS scheme; cf. Sect. 6.2.2) together with audio content information to predict the semantic labels of the CAL-500 set [480]. Other approaches exploit similar contextual sources [87, 206]. Turnbull et al. show that these individual sources already perform well in text retrieval tasks, and depending on the type of tag, different sources have both their strengths and weaknesses. For example, they find that the best retrieval results for the tag "jazz" are achieved with web-mined tags whereas the best results for "hip-hop" are achieved using MFCCs. Since jazz is a musically distinguishable concept, this shows that the chosen acoustic features are not capable of modeling its characteristics and that contextual features are instead necessary. Consequently, it is shown that combinations of the individual rankings outperform the single sources. The aspect of combining content-based and context-based information will be further investigated in Sect. 6.4.3.

Auto-tag-based indexing typically outperforms unsupervised indexing approaches such as pseudo document indexing and RRS due to the existence of a learning target and the possibility to optimize parameters in a supervised manner. On the other hand, unsupervised indexing approaches operate on an unrestricted, open indexing vocabulary that enables also to handle queries that address a priori neglected characteristics. For auto-tagging, the limiting factor is that the set of indexing terms needs to be defined first and remains fixed.

6.3 Similarity Based on Co-occurrences

Using the text-based methods presented so far, we have always constructed a feature space and estimated similarity as a function of the distance of the data points in this feature space. Instead of constructing feature representations for musical entities, the work reviewed in this section follows an immediate approach to estimate similarity. In principle, the idea is that the occurrence of two music pieces or artists within the same context is considered to be an indication for some sort of similarity.

As we have discussed before, one such contextual data source is related web pages. Determining and using such music-related pages as a data source for MIR tasks was likely first performed by Cohen and Fan [78] who automatically inspect web pages obtained by querying web search engines for co-occurrences of artist names. For all artists in the collection, the top 100 search result URLs for the query *"artist name"* are collected. Only URLs that occur at least twice are retained as these must contain at least two different artist names. The resulting HTML pages are parsed for lists according to their markup structure, and all plain text content with minimum length of 250 characters is further analyzed for occurrences of entity names. Co-occurrences of two entity names within the same list, i.e., within a very narrow context, are considered evidence for similarity and used for artist recommendation in a collaborative filtering approach (see Chap. 8).

A similar approach that broadens the context for co-occurrences to full web pages is presented in [401, Chap. 3]. First, for each artist a_i, the top-ranked web pages returned by the search engine are retrieved. Subsequently, all pages fetched for artist a_i are searched for occurrences of all other artist names a_j in the collection. The number of page hits represents a co-occurrence count $coco(a_i, a_j)$, which equals the document frequency of the artist name "a_j" in the web pages associated with artist a_i. Relating this count to $co(a_i)$, the total number of pages successfully fetched for artist a_i, a symmetric artist similarity function is constructed as shown in Eq. (6.13).

$$sim_{wp\_cooc}(a_i, a_j) = \frac{1}{2} \cdot \left(\frac{coco(a_i, a_j)}{co(a_i)} + \frac{coco(a_j, a_i)}{co(a_j)} \right) \tag{6.13}$$

Employing this method, the number of issued queries grows linearly with the number of artists in the collection. A disadvantage of this method is that particularly for large and diverse artist collections, most entries remain unpopulated. To avoid these sparsity issues, another category of co-occurrence approaches does not actually retrieve co-occurrence information but relies on result counts returned for search engine queries. Formulating a conjunctive query made of two artist names and retrieving the result count estimate from a search engine can be considered an abstraction of the standard approach to co-occurrence analysis. This also minimizes web traffic by restricting the search to display only the top-ranked page if the used search engine offers such an option. However, a shortcoming of such an approach is that creating a complete similarity matrix has quadratic computational complexity in the number of artists. It therefore scales poorly as the number of queries that has to be issued to the search engine grows quadratically with the number of artists in the collection.

Zadel and Fujinaga [531] propose a method that combines two web services to extract co-occurrence information and consecutively derive artist similarity. Given a seed artist a_i as input, their approach retrieves a list of potentially related artists a_j from the *Amazon* web service *Listmania!*. Based on this list, artist co-occurrences are derived by querying a search engine and storing the returned page counts of artist-specific queries. More precisely, *Google* is queried for "a_i" to extract the result page count $pc(a_i)$ and for "a_i" "a_j" to extract the combined result page count $pc(a_i, a_j)$, i.e., the total number of web pages on which both artists co-occur. Thereafter, the so-called relatedness of a_i to each a_j is calculated as the ratio between the combined result count and the minimum of the single page counts of both artists; cf. Eq. (6.14). The minimum is used to account for different popularities of the two artists.

$$sim_{pc\_min}(a_i, a_j) = \frac{pc(a_i, a_j)}{\min\left(pc(a_i), pc(a_j)\right)} \tag{6.14}$$

Recursively extracting artists from *Listmania!* and estimating their relatedness to the seed artist via *Google* result counts allow for an incremental construction of lists of similar artists and population of the similarity matrix. Nevertheless, while this approach focuses on promising candidates for similarity estimation, it is biased towards the candidates given by the web service in the first step and will therefore also not produce a complete similarity matrix. In order to achieve this goal, all pairwise similarities need to be estimated by sending queries that cover all pairwise combinations of artists [407]. This additional effort comes with the benefit of additional information since it also allows for predicting which artists are not similar. Analogous to Eq. (6.13), the similarity of two artists is defined as the conditional probability that one artist is found on a web page that mentions the other artist. Since the retrieved page counts for queries like *"a_i" music* or *"a_i" "a_j" music* indicate the relative frequencies of this event, they are used to estimate the conditional probability. Equation (6.15) gives a formal representation of the symmetrized similarity function that strongly resembles Eq. (6.13).

$$sim_{pc\_cp}(a_i, a_j) = \frac{1}{2} \cdot \left(\frac{pc(a_i, a_j)}{pc(a_i)} + \frac{pc(a_i, a_j)}{pc(a_j)} \right) \tag{6.15}$$

Note that since the result count values of search engines are often only coarse estimations of the true number that can vary due to the distributed architecture of modern search engines and caching strategies, these numbers are often not very reliable and may also lead to the logically impossible situation of a joint artist query (the more constraint query) having a higher result count than one of the individual queries (the more general query). Thus, practically, we need to define $pc(a_i) = max_{j \in A} pc(a_i, a_j)$ with $pc(a_i, a_i) = pc(a_i)$.

Another aspect of the conditional probabilities concerns artist prototypicality. Artists that frequently occur on web pages known to mention other artists must be of a certain relevance. This can be derived by skipping the symmetrization step of the similarity function in Eq. (6.15) and just calculating the conditional probabilities for every pair of artists. In the resulting asymmetric similarity matrix, it can be estimated how often other artists' pages refer to the artist under consideration and how often pages belonging to that artist refer to another artist. The ratio between these counts relates to the prototypicality of the artist [408]. Using genre information, also distortion effects, such as overestimation of prototypicality due to artist names consisting of common speech terms, can be accounted for, for instance, by applying a TF·IDF-like weighting scheme [409].

Figure 6.5 shows a similarity matrix based on conditional probabilities obtained from *Yahoo!* result counts processed using Eq. (6.15). The average intra-/inter-genre similarity ratios for this approach are the highest observed of all methods (mean: 8.651, median: 5.402), with the mean pushed up by the high value of Classical (22.807). Due to extreme outliers (and false assessments, i.e., Modeselektor's similarity to Jay-Z and Maroon 5, Snow's high similarity to Kiss), the visualization is less expressive. However, we can still see a tendency for higher values at the bottom right and within Electronic. The high similarity between Kiss and Snow can be attributed to the fact that both have common speech names.

With search engine result counts being not highly reliable, this simple approach can suffer from heavy distortions. On the other hand, it permits a fast, simple, and mostly accurate estimation of similarity values without the need for complex frameworks and infrastructure. However, the most severe shortcoming of the presented approaches is that, for creating a complete similarity matrix, they require

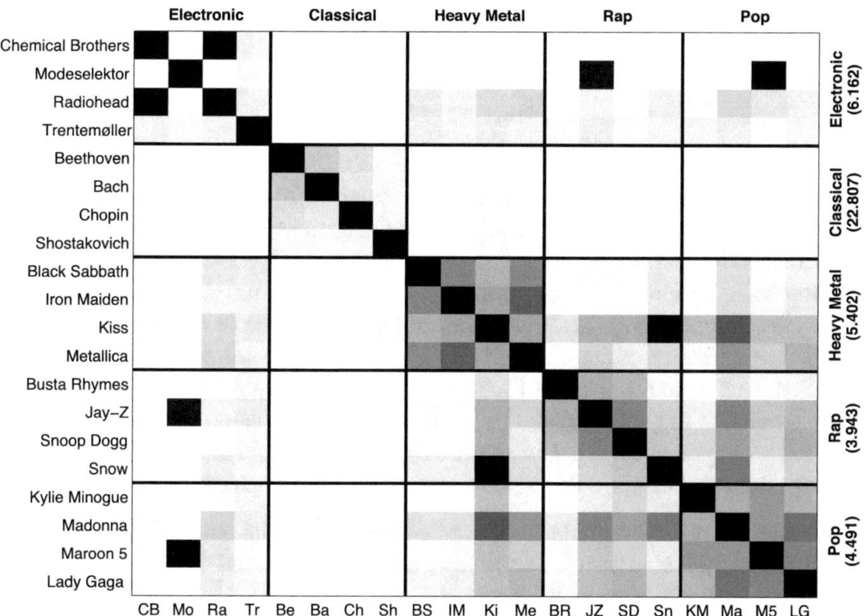

Fig. 6.5 Visualization of the similarity matrix representing conditional probabilities computed from *Yahoo!* search engine result counts

a number of search engine requests that is quadratic in the number of artists. These approaches therefore scale poorly to real-world music collections.

6.4 Combination with Audio Content Information

With context-based and content-based features covering orthogonal dimensions of music information, the logical next step is to combine these in order to obtain a holistic view and to mutually compensate for individual limitations. In this section, we discuss several methods to combine context with content information, either by creating combined similarity measures, by using context data to constrain content similarity, or by joining information from different sources for automatic labeling.

6.4.1 Combined Similarity Measures

Having a set of N range- or z-score-normalized similarity matrices $S_{1...N}$ obtained from N different sources (which can include multiple variants of content similarity and context similarity), the most straightforward way to combine these is a linear combination (e.g., [26]).

$$S_{comb} = \frac{1}{N} \cdot \sum_{i=1}^{N} \alpha_i S_i \tag{6.16}$$

where all α_i are typically empirically determined weighting factors with $\sum_{i=1}^{N} \alpha_i = 1$. This is similar to the fusion of different aspects of acoustic similarity as discussed in Sect. 3.3.5

Nonlinear combinations of similarity measures are usually obtained from a supervised learning step, i.e., given several representations or similarity values, an optimal combination of these should be found in order to approximate a given target function. McFee and Lanckriet [305] investigate different data sources in order to learn similarity judgments given by users and made available in the *aset400* data set [121]. They include five data sources representing different aspects of an artist, namely, timbre, chroma, auto-tags, social tags, and biographical text:

- *Timbre*: An artist is summarized by a GMM over all delta-MFCCs extracted from all songs by the artist (cf. Sect. 3.2.4). Similarity between GMMs is calculated using a probability product kernel.
- *Chroma*: An artist is summarized by a single Gaussian model over all pitch class profiles (PCP) extracted from all songs by the artist. Similarity between models is calculated using the KL divergence (cf. Sect. 3.2.5).
- *Auto-tags*: Semantic tags are predicted for each song using an auto-tagging system [483]. An artist is described by averaging tag probabilities across all of the artist's songs. Similarity is again calculated using a probability product kernel.
- *Collaborative tags*: TF·IDF vectors are extracted from the top 100 *Last.fm* tags for each artist and compared by cosine similarity.
- *Biography*: TF·IDF vectors are extracted from artist biographies. These are also compared by cosine similarity.

To learn a low-dimensional embedding of the high-dimensional feature space such that Euclidean distances in the low-dimensional space match subjective human similarity judgments, *partial order embedding* (POE) is applied to this set of similarity functions. The resulting unified similarity measure is suggested for music artist recommendation and similarity visualization. Note that learning robust metrics from multimodal sources is further discussed in Sect. 8.5.2, where also aspects of user information are included.

One alternative to adaptations on the level of similarity are *early fusion* approaches that combines individual feature representations before calculating similarities on this joint representation. For instance, McKay and Fujinaga [311] examine such combinations of content-based, symbolic, and context-based features for genre classification. Such an early combination can be beneficial also in context-based retrieval to allow indexing of music entities for which no associated information could be found [223]. This could be achieved, e.g., by creating pseudo document term vectors based on documents associated with acoustically similar pieces. For all other tracks, the same strategy can be used to emphasize terms that are typical among acoustically similar tracks. Pseudo document term vectors can be created by combining (or reweighing) term vectors using, e.g., a simple Gauss weighting over the $s = 10$ most similar tracks (for which contextual information could be derived) as given by an audio similarity measure. The audio-reweighed weight of term t for music piece a_i can hence be defined as given in Eq. (6.17).

$$w_{awt}(t, a_i) = \sum_{j=1}^{s} \frac{1}{\sqrt{2\pi}} e^{-\frac{(j/2)^2}{2}} \cdot w(t, a_j), \tag{6.17}$$

Note that audio-weighed term vectors again need to be normalized.

Late fusion approaches, on the other hand, combine the outputs of classifiers trained on data from different modalities. Whitman and Smaragdis [513] use audio-based and web-based genre classification for the task of style detection on a set of five genres with five artists each. In that particular setting, combining the predictions made by both methods linearly yields perfect overall prediction for all test cases.

6.4.2 Contextual Filtering

Different sources can be combined by using one as a filter for the other. For instance, context information can be used to restrict audio similarity to a subset of candidates for retrieval. This can be motivated by a need to avoid strong outliers or unexpected retrieval results that are occurring with audio-only approaches.

Figure 6.6 shows an example of a web-filtered similarity matrix for the toy music data set. First, the task of finding artists similar to the seed artist is carried out by selecting all artists that have a higher similarity value than the 70 percentile of the seed's distribution of similarity values, i.e., selecting the six most similar artists as candidates in the given example. For calculating artist similarity, web features are used; cf. Fig. 6.1. In the next step, content-based similarity values for all tracks by all candidate artists are calculated (in our example, BLF similarity; cf. Fig. 3.19). That is, within the artist-based preselection, audio similarity defines the ranking of retrieved tracks.

Figure 6.6 reveals the asymmetry in nearest neighbor ranking (see also Sect. 3.4 for an explanation). For instance, Iron Maiden is often among the candidates for retrieval even for electronic and classical pieces, whereas Kiss often appears in the retrieved set also for rap and pop pieces. Here, the drawbacks of contextual approaches (common speech name effects) affect the preselection. However, since the final ranking is obtained by acoustic similarity, these candidates will not be top-ranked at retrieval stage. This (and the thresholding; see next) has a positive impact on the overall performance, with averaged intra-/inter-genre similarity ratios of 7.435 and 5.202 for mean and median, respectively.

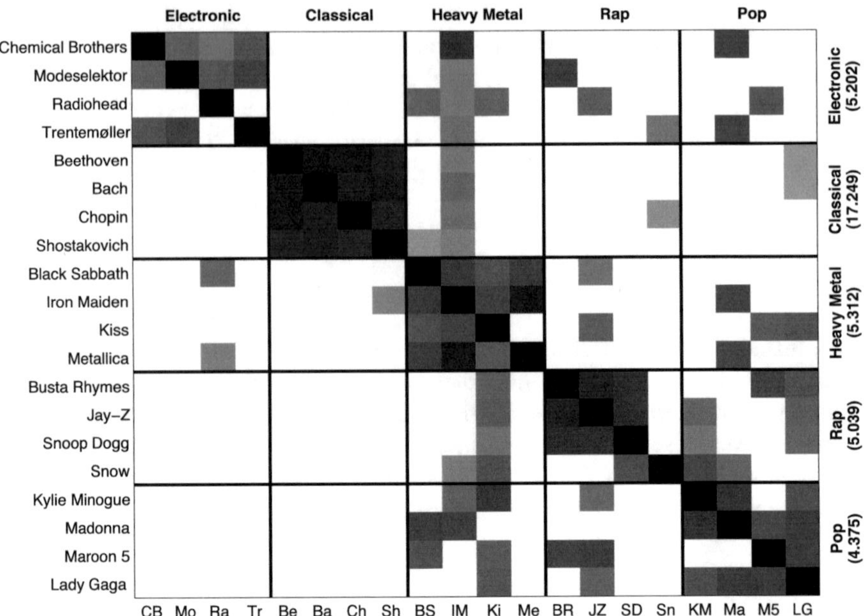

Fig. 6.6 BLF similarity matrix after filtering using web similarity; cf. Figs. 3.19 and 6.1

In the given example, besides the choice of data to use, the crucial parameter is the threshold to be used in order to build a filter. This can be either a defined similarity value or a fixed number of potential candidates. Another example shows that *self-organizing maps* (SOM) [235] can be used instead of a static thresholding [221, 363]. First, web-term-based artist profiles are pre-clustered using a SOM to find promising relations between artists. Then, candidates are selected based on their placement on the SOM grid. More precisely, only audio similarities between tracks of artists that are mapped to the same or adjacent SOM unit are calculated. It is shown that, for the task of playlist generation, this not only improves the quality of playlists by reducing the number of outliers but also accelerates the creation process by reducing the number of necessary audio similarity calculations.

6.4.3 Combined Tag Prediction

Finally, we want to briefly highlight the combination of sources in the prediction of tags. The setting is the same as described in Sects. 4.2 and 6.2.3; however, data from different modalities, explicitly content and context, is jointly exploited to improve accuracy.

As discussed in Sect. 4.3.2, Laurier et al. [252] combine lyrics and audio features for emotion tagging. To this end, they train one binary classifier for each class of emotions and can therefore tag each song with multiple emotion labels. As features, Laurier et al. extract timbral, rhythmic, tonal, and temporal audio descriptors, as well as LSI-compressed TF·IDF vectors from lyrics. For feature fusion, both early and late fusion approaches are investigated with early fusion resulting in the best predictions.

Aucouturier et al. [17] combine timbre similarity with meta-data information to improve the quality of classification according to certain tag categories like genre, instrument, mood, or listening situation. In contrast to the work by Laurier et al., they train all binary meta-data tag classifiers simultaneously and exploit the correlations between tags to mutually reinforce the individual classifiers. This happens in an iterative way by feeding classification decisions made on easier tags as new features into the classification process of more difficult tags. Accounting for the mutual dependencies of musical categories in the described manner results in improved accuracy, particularly for culturally and subjectively motivated categories.

Turnbull et al. [484] investigate social and web-mined tags with MFCC and chroma features with three different combination algorithms. Two of these algorithms, namely, calibrated score averaging (CSA) and RankBoost, operate on the relevance rankings obtained from the individual modalities and joins these ranking into a single ranking based on parameters learned in a training phase. The third, kernel combination support vector machine (KC-SVM), optimizes the weights for a linear combination of the individual similarity matrices (cf. Sect. 6.4.1) while learning the largest margin boundaries for tag class separation.

Experiments show that chroma vectors perform worst of all features and that the combination approaches benefit from omitting them. With regard to the different combination approaches, none is significantly better than the others including a hypothetical combination that automatically selects the best single source for prediction of each tag. However, prediction of the majority of tags is improved by one of the combination algorithms. The authors conclude that another combination step that fuses the outputs of the different combined rankings could further improve performance.

6.5 Stylistic Analysis and Similarity

The existence of rich textual context data renders further analysis and representations beyond simple term profiles possible. To round up this chapter on contextual data analysis and feature extraction, we want to highlight some directions of stylistic analysis using text processing methods. Particularly lyrics are a rich source of stylistic information since they immediately represent artistic output (content), rather than external interpretation or reception of artistic output.

Mahedero et al. [291] demonstrate the usefulness of lyrics for four important tasks: language identification, structure extraction (i.e., recognition of intro, verse, chorus, bridge, outro, etc.), thematic categorization, and similarity measurement.

Using standard TF·IDF vectors as song representations, exploratory experiments indicate some potential for cover version identification and plagiarism detection.

Mayer et al. [302] analyze lyrics for rhyme and statistical properties to improve lyrics-based genre classification. To extract rhyme features, lyrics are transcribed to a phonetic representation and searched for different patterns of rhyming lines (e.g., AA, AABB, ABAB). Features consist of the number of occurrences of each pattern, the percentage of rhyming blocks, and the fraction of unique terms used to build the rhymes. Statistical features are constructed by counting various punctuation characters and digits and calculating typical ratios like average words per line or average length of words. Classification experiments show that the proposed style features and a combination of style features and classical TF·IDF features outperform the TF·IDF-only approach.

Hussein and Brown [178] analyze lyrics for rhyming style information to automatically identify (possibly imperfect) internal and line-ending rhymes. In continuation of this work [179], the proposed high-level rhyme features can be used to identify rap artists based on the stylistic information extracted from the lyrics.

Singhi and Brown [449] compare stylistic features of lyrics, poetry, and articles. Based on adjectives, they investigate whether specific types of expressions are specific to one of the three types of texts. To this end, they estimate the probability distribution of synonymous adjectives for each text type. Given the adjectives of a text, this can be used to classify the genre. Their experiments show a success rate of 67 % for lyrics, 57 % for poetry, and 80 % for articles. Furthermore, they find that adjectives likely to be used in lyrics are easier to rhyme than adjectives likely to be used in poetry. The same method can be used to distinguish “poetic” from “non-poetic” artists. For instance, the lyrics by Bob Dylan and Ed Sheeran are often classified as poetry, whereas lyrics by Drake, Led Zeppelin, and Bryan Adams are not.

6.6 Summary

In this chapter, we dealt with text IR methods to process data retrieved from contextual sources described in Chap. 5 and model this data in order to facilitate calculation of music-context-based similarity measures and to complement content-based approaches. In order to calculate similarities from textual data retrieved from web pages or collaborative tags, in Sect. 6.1, we discussed standard approaches, in particular the vector space model, the TF·IDF weighting scheme, and latent space indexing methods. Based on these approaches, we presented indexing methods and functions for building text-based music search engines in Sect. 6.2. Section 6.3 dealt with similarity calculation by analyzing co-occurrences of identifiers in different data sources, as well as approximations of co-occurrences on web pages by exploiting web search engine result count estimates. Section 6.4 outlined ways to integrate the extracted similarity information with content-based methods such as the ones presented in Part I. We concluded the methodological part of the chapter by

Table 6.3 Average intra-/inter-genre similarity ratios for different approaches presented in this chapter

Approach	Figure	Mean	Median
Web page VSM-Cosine	6.1	3.447	2.957
Tag SVD-Euclidean	6.2	5.942	5.961
Search engine-Co-oc	6.5	8.651	5.402
BLF (web-filtered)	6.6	7.435	5.202

Considered functions to compute a summarization are arithmetic mean and median

discussing methods for stylistic text analysis, in particular in the context of lyrics, in Sect. 6.5.

Table 6.3 summarizes the overall intra-/inter-genre similarity ratios for the similarity approaches presented in this chapter. From the context-only measures, in our setting, the search engine page count co-occurrence approach performs overall best, followed by the SVD-modeled tag approach and the VSM using web pages. All of the values shown are above the values seen for content-based methods; cf. Table 3.1. While this might lead to the easy interpretation that context-based methods in general outperform content-based methods, one needs to be careful not to over-interpret these results. The main reason for that is that meta-data methods are biased, in the sense that they build upon the same cultural knowledge that is underlying the "ground truth" used for evaluation. Audio-based methods might capture other qualities that we do not account for in our evaluation. Furthermore, the context-based method operates on the level of artists, which is more robust than operating on the level of individual tracks, which is usually more heterogeneous. In contrast to the toy music data set, in a real-world setting, an artist is typically assigned more than one track. The "value" of audio information can be seen in the combined measure of web-filtered BLF similarity, which outperforms the measure based on web pages only (which is the weakest of the context-based methods but still outperforms all the audio-based methods). In fact, we can argue that this measure combines the advantages of both domains. Due to the context-based component, the measure gets more "conservative," i.e., it finds the obvious cases as these are represented in cultural sources (incorporating, e.g., popularity biases). At the same time, due to the audio-based component, potential errors made by the context-based methods are less severe, as the final values within the constrained selection stem from acoustic properties. Hence, such a combination typically can avoid outliers from both domains.

6.7 Further Reading

As the methods in this chapter pertain to classic text IR, more detail on all the methods discussed can be found in dedicated IR books. In particular, we recommend to consult the books by Manning, Raghavan, and Schütze [297]; Büttcher, Clarke, and Cormack [53]; and Baeza-Yates and Ribeiro-Neto [20]. In addition to the basic methods presented here, other representation and retrieval models popular in text IR (and applicable to MIR as well) are discussed. The covered topics comprise probabilistic retrieval, language models, classification and clustering techniques, meta-learning, parallel and distributed retrieval, inverted indices construction, relevance feedback, query expansion, and system evaluation.

Part III
User-Centric MIR

Part III covers chapters on listener-centered and user-context-aware MIR. Next to aspects of multimodal similarity and retrieval, this includes discussions of user needs beyond accuracy, such as diversity and serendipity, factors of user- and context-awareness, adaptive similarity measures, and collaborative recommendation methods.

Chapter 7 is devoted to sources of listener-centered information. More precisely, we address users' interaction traces with music, which can be gathered from social networks, ratings, playlists, listening histories, and others. Furthermore, aspects of the listening context are discussed, including the listener's location and activity as well as time and weather. Since music recommendation is one of the major application areas of this type of data, Chap. 8 covers corresponding collaborative similarity and recommendation techniques. Hybrid methods are discussed both in Chaps. 6 and 8, where the former summarizes hybrids including at least two aspects of content and music context information, whereas the latter covers methods that combine user-centric factors with either content or music context aspects as well as methods that integrate all three aspects.

Chapter 7
Listener-Centered Data Sources and Aspects: Traces of Music Interaction

While Chaps. 5 and 6 addressed contextual factors of the music items themselves, this chapter targets the interaction between users and music items. We first describe how respective data can be organized in terms of abstraction level. Subsequently, we detail data acquisition from social networks, ratings, playlists, listening histories, and others. Furthermore, aspects of the user context are discussed, including the listener's location and activity as well as time and weather, which are typically acquired through sensor data from smart phones and web services, respectively. Users' connections in social networks are also discussed in this context. We round off the chapter by analyzing some factors of user intention when listening to music. How the discussed interaction traces and aspects of the user context and user intention can be exploited for music similarity and music recommendation is the topic of Chap. 8.

7.1 Definition and Comparison of Listener-Centered Features

We categorize listener-centered aspects that reflect or influence music preference into (1) *interaction traces* between users and music items and (2) the *listener context*. While the former expresses explicit or implicit relationships between a listener and a music item, the latter refers to factors external to and usually independent of the listener, for instance, time, weather, location, or surrounding people. Examples of an explicit relationship between a listener and an item include a particular item rating or selection of a song to play. An implicit relationship can be established, for example, by uninterrupted playback of a track. However, such implicit feedback should be treated with care. While playing a song to completion may be interpreted as implicit positive feedback, it might also just indicate that the user has become disengaged or decreased the volume.

P. Knees, M. Schedl, *Music Similarity and Retrieval*, The Information Retrieval Series 36, DOI 10.1007/978-3-662-49722-7_7

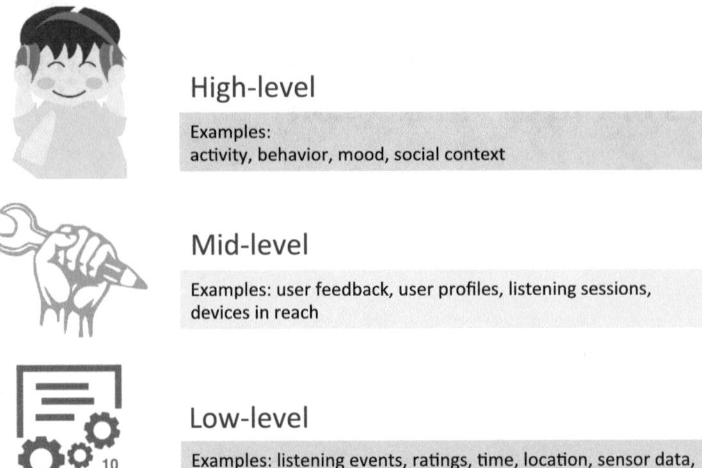

Fig. 7.1 The different levels of feature abstraction for user data

Both categories of listener-centered aspects exhibit to a certain degree an interrelationship. For example, the current weather condition (listener context) may well have an impact on the rating a listener assigns to an item (interaction trace). We hence refrain from discussing interaction traces and context strictly separately and instead provide a joint analysis.

As shown in Fig. 7.1, listener-centered aspects can again be categorized according to their level of abstraction. Low-level features include singular listening events (interaction trace) or raw sensor data gathered, for instance, from smart phones (context). Semantically more meaningful, and therefore to be found on the mid-level, is information about a user's listening session or playlist (interaction trace) or other devices in reach or connected to the user's smart phone (context). Examples of high-level aspects include the user's musical activity (interaction trace) or his or her social context, e.g., being together with family, colleagues, or alone.

Similar to music context features discussed in the previous two chapters, listener-centered data can be gathered from a variety of sources. The ones we are dealing with in this book are compared in Table 7.1, according to various dimensions. All data sources suffer from the problem of incomplete or erroneous meta-data, for instance, typos in postings on social networks and microblogs or tracks shared on peer-to-peer (P2P) networks without ID3 tags assigned. While playlist and rating data as well as data gathered from microblogs and social networks is commonly available locally on the user device or can easily be accessed via APIs, P2P networks typically do not provide APIs to fetch user and collection data. Hence, accessing this kind of data source requires dedicated crawlers. In terms of the type of relationship or feedback, playlists, ratings, and music collections shared on P2P networks establish an explicit relationship between users and music items.

Table 7.1 Overview of different listener-centered sources

	Playlists	*Ratings*	*Microblogs*
Source	Local, API	Local, API	API
Community required	No	No	Yes
Relationship	Explicit	Explicit	Implicit, explicit
Level	Track	Track	Artist, track
Specific bias	Low	Low	High
	P2P nets	*Social nets*	*Sensor data*
Source	crawler	API	Smart devices
Community required	Yes	Yes	No
Relationship	Explicit	Implicit, explicit	Implicit, explicit
Level	Track	Artist, track	Artist, track
Specific bias	High	High	No

While such an explicit relationship can also be found on microblogs or social networks (e.g., a posting *"#nowplaying The Fox (What Does The Fox Say?) by Ylvis"* or a "Like" given to a band on *Facebook*), these data sources further provide implicit information, such as an indication of music taste in a user's profile description or a user's geo-tag indicating the attendance of a particular concert (together with the contextual aspect of time). Music items added to playlists, given ratings to, or shared on P2P networks are almost always tracks, whereas user–item relationships in microblogs and social networks are also established on the level of artists. Data acquired from microblogs, peer-to-peer networks, and social networks is biased towards the people that use these services, which are typically not representative of the overall population. In contrast, creating playlists and—to a lesser extent—assigning ratings are not bound to a specific group of people or community, as a wide range of applications come with these functionalities and they are used by almost every music listener.

In the following, we describe how to gather listener-centered data from a variety of sources and we connect them to the respective methods for music similarity computation, music retrieval, and music recommendation, which are detailed in the subsequent chapter.

7.2 Personal Collections and Peer-to-Peer Network Folders

In order to learn about music preference of users, access to their **personal music collections** should provide information of high quality. Moreover, access to the whole music consumption ecosystem of a listener, spanning multiple devices—stationary as well as portable—could even allow for finer distinction of subsets of the collection typically intended for different purposes. While full knowledge about users' personal collections can therefore be highly valuable, the drawback of this source is that it is usually only accessible to companies such as *Apple*, *Gracenote*, *Amazon*, or *Google* who are granted access by users of their products, for instance, through applications like *iTunes*[1] or cloud-based services like *Amazon Cloud Player*[2] or *Google Play Music*.[3] From an academic point of view, this data is therefore very difficult, if not impossible, to obtain. The same argument holds for methods that analyze data traffic of music downloads through server logs, which reveal information on users' tastes but are again publicly unavailable [78, 537].

Fortunately, a view of, or access to, some personal collections (or at least their shared parts) is possible through **peer-to-peer (P2P) networks**. Users of such networks share all kinds of material and media, ranging from software to movies to books and, of course, music. They frequently even share their entire music repository, in which case P2P networks represent a good source of information about personal collections. P2P networks have been seeing a decline in usage over the past couple of years, not least due to freely available music streaming services, such as *Spotify* or *Apple Music*. Furthermore, exploiting P2P networks as a data source is obviously limited to capture only behavior of users of such networks. The frequently expressed requirement of anonymity for P2P network users does not contribute to facilitate the task. Nevertheless, we would like to briefly discuss them as they have been used to a certain extent in co-occurrence and graph-based music similarity and retrieval methods; cf. Sects. 8.1 and 8.2.

Although providers of P2P networks usually do not offer public APIs to access shared content and meta-data, content provision and sharing typically relies on open protocols, such as the *Gnutella* protocol [384], which can be used to discover users and items and retrieve information on both. To this end, Koenigstein et al. [232] and Shavitt et al. [446] describe a crawler that first discovers the network topology of the *Gnutella* network in a breadth-first manner, i.e., it detects active users and their connections (downloaders and uploaders) and queues them in a list. Another component, the browser, then retrieves the active users from this list and extracts meta-data about their shared music content, both from file names and ID3 tags. However, similar to music items mined from other data sources, this meta-data frequently suffers from noise and incompleteness. Matching methods, such as those detailed in Chap. 5 and Sect. 7.3, should thus be employed for data cleansing.

7.3 Listening Histories and Playlists

The most obvious source for listening-related data is listening events. Each listening event is typically assigned a time stamp. Its simplest representation is a triple *<artist, track, timestamp>*. The entirety of these events for a given user is

[1] https://www.apple.com/itunes/.

[2] http://www.amazon.com/gp/dmusic/marketing/CloudPlayerLaunchPage.

[3] https://play.google.com/music/listen.

commonly referred to as his or her listening history. In contrast to listening histories, playlists are explicitly generated sequences of songs. When creating such a playlist, the user has a certain intention in mind, for instance, assembling a mix for a romantic dinner or for a Saturday night party. A time stamp is in general not necessary.

One of the most frequently used sources to acquire possibly large amounts of **listening histories** is again *Last.fm*. They provide an API function[4] which can be used to gather the complete listening history of a given user, provided the user under consideration always "scrobbles," which is the *Last.fm* term for letting *Last.fm* know which track is played at a certain moment in time. Details on how to use this API function are provided on the referred web page. To give an example, Fig. 7.2 shows parts of the XML response to the RESTful query that returns the listening history of user *dampfwalze80*. In addition to artist, album, and track names, *MusicBrainz* identifiers, links to images of album covers in different sizes, and listening time both in human-readable and Unix time format are provided.[5]

Another data source to acquire listening histories is **microblogs**. The use of microblogging services through which users can share short text messages has been strongly increasing during the past couple of years. *Twitter*[6] is currently the most popular of these services on a worldwide scale, but a variety of microblogging platforms that are very popular in certain regions also exist, for example, *Sina Weibo* in China.[7] Both platforms count hundreds of millions of users at the time of writing. More general details on microblogs were already given in Sect. 5.4.3.

Since many users share their music listening habits in microblogs, they provide a valuable data source for gathering listening histories, which can eventually be exploited to infer music similarity. In particular, the techniques of co-occurrence analysis and vector space model are applicable (cf. Sect. 8.1). Since users do not only post listening-related messages, the analysis of their other messages via natural language processing and related techniques can uncover additional facts, which can again be connected to the music taste of the user. For instance, by applying sentiment analysis or opinion mining [278] on text surrounded by an artist name, we can infer whether the user likely has a positive or negative attitude towards the artist. One thing that always needs to be borne in mind, however, when using such microblogging data is that it is heftily biased towards the community of microbloggers.

As for data acquisition from microblogs for our purposes of extracting listening events, both *Twitter*[8] and *Sina Weibo*[9] offer convenient APIs to automatically access many functions, among others, read and post messages, get the friends of a user, or perform location-based search. In the following, we will focus our discussion on

[4] http://www.last.fm/api/show/user.getRecentTracks.

[5] Unix time is defined as the number of seconds that have elapsed since midnight UTC, Thursday, January 1, 1970.

[6] http://www.twitter.com.

[7] http://www.weibo.com.

[8] https://dev.twitter.com.

[9] http://open.weibo.com/wiki/API%E6%96%87%E6%A1%A3/en.

```
<lfm status="ok">
<recenttracks user="dampfwalze80" page="1" perPage="10" totalPages="253" total
     ="2529">

<track>
<artist mbid="fa34b363-79df-434f-a5b8-be4e6898543f">Ludovico Einaudi</artist>
<name>Divenire</name>
<streamable>0</streamable>
<mbid>625e9b7f-1ae5-47d1-89b9-5dfc6c4d7f6f</mbid>
<album mbid="931d4cfa-53c6-41cd-8ab0-a2917a7c1f3e">Divenire</album>
<url>http://www.last.fm/music/Ludovico+Einaudi/_/Divenire</url>
<image size="small">http://userserve-ak.last.fm/serve/34s/36086307.png</image>
<image size="medium">http://userserve-ak.last.fm/serve/64s/36086307.png</image
     >
<image size="large">http://userserve-ak.last.fm/serve/126/36086307.png</image>
<image size="extralarge">http://userserve-ak.last.fm/serve/300x300/36086307.
     png</image>
<date uts="1375700713">5 Aug 2013, 11:05</date>
</track>

<track>
<artist mbid="b2d122f9-eadb-4930-a196-8f221eeb0c66">Rammstein</artist>
<name>Du hast</name>
<streamable>0</streamable>
<mbid>6817afbe-dc80-4b26-84d5-c02e4d1e6683</mbid>
<album mbid="">Music From And Inspired By The Motion Picture The Matrix</album
     >
<url>http://www.last.fm/music/Rammstein/_/Du+hast</url>
<image size="small">http://userserve-ak.last.fm/serve/34s/51478675.jpg</image>
<image size="medium">http://userserve-ak.last.fm/serve/64s/51478675.jpg</image
     >
<image size="large">http://userserve-ak.last.fm/serve/126/51478675.jpg</image>
<image size="extralarge">http://userserve-ak.last.fm/serve/300x300/51478675.
     jpg</image>
<date uts="1373525974">11 Jul 2013, 06:59</date>
</track>

...

</recenttracks>
</lfm>
```


Fig. 7.2 A sample XML response obtained via the *Last.fm* API function *user.getRecentTracks*

Twitter, because most of the content on *Sina Weibo* is in Chinese, which might present a serious obstacle to most readers when given examples. For gathering data on a larger scale, their Streaming API[10] should be considered. Even though it returns a few tens of tweets every second, this is just a tiny snapshot of all public tweets posted. The freely available version of the Streaming API restricts the amount of tweets to approximately 1 % of all available ones. Given the several millions of tweets shared every day, this percentage still yields a considerable amount of user-generated data. Using the *statuses/filter* function[11] seems the most appropriate way for the purpose of constructing a corpus of microblogs that refer to listening events. Obvious filter terms include *music*, *#nowplaying*, or *#listeningto*. Figure 7.3 gives an example of the JSON response returned by the *statuses/filter* function, using the

[10]https://dev.twitter.com/streaming.

[11]https://stream.twitter.com/1.1/statuses/filter.json.

```
{"created_at":"Thu Jan 16 20:07:32 +0000 2014",
 "id":423909402475630593,
 "id_str":"423909402475630593",
 "text":"#NowPlaying Psy - Gangnam Style. It this doesn't make you want to get up and dance then you must not
        have a pulse lol",
 "source":"web",
 "truncated":false,
 "in_reply_to_status_id":null,
 "in_reply_to_status_id_str":null,
 "in_reply_to_user_id":null,
 "in_reply_to_user_id_str":null,
 "in_reply_to_screen_name":null,
 "user":{"id":32216543,
         "id_str":"32216543",
         "name":"Carmen",
         "screen_name":"WrigleyCub220",
         "location":"Chicago, IL",
         "url":"http:\/\/www.facebook.com\/FieldzOBarley",
         "description":"Just a girl that loves beer, cars, music & movies. Trying every beer I can. Check out
               my FB page to see how many and what ones I have had!!",
         "protected":false,
         "followers_count":662,
         "friends_count":907,
         "listed_count":12,
         "created_at":"Fri Apr 17 00:12:44 +0000 2009",
         "favourites_count":36,
         "utc_offset":-21600,
         "time_zone":"Central Time (US & Canada)",
         "geo_enabled":false,
         "verified":false,
         "statuses_count":36722,
         "lang":"en",
         "contributors_enabled":false,
         "is_translator":false,
         "profile_background_color":"000505",
         "profile_background_image_url":"http:\/\/a0.twimg.com\/profile_background_images
               \/378800000163857825\/nT-oOmCi.jpeg",
         "profile_background_image_url_https":"https:\/\/si0.twimg.com\/profile_background_images
               \/378800000163857825\/nT-oOmCi.jpeg",
         "profile_background_tile":true,
         "profile_image_url":"http:\/\/pbs.twimg.com\/profile_images\/378800000791649597\/
               c5d22dd3c4b0011f53bb6df998ca2dfb_normal.jpeg",
         "profile_image_url_https":"https:\/\/pbs.twimg.com\/profile_images\/378800000791649597\/
               c5d22dd3c4b0011f53bb6df998ca2dfb_normal.jpeg",
         "profile_banner_url":"https:\/\/pbs.twimg.com\/profile_banners\/32216543\/1388182191",
         "profile_link_color":"1259E6",
         "profile_sidebar_border_color":"000000",
         "profile_sidebar_fill_color":"DDFFCC",
         "profile_text_color":"333333",
         "profile_use_background_image":true,
         "default_profile":false,
         "default_profile_image":false,
         "following":null,"follow_request_sent":null,
         "notifications":null
        },
 "geo":null,
 "coordinates":null,
 "place":null,
 "contributors":null,
 "retweet_count":0,
 "favorite_count":0,
 "entities":{"hashtags":[{"text":"NowPlaying","indices":[0,11]}],
             "symbols":[],
             "urls":[],
             "user_mentions":[]},
 "favorited":false,
 "retweeted":false,
 "filter_level":"medium",
 "lang":"en"
}
```


Fig. 7.3 A snippet of the JSON response (one tweet) obtained via the *Twitter* API function *statuses/filter* with the filter term *#nowplaying*

text filter *#nowplaying*. Since the figure shows only one tweet, you can see that the offered information is quite detailed. In addition to the actual message (`text`), each tweet comes with a unique identifier (`id`), a creation date (`created_at`), whether it was posted on the web or using a mobile app (`source`), whether it was a reply to another tweet (`in_reply_to_*`), geographical information (`geo`,

```
<playlists user="RJ">
<playlist>
<id>2614993</id>
<title>Doglist</title>
<date>2008-05-22T09:24:23</date>
<size>18</size>
<duration>3812</duration>
<creator>http://www.last.fm/user/RJ</creator>
<url>http://www.last.fm/user/RJ/library/playlists/1k1qp_doglist</url>
</playlist>
</playlists>
```


Fig. 7.4 A snippet of the XML response obtained via the *Last.fm* API function *user.getPlaylists*

`coordinates`, `place`), language (`lang`), and statistics such as how frequently it was retweeted by another user (`retweet_count`). Furthermore, a variety of information about the tweet's author is given (`user`).

Due to the free-form text used in tweets, they frequently contain typos, and users further employ different patterns to indicate artist and track names, e.g., *[track name] by [artist name]* or *[artist name] - [track name]*. Even if the tweet contains the music-related filter term, we thus still need to map it to a proper artist or track. To this end, the gathered tweets are usually indexed by identifying the most frequently used patterns and matching the resulting prospective artist and track names to a music database, such as *MusicBrainz*. In a quantitative study of several million tweets, we have shown that approximately 20–25 % of tweets filtered by the music-related terms *#nowplaying*, *#np*, *#itunes*, *#musicmonday*, and *#thisismyjam* can be matched to existing artists and tracks [168].

As for creating a corpus of **playlists**, again *Last.fm* offers an API call to fetch the playlists defined by a given user.[12] Figure 7.4 shows an XML snippet returned in response to a query for the playlists of user *RJ*. While this API function returns only meta-data about the user's playlist, the link provided in the `<url>` tag directs to a web page containing the actual titles in the playlist. Another service to share playlists is *Art of the Mix*.[13] Although they do not offer an API, each mix is assigned a unique identifier, and convenient retrieval functions are available, e.g., to search for mixes[14] and to access a mix's corresponding web page.[15] In addition to the mix name, actual tracks in the mix, submission date, and format (e.g., cassette or CD), a category (e.g., genre) and comments can be provided by the uploader.

Both listening histories and explicit playlists can be used to derive a music similarity measure based on the vector space model or on co-occurrence analysis (cf. Sect. 8.1). Furthermore, if user ratings are provided per listening event or per track or artist, techniques for rating prediction can be employed, which is the most frequent basis for building recommender systems (cf. Sect. 8.4). In addition,

[12]http://www.last.fm/api/show/user.getPlaylists.

[13]http://www.artofthemix.org.

[14]http://www.artofthemix.org/FindaMix/FindMix2.aspx?q=[query].

[15]http://www.artofthemix.org/FindAMix/getcontents2.aspx?strMixID=[id].

listening histories can be exploited to model listening sessions, i.e., sequences of listening events with at most very short breaks between tracks (cf. Sect. 8.3). In contrast to playlists, such listening sessions are not intentionally created.

7.4 User Ratings

A core component of many recommendation algorithms is rating prediction, i.e., given a corpus of triples *<user, item, rating>*, to predict the rating of a user for an unseen item (Sect. 8.4). Such user ratings are frequently given either on a binary scale (like/dislike) or on a five-point scale. Although, or perhaps because, user ratings play a crucial role in recommender systems, in the music domain, services that allow to access rating data are not very common. Ratings for music items are typically available only as a subset of a more general domain, e.g., by restricting the space of items on *Amazon*, *Epinions*, or *YouTube* to the music domain. In the case of *YouTube*, a dedicated function to obtain ratings through their API exists.[16] As many of the shared videos are music clips, this API can be used to some extent to acquire music-related user ratings. However, such ratings are given to the entire video. For this reason, we do not know the intention of the raters, i.e., whether their ratings are targeted at the video or the audio.

Another possibility to use music rating data is provided by the *Yahoo! Music*[17] data set [113]. It represents one of the largest music rating data sets currently available for research, including 262,810,175 ratings of 624,961 music items by 1,000,990 users and spanning the time period from 1999 to 2010. Ratings are provided on the granularity levels of tracks, albums, artists, and genres. They are partly given on a five-point Likert scale, partly on an integer scale between 0 and 100. However, all identifiers of tracks, albums, and artists are anonymized, making a meaningful connection to content and context features infeasible. Like typical data sets of user ratings, the *Yahoo! Music* set is highly sparse, i.e., 99.96 % of items in the user–item matrix are zero. This number is even higher than for common movie recommendation sets. For instance, the very popular *Netflix* set shows 98.82 % sparsity [113]. This high sparsity renders the rating prediction and recommendation tasks very challenging; cf. Sect. 8.4.

The particularly high sparsity in music rating data is the result of a major difference between ratings in the movie and music domain. While the common duration of a movie is between 90 and 120 min, a song typically lasts only 3–5 min. Since rating an item requires active user interaction, the time needed to rate a movie is substantially shorter than the time needed to consume it. This ratio, however, is far more unfavorable when the items under consideration are music pieces. In consequence, many users do not even rate small parts of their music collections

[16] https://developers.google.com/youtube/2.0/reference#youtube_data_api_tag_yt:rating.

[17] http://music.yahoo.com.

(which tend to be much larger than their movie collections), while they frequently do rate most, if not all, of their movies.

Another problem with rating data is that the scales available to raters differ between data sources. As already mentioned, some platforms, such as *Pandora* or *Last.fm*, only provide binary ratings (typically, "Thumbs up," "Like," and "Love" and "Thumbs down," "Hate," and "Ban," to express liking and disliking, respectively), while others offer more fine-grained ways to express liking. However, in the latter case, the distribution of ratings tend to be skewed towards the extremes [316]. It is thus very challenging to interpret the different rating scales and even more to integrate rating data that use different scales.

7.5 Modeling User Context

The data sources discussed so far provide explicit information on the listeners' interaction traces with music items. However, from the data used, only a narrow aspect of user behavior can actually be observed, while comprehensive models of user behavior and intention are what is typically desired by researchers in user-aware recommendation and retrieval. As a consequence, we need to be aware that the models created based on these sources represent the observed data but might not necessarily be related to the real user context. Needless to say, this challenges many of the assumptions made regarding the modeling of users.

In this section, we address the problem of describing important aspects of the user context, using dedicated data sources. First, we discuss approaches that make use of sensor data in order to obtain a closely measured and more realistic context hypothesis of the *environmental context*. Second, we discuss social networks and user connections to model the *social context* of a user.

7.5.1 *Sensor Data for Modeling User Context*

User context encompasses all aspects of the user's environment, which in the context of this book is connected to some form of interaction with music, from passive listening to a piece to actively deciding on rating or skipping a track.

The abundance of sensors available in today's smart devices, such as tablets, smart phones, smart watches, but also fitness trackers, opens a wealth of user-centered data, which can be exploited to build detailed user profiles. Connected with information on the user's listening habits, these profiles can support music retrieval and recommendation tasks by incorporating aspects of the listeners and their context. An overview of some of these contextual user factors is provided in general in [3, 25] and specifically for music in [424].

This kind of user-centered information can be categorized into raw sensor data, derived features, and more complex aspects learned via machine learning. Examples of the former include longitude and latitude, heart rate, or time. As for derived features, for instance, given time and location, it is quite easy to obtain weather information through web services like *AccuWeather*.[18] Machine learning, more precisely classification or regression techniques, can be applied on accelerometer and gyroscope data to predict user activity.

In the following, we would like to give the reader an idea of the types of sensor data that can be acquired from current smart devices and used to predict music taste at different levels: genre, artist, track, and mood [144]. The exploited features can be categorized and summarized as follows, where (N) indicates that the feature is numeric, whereas (C) denotes a categorical feature:

- **Time:** day of the week (N), hour of the day (N)
- **Location:** provider (C), latitude (C), longitude (C), accuracy (N), altitude (N)
- **Weather:** temperature (N), wind direction (N), wind speed (N), precipitation (N), humidity (N), visibility (N), pressure (N), cloud cover (N), weather code (N)
- **Device:** battery level (N), battery status (N), available internal/external storage (N), volume settings (N), audio output mode (C)
- **Phone:** service state (C), roaming (C), signal strength (N), GSM indicator (N), network type (N)
- **Task:** up to ten recently used tasks/apps (C), screen on/off (C), docking mode (C)
- **Network:** *mobile network*: available (C), connected (C); *active network*: type (C), subtype (C), roaming (C); *Bluetooth*: available (C), enabled (C); *Wi-Fi*: enabled (C), available (C), connected (C), BSSID (C), SSID (C), IP (N), link speed (N), RSSI (N)
- **Ambient:** mean and standard deviation of all attributes: light (N), proximity (N), temperature (N), pressure (N), noise (N)
- **Motion:** mean and standard deviation of acceleration force (N) and rate of rotation (C); orientation of user (N), orientation of device (C)
- **Player:** repeat mode (C), shuffle mode (C), *sound effects*: equalizer present (C), equalizer enabled (C), bass boost enabled (C), bass boost strength (N), virtualizer enabled (C), virtualizer strength (N), reverb enabled (C), reverb strength (N)

A detailed elaboration on how to acquire these features is out of the scope of this book. For this, the reader is referred to the developer web pages of the respective providers, the major ones being *Google*'s *Android*[19] and *Apple*'s *iOS*[20] where current API definitions can be found.

Approaches to **context-aware music retrieval and recommendation** differ significantly in the degree to which the user context is accounted for, i.e., which context features are exploited. Many approaches rely solely on one or a few aspects,

[18]https://api.accuweather.com.

[19]https://developer.android.com/guide/index.html.

[20]https://developer.apple.com/devcenter/ios.

of which temporal and spatial features are the most prominent ones. For instance, Cebrian et al. [60] exploit temporal features; Lee and Lee [260] incorporate weather conditions and listening histories. Some works target mobile music consumption, typically matching music with the current pace of the user while doing sports, e.g., Moens et al. [320], Biehl et al. [37], Elliott and Tomlinson [120], Dornbush et al. [109], and Cunningham et al. [92]. To this end, the user's location or heartbeat is monitored. Elliott and Tomlinson [120] focus on the particular activities of walking and running. The authors present a system that adapts music to the user's pace by matching the beats per minute of music tracks with the user's steps per minute. Additionally, the system uses implicit feedback by estimating the likelihood of a song being played based on the number of times the user has previously skipped the song. In similar research, de Oliveira and Oliver [99] compare the user's heart rate and steps per minute with music tempo to moderate the intensity of a workout.

Among the works that incorporate a larger variety of user aspects is the *NextOne* player by Hu and Ogihara [192]. It models the music recommendation problem under five perspectives: music genre, release year, user preference, "freshness"—referring to old songs that a user almost forgot and that should be recovered (cf. Sect. 7.6)—and temporal aspects per day and week. These five factors are then individually weighted and aggregated to obtain the final recommendations. Aiming at modeling the listener's context as comprehensively as possible, a recent work of ours presents the *Mobile Music Genius* (MMG) player, which gathers a wide range of user-context attributes during music playback, e.g., time, location, weather, device- and phone-related features (music volume), tasks (running on the device), network, ambient (light, proximity, pressure, noise), motion (accelerometers, orientation), and player-related features (repeat, shuffle, sound effects) [421]. MMG then learns relations (using a C4.5 decision tree learner) between these $\sim$100-dimensional feature vectors and meta-data (genre, artist, track, and mood are considered) and uses these learned relations to adapt the playlist on the fly when the user's context changes by a certain amount. An in-depth analysis of the data collected with the MMG application reveals that predicting music preference from contextual sensor data works to a certain extent on the level of artists and genres but does not work well to predict individual tracks or moods [144]. Furthermore, the most relevant features for predicting music preference are identified as time, location, weather, and apps running (tasks).

Finally, some approaches do not directly relate computational features of the user context to music preference, but infer higher-level user aspects, such as activity [508] or emotional state [57, 352], in an intermediate step. For instance, Park et al. [352] exploit temperature, humidity, noise, light level, weather, season, and time of day to classify the emotional state of the user into the categories depressed, content, exuberant, and anxious/frantic. Given explicit user preferences on which kind of music (according to genre, tempo, and mood) a listener prefers in which emotional state, and to what degree (on a five-point Likert scale), the proposed system recommends music that should fit the listener's current affective state. Wang et al. [508] present a music recommender system for mobile devices, which monitors time of day, accelerometer data, and ambient noise and predicts one of the following user activities: running, walking, sleeping, working, or shopping. Similar to Park et al.'s work [352], Wang et al.'s system then requires music tracks that are labeled for use with certain activities in order to effect activity-aware recommendations.

7.5.2 Social Networks and User Connections

Although microblogging services discussed previously in Sect. 7.3 typically also qualify as social networks, in this section, we focus on the additional possibility to extract information from the connections between users. These connections enable graph-based approaches to model music similarity and support music recommendation tasks (cf. Sect. 8.2). In typical social networks, users are either connected in a directed or an undirected way. The former refers to the fact that users can follow others without being followed by the others themselves. Such user relationships are common, for instance, in *Twitter* and *Spotify*. Networks that foster undirected or reciprocal connections include *Facebook* and *Last.fm*.

Friendship relations in the context of music, but also other domains, can be acquired through various web services. In *Spotify*, extracting followers or followed users of an arbitrary user is not straightforward. Even though they offer an API function[21] to check if the current user follows a set of other users or artists, it is not possible to directly retrieve a list of users connected to a given seed. The music network *Soundcloud*[22] offers more convenient API functions[23] for retrieving relations between users, in particular, lists of followings and followers[24] of the user. Probably the most convenient API function to obtain user connectivity data is provided by *Last.fm*. Based on a set of seed users, an API call[25] can be used to gather this kind of data and eventually construct a friendship graph. Figure 7.5 shows a response in JSON format that outputs the top two friends of user *MetalKid* together with several of their characteristics.

[21]The endpoint */v1/me/following/contains*, see https://developer.spotify.com/web-api/check-current-user-follows.

[22]https://soundcloud.com.

[23]https://developers.soundcloud.com/docs/api/reference#users.

[24]The endpoints */users/{id}/followings* and */users/{id}/followers*, respectively.

[25]http://www.last.fm/api/show/user.getFriends.

```
{friends:
    {user: [{name: "chrisjacksonftw",
            realname: "chris jackson",
            image: [{#text: "http://userserve-ak.last.fm/serve/34/61955395.jpg", size: "small"},
                    {#text: "http://userserve-ak.last.fm/serve/64/61955395.jpg", size: "medium"},
                    {#text: "http://userserve-ak.last.fm/serve/126/61955395.jpg", size: "large"},
                    {#text: "http://userserve-ak.last.fm/serve/252/61955395.jpg", size: "extralarge"}
                   ],
            url: "http://www.last.fm/user/chrisjacksonftw",
            id: "10344536",
            country: "UK",
            age: "22",
            gender: "m",
            subscriber: "0",
            playcount: "55916",
            playlists: "0",
            bootstrap: "0",
            registered: {#text: "2008-04-28 16:46", unixtime: "1209401178"},
            type: "user"},
           {name: "Betulas",
            realname: "Aaron Barefoot",
            image: [{#text: "http://userserve-ak.last.fm/serve/34/49651755.jpg", size: "small"},
                    {#text: "http://userserve-ak.last.fm/serve/64/49651755.jpg", size: "medium"},
                    {#text: "http://userserve-ak.last.fm/serve/126/49651755.jpg", size: "large"},
                    {#text: "http://userserve-ak.last.fm/serve/252/49651755.jpg", size: "extralarge"}
                   ],
            url: "http://www.last.fm/user/Betulas",
            id: "11616303",
            country: "UK",
            age: "26",
            gender: "m",
            subscriber: "0",
            playcount: "10082",
            playlists: "0",
            bootstrap: "0",
            registered: {#text: "2008-08-04 00:32", unixtime: "1217809965"},
            type: "user"}
           ],
     @attr: {for: "MetalKid",
            page: "1",
            perPage: "2",
            totalPages: "8",
            total: "16"
           }
    }
}
```


Fig. 7.5 A sample JSON response obtained via the *Last.fm* API function *user.getFriends*

7.6 Factors of User Intentions

As we have seen, there exists a variety of data sources for user–music interaction traces and user context features. However, there are a few other important user-centric factors that should be considered when elaborating music recommendation algorithms. These factors relate to the perceived quality and usefulness of music recommendations and go beyond the notion of similarity. They can be regarded as closely connected to the intention of the user when accessing a music collection. Some of these user-aware factors are summarized in the following; cf. [42, 271, 418, 536]:

- **Diversity:** Even though the items suggested by a music recommender or the results of a query to a music retrieval system should contain music items similar to the input, they should also exhibit a certain amount of diversity. This is evidenced by studies that have identified purely similarity-based recommendation as a shortcoming that leads to decreased user satisfaction and monotony [63, 456, 535]. On the other hand, inclusion of diversity must not significantly harm the accuracy of the approaches, i.e., recommended items still need to be relevant [495], but should not be too similar to each other “in order to provide the user with optimal coverage of the information space” [456]. A simple example that reveals the importance of diversity is the well-known “album effect” [514], i.e., due to same recording settings; tracks on the same album usually show a higher level of audio similarity than other tracks, even by the same artist. In many music recommendation use cases, however, the user does not want to listen to all tracks by an artist or on an album, as this could have been achieved by a simple meta-data-based query. To alleviate this issue, some systems filter results from the same album or even by the same artist. Others offer a parameter to avoid repetitions of the same artist within a number of consecutive songs.

- **Familiarity/Popularity and Hotness/Trendiness:** Familiarity or popularity measures how well-known a music item—typically an artist—is among a given population, whereas hotness or trendiness relates to the amount of buzz or attention the item is currently attracting. While popularity has a more positive connotation than familiarity, the terms are frequently used interchangeably in the literature. According to the temporal dimension, popularity can be regarded as a longer-lasting characteristic, whereas hotness usually relates to recent appreciation, which often lasts for a shorter period of time. As an example, The Rolling Stones are certainly popular, whereas Lady Gaga and Psy rank higher in terms of hotness in the beginning of the 2010s.
- **Novelty/Freshness:** Music recommenders that keep on suggesting tracks, albums, or artists already known by the user under consideration will likely not satisfy them, even if the recommended items are perfectly suited in terms of similarity and diversity. An exception might be the recommendation of new releases by a liked artist. For this reason, recommending at least a certain amount of items that are novel to the user is vital [63]. Freshness might also refer to the last time a user listened to a certain piece or artist. Even if the item has been listened to before by the user some time ago, it may be a valuable and fresh recommendation if the user has forgotten about the item and rediscovers it by means of the music recommender.
- **Serendipity:** Eventually, all of the user-centric aspects discussed above contribute to the property of serendipity. It refers to the fact that a user is surprised in a positive way, because she discovered an interesting item she did not expect or was not aware of. Therefore, providing serendipitous recommendations is an important requirement for any (music) recommender system [61, 536]. In order to affect recommendations of this kind, a wide variety of information needs to be incorporated into the recommendation process. In particular, the system should possess decent knowledge about the listeners, including but not limited to their full listening history, personal and situational characteristics, and both general and context-dependent music preferences. To give an example, a fan of German medieval folk metal might be rather disappointed and bored if the system recommends the band Saltatio Mortis, which is well known for this particular style. In that case, the recommendation is certainly not serendipitous. In contrast, for a user who occasionally enjoys folk and metal music and has some

knowledge of the German language, this band may represent a serendipitous recommendation. However, how to integrate these varieties of information to provide recommendations as serendipitous as possible is still an unsolved research problem.

7.7 Summary

In this chapter, we have seen that there exists quite some variety of data sources for *user interaction traces with music* and *listener context*. We reviewed the most important and most frequently exploited ones and compared them to each other. In particular, we discussed personal music collections and peer-to-peer networks. The former, while offering highly valuable indications of user preference and consumption behavior, is typically only available at large to companies. The latter suffers from anonymity of users and is therefore of limited value for personalized recommendation. We continued with listening histories and playlists, which can be acquired through APIs of dedicated platforms, including *Last.fm* and *Spotify*. Listening histories can, to some extent, also be extracted from microblog posts on platforms like *Twitter* or *Sina Weibo*. From time-stamped data provided in listening histories, information about listening sessions can be derived too. While more prominent in the movie domain, there also exist resources that provide rating information of music items by users. For instance, *Last.fm* and *Pandora* offer binary ratings (like vs. dislike), while more precise ratings on a Likert scale (e.g., 1–5 stars) can be extracted for music items from *Amazon* or *YouTube*, among others. While such user ratings are the most common pieces of information for recommender systems in general, for music recommendation, they are less popular due to reasons including data sparsity, different rating scales, and skewed rating distributions.

We further reviewed aspects of the user context, either derived from sensor data (environmental context) or extracted from user connections in social networks (social context). Recommendation or retrieval approaches based on the former kind of data typically exploit sensors of smart devices, while user connections are frequently gathered through *Soundcloud*, *Twitter*, or *Facebook*. Some systems directly learn relationships between context data and music preference, while others first infer higher-level user aspects such as activity or emotional state and subsequently relate these aspects to music preference.

All discussed data sources come with advantages and disadvantages, some aspects of which are ease of data acquisition (convenient API provided, for instance, by *Last.fm* vs. requirement to build a sophisticated crawler in case of P2P networks), specific biases (e.g., community bias or popularity bias), or data level (track vs. artist), detail (pieces of information), and volume (sensor data collected by users of a specific smart phone app vs. large-scale data on listening events provided by *Last.fm* or extracted from *Twitter*). The choice for a particular data source thus highly depends on the addressed problem or intended system. For researchers and practitioners who are new to user-aware music recommendation, we believe that

Last.fm is a good starting point for data acquisition due to their extensive API functionality.

We concluded the chapter by a discussion of aspects that capture user intention to some extent and can be regarded as desired properties for results or recommendations of user-aware music retrieval or recommender systems. In addition to the major concept discussed in this book, i.e., similarity, such aspects include diversity, familiarity/popularity, hotness/trendiness, novelty/freshness, and serendipity, the positive surprise when discovering previously unknown or unexpected music. While a lot of work is being devoted to build music recommender systems whose recommendations fulfill these requirements, as of now, we are not aware of any system that successfully addresses them all.

7.8 Further Reading

Since the major part of this chapter dealt with data sources and how to acquire music-related information from these, the most important additional material constitutes the documentation of the data providers on how to access their services and build applications to gather the desired data. The following table summarizes the URLs to the respective APIs or SDKs of major platforms and services.

Provider/platform	URL
Art of the Mix	http://www.artofthemix.org/aotm.asmx
Discogs	http://www.discogs.com/developers
Echonest	http://developer.echonest.com
Last.fm	http://www.last.fm/api
MusicBrainz	https://musicbrainz.org/doc/Developer_Resources
Soundcloud	https://developers.soundcloud.com/docs/api/guide
Spotify	https://developer.spotify.com/web-api
YouTube	https://developers.google.com/youtube/v3/docs
Twitter (REST API)	https://dev.twitter.com/rest/public
Twitter (Streaming API)	https://dev.twitter.com/streaming/overview
Sina Weibo	http://open.weibo.com/wiki/API%E6%96%87%E6%A1%A3/en
Apple iOS	https://developer.apple.com/devcenter/ios
Google Android	https://developer.android.com/guide

An alternative to acquiring the data yourself is the use of existing data sets. Since a detailed review of the available collections is out of the book's scope, we would like to point the interested reader to a book chapter on music recommender systems, which was recently published in the second edition of the *Recommender Systems Handbook* [424]. Next to an introduction to this kind of recommender systems, a discussion of state-of-the-art approaches to content-based, contextual, hybrid recommendation and automatic playlist generation, we review in that chapter data sets and evaluation strategies for music recommender systems.

Chapter 8
Collaborative Music Similarity and Recommendation

In this chapter, we illustrate how the data on interaction traces between users and music items, whose acquisition we discussed in the previous chapter, can be used for the tasks of determining similarities between music items and between users. We particularly focus on the topic of music recommendation since it is presumably the most popular task carried out on this kind of data. We first start with a discussion of methods to infer music similarity via co-occurrence analysis, i.e., define similarity via information on which items are listened to together by users (Sect. 8.1). Subsequently, Sect. 8.2 shows how to exploit the graph structure of artist and of user networks, who are connected on various social media platforms. We then describe a recently popular technique—*latent Dirichlet allocation* (LDA)—and its application to listening sessions in order to recommend items or playlists (Sect. 8.3). As many (music) recommender systems rely on predicting ratings of items unseen by the user under consideration, Sect. 8.4 deals with the topic of rating prediction. Eventually, Sect. 8.5 illustrates how to integrate music item based features and user-based interaction traces to build hybrid systems.

Many recommender systems approaches, including the ones presented in this chapter, rely on *collaborative filtering* (CF) techniques [146]. From a general point of view, collaborative filtering refers to the process of exploiting large amounts of collaboratively generated data (interaction traces between users and items) to filter items irrelevant for a given user from a repository, for instance, of music tracks. The aim is to retain and recommend those items that are likely a good fit to the taste of the target user [383]. Collaborative filtering is characterized by large amounts of users and items and makes heavy use of users' taste or preference data, which expresses some kind of explicit rating or implicit feedback (e.g., skipping a song).

P. Knees, M. Schedl, *Music Similarity and Retrieval*, The Information Retrieval Series 36, DOI 10.1007/978-3-662-49722-7_8

8.1 Similarity Estimation via Co-occurrence

When it comes to computing music similarity from data on user–item relationships, without having access to explicit rating data, co-occurrence-based approaches are among the most widely applied ones. Such methods assume that items frequently listened to in the same context, for instance, a music playlist, are more similar than items without that property. The basic technique of co-occurrence analysis has already been introduced in Sect. 6.3, to infer similarities from documents representing a music item, for instance, web pages about an artist. Here we focus on its application to data that directly connects listening events to users. In particular, user-centric sources exploited via co-occurrence analysis include microblogs, explicit listening histories or playlists (cf. Sect. 7.3), and shared folders in P2P networks (cf. Sect. 7.2).

Reviewing the literature on co-occurrence-based music similarity computation, we propose an integrated framework by defining a family of similarity functions between music items a_i and a_j as the product shown in Eq. (8.1), where $c(a_i, a_j)$ is a scoring function and $p(a_i, a_j)$ is an optional factor to correct exorbitant item popularity [423].[1] Six variants of scoring functions shown in Eqs. (8.2)–(8.7) were identified, where $cooc_{a_i,a_j}$ denotes the frequency of co-occurrence of items a_i and a_j, i.e., the number of users who listen to both a_i and a_j, and occ_{a_i} refers to the total number of users who listen to item a_i. Note that the simplest variant [Eq. (8.2)] yields a similarity measure which is typically asymmetric as it reflects the relative frequency of item a_j in the context of artist a_i but not vice versa.

$$sim_{cooc}(a_i, a_j) = c(a_i, a_j) \cdot p(a_i, a_j) \tag{8.1}$$

$$c_1(a_i, a_j) = \frac{cooc_{a_i,a_j}}{occ_{a_i}} \tag{8.2}$$

$$c_2(a_i, a_j) = \frac{cooc_{a_i,a_j}}{\min(occ_{a_i}, occ_{a_j})} \tag{8.3}$$

$$c_3(a_i, a_j) = \frac{cooc_{a_i,a_j}}{\max(occ_{a_i}, occ_{a_j})} \tag{8.4}$$

$$c_4(a_i, a_j) = \frac{cooc_{a_i,a_j}}{\frac{1}{2} \cdot (occ_{a_i} + occ_{a_j})} \tag{8.5}$$

$$c_5(a_i, a_j) = \frac{cooc_{a_i,a_j}}{occ_{a_i} \cdot occ_{a_j}} \tag{8.6}$$

$$c_6(a_i, a_j) = \frac{cooc_{a_i,a_j}}{\sqrt{occ_{a_i} \cdot occ_{a_j}}} \tag{8.7}$$

[1] a_i and a_j refer to artists, but the approaches can also be applied to other kinds of music items.

The use of a popularity correction term $p(a_i, a_j)$ is motivated by the fact that very popular items commonly distort co-occurrence-based similarity measures due to the popularity bias (cf. Sect. 5.1.2). To mitigate the problem, Whitman and Lawrence (and later other authors) propose in [512] the use of a popularity correction weight, as shown in Eq. (8.8), where $\max_k occ_{a_k}$ indicates the maximum frequency of any item k in the whole data set. Such a popularity correction downweights similarities between items of highly differing popularity.

$$p(a_i, a_j) = 1 - \frac{\left|occ_{a_i} - occ_{a_j}\right|}{\max_k occ_{a_k}} \tag{8.8}$$

It is noteworthy that such co-occurrence-based similarity computation is not unlike calculating similarities in the vector space model (cf. Sect. 6.1.1), where weight vectors are, in this case, constructed on a dictionary containing the names of the music items under consideration, such as artist or track names. To illustrate this, we interpret the set of playlists or listening histories which contain a certain music item a_i as a document. Computing a term weight vector representation by counting the number of occurrences of all other items $a_j \in A \backslash a_i$ in the document of a_i can be considered calculating term frequencies of item names a_j, from which, similar to the VSM approach, similarities are derived. In this case, A denotes the set of all items.

Experiments on a collection of several millions of listening events extracted from **microblogs** show best performance for the variant in Eq. (8.7) when comparing Eqs. (8.2)–(8.7) with and without the use of popularity correction [423]. In the latter case, $p(a_i, a_j)$ is just set to 1. Surprisingly, the popularity correction term turns out to be negligible in these experiments. This could be caused by the fact that the listening events are biased towards popular artists, resulting in rather small differences between popularity, hence having little influence on the numerator in Eq. (8.8). It can further be shown that co-occurrence-based music similarity computation outperforms methods based on TF·IDF weights in the vector space model, at least when using microblogs as data source and artists as music items. Zangerle et al. [532] show that co-occurrence analysis of listening histories inferred from microblogs also works considerably well on the level of songs, even when taking into account only absolute numbers of co-occurrences between songs to approximate similarity.

An earlier approach that makes use of co-occurrence analysis is presented by Pachet et al. [340]. The authors utilize radio station **playlists** and compilation CDs as data sources and extract co-occurrences between artists and between tracks. They employ a symmetric version of Eq. (8.2), similar to Eq. (6.15). A shortcoming of this type of co-occurrence formulations is that they fall short of capturing transitive relationships, i.e., indirect links, between items. For example, given three items a_i, a_j, and a_k, let us assume that a_i and a_j frequently co-occur and so do a_j and a_k. However, in all co-occurrence formulations discussed so far, the presumable high similarity between a_i and a_k is not reflected.

To account for such indirect links, Pachet et al. propose to construct co-occurrence vectors of items a_i and a_j and compute Pearson's correlation coefficient between these, according to Eq. (3.18). Similar to what was previously said about interpreting co-occurrence analysis as a special case of computing similarities in the VSM, a co-occurrence vector of item a_i contains, for each item in the collection, the number of co-occurrences with a_i.

Research on co-occurrence-based similarity that exploits shared folder data from **P2P networks** includes [33, 284]. Both extract data from the P2P network *OpenNap*[2] to derive what can be considered a social music similarity measure. Logan et al. [284] and Berenzweig et al. [33] gather meta-data for approximately 3200 shared music collections, yielding a total of about 175,000 user–artist relationships. The authors subsequently compare similarities inferred from artist co-occurrences in the shared folders, expert opinions from *allmusic*, playlist co-occurrences from *Art of the Mix*, data gathered via a web survey, and MFCC audio features. For this purpose, they investigate the top N most similar artists to each seed artist, according to each of the five data sources, and calculate the average overlap between these. They find that the co-occurrence information from *OpenNap* and from *Art of the Mix* overlap to a high degree, the experts from *allmusic* and the participants of the web survey show a moderate agreement, and the content-based audio similarity has a low agreement with the other sources.

8.2 Graph-Based and Distance-Based Similarity

In contrast to considering only direct co-occurrences (as we do in [423]) or a one-step transitive approach (as Pachet et al. do in [340]), which we discussed in the previous section, graph-based methods consider parts of or even the entire network formed by connections between artists, between users, or between both. Please note that we will use the terms "graph" and "network" interchangeably in the following. Regardless of the used term, nodes represent users or items, which are connected via an edge if they appear in the same context (e.g., songs in the same playlist) or are regarded as being known to each other (e.g., friendship relations in a social network). Such networks are typically built on data mined from listening histories or playlists (cf. Sect. 7.3), from social networks (cf. Sect. 7.5.2), or from P2P networks (cf. Sect. 7.2). Information on the closeness of two users, two artists, or two items a_i and a_j is then inferred by counting the number of intermediate objects when traversing the graph from a_i to a_j. Typically, respective approaches take as distance measure the shortest path between the two items. But also the number of shortest paths connecting a_i and a_j is sometimes considered, to account for item pairs that are connected by more than one path of minimum length.

[2]http://opennap.sourceforge.net.

An early work that considers artist graphs built from **P2P networks** and **expert opinions** is [121] by Ellis et al. They mine *OpenNap* to extract about 400,000 user-track relations that cover about 3000 unique artists. In addition to *OpenNap*, the authors analyze expert similarity judgments gathered from *allmusic*. These judgments are made by music editors who manually assign a list of similar artists to each target artist. To take indirect links along these similarity judgments into account, Ellis et al. propose a transitive similarity measure, which they call Erdös distance. The distance $d_{Erd}(a_i, a_j)$ between two artists a_i and a_j is computed as the minimum number of intermediate nodes required to form a path between the two. Unlike simple co-occurrence-based techniques, this approach also gives information about which artists are dissimilar, i.e., those with a high minimum path length. In addition, Ellis et al. propose another distance measure, which they name *resistive Erdös*. It accounts for the fact that more than one shortest path of length l may connect a_i and a_j and in turn presumes that a_i and a_j are more similar if they are connected by several shortest paths. The resistive Erdös measure resembles the electrical resistance in a resistor network and is defined in Eq. (8.9), where the set of paths $Paths(a_i, a_j)$ from a_i to a_j are modeled as resistors whose resistance equals the path length $|p|$. Even though this measure takes into account different possibilities of reaching a_j from a_i, it does not overcome the popularity bias since a large number of paths between particularly popular artists may lower the total resistance.

$$d_{Erd\text{-}resist}(a_i, a_j) = \left(\sum_{p \in Paths(a_i, a_j)} \frac{1}{|p|} \right)^{-1} \tag{8.9}$$

More recently, Shavitt and Weinsberg approach the problem of defining music similarity from **P2P network** data [445]. They gather data on the artist and on the song level from the *Gnutella* network by collecting meta-data of shared files from more than 1.2 million users. A total of 530,000 songs are covered in their data set. Both users and songs are then jointly represented in a two-mode graph, in which a song is linked to a user if the user shares the song. The authors exploit this kind of data for recommendation. To this end, they construct a user–artist matrix V, where $v_{i,j}$ refers to the number of songs by artist a_j that user U_i shares. The items in the user–artist matrix are subsequently clustered by applying k-means [288], and artist recommendation is performed using either (1) the centroid of the cluster to which the target user U_i belongs or (2) the nearest neighbors of U_i within her cluster. In addition to artist recommendation, Shavitt and Weinsberg also perform song-level recommendation. To this end, they construct a graph in which nodes represent songs and edges between two songs are weighted by the number of users who share the two. Removing edges with relatively small weights, the size of the graph is then reduced from 1 billion to 20 million edges. To mitigate the popularity bias, the authors propose a normalized distance function as shown in Eq. (8.10), in which $u(S_i, S_j)$ represents the number of users who share songs S_i and S_j, and C_i and

C_j, respectively, denote the popularity of song S_i and S_j, measured as their total frequency in the data set.

$$d_{P2P\text{-}popnorm}(S_i, S_j) = -\log_2 \left(\frac{u(S_i, S_j)}{\sqrt{C_i \cdot C_j}} \right) \tag{8.10}$$

Applying again a variant of k-means to the song similarity matrix resulting from this distance function, the authors obtain clusters of similar songs with different popularity. Songs are eventually recommended by finding the cluster that best represents the target user U_i's shared music and recommending those songs that are most similar to any song in U_i's collection.

An extension to the co-occurrence techniques presented above further exploits information about the sequence of items in **playlists**. This approach can be considered graph-based when items are represented as nodes and edges are assigned a weight corresponding to the closeness in the playlist of the two items they connect. To this end, Baccigalupo et al. [18] investigate more than one million playlists provided by the community of the (in the meantime discontinued) web service *MusicStrands*. The authors then consider the 4000 most popular artists in the corpus of playlists, where popularity is defined as the number of playlists in which each artist occurs. Baccigalupo et al. assume two artists being more similar if they occur consecutively in a playlist, in contrast to artists who occur farther away. To reflect this assumption in a distance measure, the authors define one as $d_h(a_i, a_j)$, which counts the frequency of co-occurrences of songs by artist a_i with songs by artist a_j at a distance of h in all playlists. The parameter h thus defines the number of songs between any occurrence of a song by a_i and any occurrence of a song by a_j. The overall distance between artists a_i and a_j is eventually defined as shown in Eq. (8.11), where co-occurrences at different distances are weighted differently, using a weighting term β_h. More precisely, co-occurrences at distance 0, i.e., two consecutive songs by artists a_i and a_j, are given a weight of $\beta_0 = 1$. Co-occurrences at distance 1 and 2 are weighted, respectively, by $\beta_1 = 0.8$ and $\beta_2 = 0.64$.

$$d_{playlist}(a_i, a_j) = \sum_{h=0}^{2} \beta_h \cdot \left[d_h(a_i, a_j) + d_h(a_j, a_i) \right] \tag{8.11}$$

The authors also propose a normalized version to alleviate the popularity bias. It is shown in Eq. (8.12), where $\widehat{d_{playlist}(a_i)}$ denotes the average distance between a_i and all other artists, i.e., $\frac{1}{|A|-1} \cdot \sum_{a_j \in A \setminus a_i} d_{playlist}(a_i, a_j)$, A being the set of all artists.

$$d_{pl\text{-}norm}(a_i, a_j) = \frac{d_{playlist}(a_i, a_j) - \widehat{d_{playlist}(a_i)}}{\left| \max \left(d_{playlist}(a_i, a_j) - \widehat{d_{playlist}(a_i)} \right) \right|} \tag{8.12}$$

Graph-based analysis aiming at gauging music similarity can also be performed on user data mined from **social networks**. Among others, Fields et al. [126] exploit artist-related data from *MySpace*.[3] They build an artist network by exploring the "top friends" of a given seed artist and performing a breadth-first search, including all artist nodes until a depth of 6. In turn, they construct a graph of about 15,000 nodes. They convert the directed graph to an undirected one by maintaining only edges that are reflexive, i.e., edges between artists a_i and a_j who mutually occur among the other's top friends. This reduces the number of edges from approximately 120,000–90,000. Fields et al. further extract audio features and compute similarities for all songs made available by each artist on their *MySpace* page. MFCCs (cf. Sect. 3.2.1) are used as audio features, on which GMMs are modeled, and similarities are computed via EMD (cf. Sect. 3.2.4). The authors investigate whether artist similarity defined via the shortest-path distance in the network (cf. Ellis et al. [121]) correlates with content-based similarity between the respective songs shared by the artists. They find no evidence to support this hypothesis. By applying several community detection algorithms to this *MySpace* graph, Jacobson et al. [198] find that the artist network reveals groups of artists who make music of similar genres. They quantify this showing that the genre entropy within the identified groups is lower than within the overall network.

In another work that exploits user connections in a social network, Mesnage et al. [316] explore a social recommendation strategy based on users' friendship relationships on *Facebook*. The authors first gather a set of the most popular tracks from about 300,000 *Last.fm* users and connect these to *YouTube* music videos used in the subsequent experiment. They implement a *Facebook* application dubbed "Starnet," which they use to investigate different recommendation strategies. In particular, Mesnage et al. propose a social recommendation method that recommends tracks given a rating of at least 3 (on a five-point Likert scale) by friends of the target user but at the same time have not been rated yet by the target user. The target user is then exposed to a *YouTube* video of the track and has to rate the recommendation. Restricting these social recommendations to items that are given at least 3 stars by the user's friends can be interpreted as social diffusion of items through the user network. For instance, if Alice is presented a random track and rates it with 4 stars, it will be diffused to her friends. If Bob is a friend of Alice but gives the track only a rating of 1, diffusion stops as the track will not be recommended to friends of Bob, unless they are also friends to people who rate the track highly. Mesnage et al. compare their social recommender to two baselines: (1) recommending random tracks without any rating and (2) randomly selecting tracks from users not befriended with the target user but rated with more than 2 stars. In a study including 68 participants, the authors show that the second baseline outperforms the first one and that users prefer the social recommendations over both baselines.

[3] http://www.myspace.com.

8.3 Exploiting Latent Context from Listening Sessions

We already discussed in Sect. 6.1.2 the technique of latent semantic indexing (LSI), which is used to unveil latent semantic associations of terms by means of co-occurrence analysis. A probabilistic method with the same goal is LDA, which we introduce in the following section. It belongs to the class of topic modeling approaches, which are statistical methods for detecting abstract or latent "topics" in documents and has frequently been used in text processing and mining.

Research that employs LDA for music clustering and recommendation tasks typically uses **listening histories** or **listening sessions** (cf. Sect. 7.3) as input to LDA. The purpose of applying LDA in this context is commonly to identify music items that co-occur in listening sessions and belong to a certain concept or topic, which may, for instance, correspond to a genre or style but also to something less obvious like pop music of the 2010s featuring female singers.

Next, we first lay the theoretical foundation of LDA in Sect. 8.3.1. Subsequently, we report in Sect. 8.3.2 on a small case study, in which we show how LDA can be used to detect artist topics based on large-scale listening data mined from *Last.fm*. Section 8.3.3 then presents other literature in which LDA is used for music recommendation.

8.3.1 Latent Dirichlet Allocation

LDA as proposed by Blei et al. [41] is a generative model in which each document is modeled as a multinomial distribution of topics and each topic is modeled as a multinomial distribution of words. The generative process begins by choosing a distribution over topics $\mathbf{z} = (z_{1:K})$ for a given document out of a corpus of M documents, where K represents the number of topics. Given a distribution of topics for a document, words are generated by sampling topics from this distribution. The result is a vector of N words $\mathbf{w} = (w_{1:N})$ for a document. LDA assumes a Dirichlet prior distribution on the topic mixture parameters Θ and Φ, to provide a complete generative model for documents. Θ is an $M \times K$ matrix of document-specific mixture weights for the K topics, each drawn from a Dirichlet prior, with hyperparameter α. Φ is a $V \times K$ matrix of word-specific mixture weights over V vocabulary items for the K topics, each drawn from a Dirichlet prior, with hyperparameter β.

Figure 8.1 illustrates the process. The inner plate N shows the modeling of topics $\mathbf{z}$ as a distribution over N words $\mathbf{w}$. Words are further dependent on a Dirichlet distribution (β), from which they are drawn. The outer plate shows the modeling of the M documents in the corpus. The mixture weights Θ that describe each document as a distribution over topics are again assumed to be Dirichlet distributed (α).

There are two main objectives of LDA inference: (1) find the probability of a word given each topic k: $p(w = t|z = k) = \phi_k^t$ and (2) find the probability of a topic given each document m: $p(z = k|d = m) = \theta_m^k$. Several approximation techniques

Fig. 8.1 Graphical model of latent Dirichlet allocation (LDA)

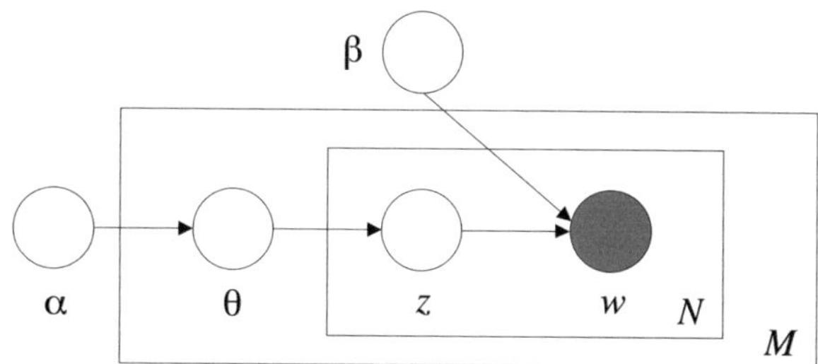

have been developed for inference and learning in the LDA model [41, 161]. A very common technique is Gibbs sampling [161], which results in Eq. (8.3.1), where n_k^t and n_m^k are the number of times word $w = t$ and document $d = m$ have been assigned to topic $z = k$, respectively, and $n_k = \sum_{t=1}^{V} n_k^t$, and $n_m = \sum_{k=1}^{K} n_m^k$.

$$\phi_k^t = \frac{n_k^t + \beta}{n_k + V\beta}, \qquad \theta_m^k = \frac{n_m^k + \alpha}{n_m + K\alpha}, \tag{8.13}$$

From a trained LDA model, similarities between music items can be easily computed by employing a similarity function to the topic vectors of pairs of items. When it comes to evaluation, most figures of merit from retrieval and machine learning can be used (cf. Sect. 1.5). In addition, LDA and other topic models are frequently evaluated in terms of their ability to generalize to unseen data [170]. A common performance measure for this purpose is **perplexity**. In the context of topic modeling, perplexity measures how well the topic model learned from a training corpus generalizes to a set of unseen documents in a test corpus. The lower the perplexity of a model, the better its predictive power. Perplexity is defined as the reciprocal geometric mean of the likelihood of a test corpus given a model $\mathcal{M}$, as shown in Eq. (8.14), where N_m is the length of document m, w_m is the set of unseen words in document m, and M is the number of documents in the test set.

$$Perp = \exp\left[-\frac{\sum_{m=1}^{M} \log p(w_m|\mathcal{M})}{\sum_{m=1}^{M} N_m}\right], \tag{8.14}$$

8.3.2 *Case Study: Artist Clustering from Listening Events*

To illustrate the use of LDA for the task of detecting clusters (or topics) of artists from listening co-occurrences, we first acquire a corpus of listening events for a set of diverse users of *Last.fm*. Starting with the top 250 tags,[4] we retrieve the top

[4] http://www.last.fm/api/show/tag.getTopTags.

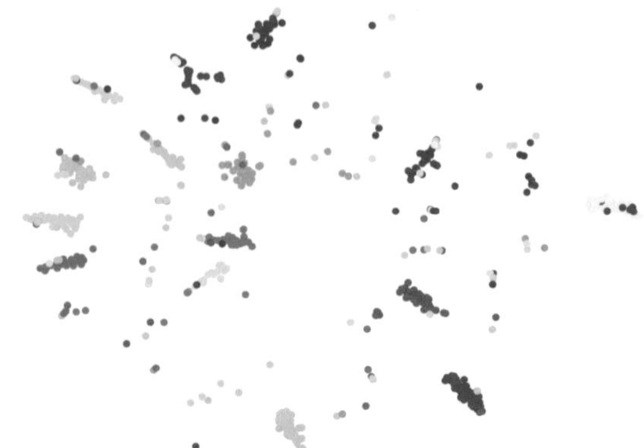

Fig. 8.2 Visualization of the top 50 artists for the most important 15 topics after applying a clustering technique, i.e., t-distributed stochastic neighborhood embedding (t-SNE). Artists belonging to different topics are depicted in *different colors*

artists for these top tags.[5] This yields a set of roughly 50,000 unique artists. We then obtain the users most frequently listening to these artists,[6] resulting in a set of about 485,000 unique users. We retrieve the listening histories of a randomly selected subset of 38,000 users.[7] This process eventually results in a set of more than 400 million listening events, involving slightly more than 300,000 unique artists.

As the LDA model is built upon co-occurrence information, we subsequently have to filter this data set by removing artists that are listened to by only one user and users that listen to only one artist. We iteratively perform these cleaning processes by artist and by user, until we reach a fixed point at which the final data set comprises 414 million listening events, involving 245,000 distinct artists and 33,400 distinct users.

This final data set is then fed into an LDA model configured to learn 50 topics. Words are represented as artists and documents as listening sessions. As the learned topic model cannot be directly visualized in an easy manner, we apply a clustering technique to the topic vectors of all artists. More precisely, we compute a t-distributed stochastic neighborhood embedding (t-SNE) [492] on the 50-dimensional feature vectors for each artist t, given by the probabilities over all topics $\phi^t_{1:K}$. The resulting embedding is visualized in Fig. 8.2, where only the top 50 artists for the most important 15 topics found by LDA are depicted to avoid visual clutter. Artists assigned to the same topic are shown in the same color. As we can see, the clusters in general correspond quite well to the identified topics. To perform a qualitative analysis, Fig. 8.3 depicts the most probable artists (defined as $\phi^t_k > 0.03$) for the most important 15 topics. Again, t-SNE clustering is employed and probabilities ϕ^t_k are linearly mapped to font size to visualize artist names. Topics are color-coded as in Fig. 8.2. The topics 25 and 42 are the most important ones with topic probabilities of $z_{25} = 0.033$ and $z_{42} = 0.031$, respectively. Taking a closer

[5]http://www.last.fm/api/show/tag.getTopArtists.

[6]http://www.last.fm/api/show/artist.getTopFans.

[7]http://www.last.fm/api/show/user.getRecentTracks.

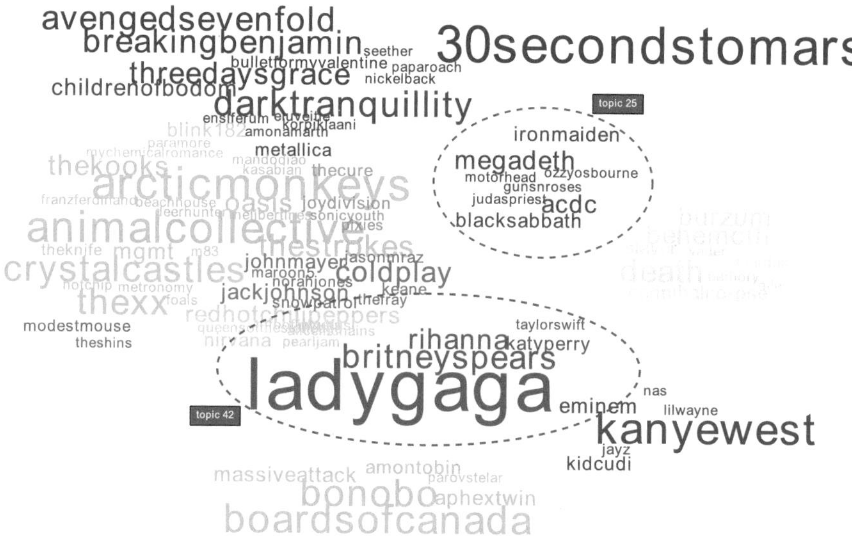

Fig. 8.3 Most probable artists given the topic k, i.e., $\phi_k^t > 0.03$, for the 15 top topics

- heavymetal, hardrock, metal, classicrock, 80s, speedmetal, thrashmetal, hairmetal, 70s, glamrock
- femalevocalists, dance, rnb, hiphop, sexy, guiltypleasure, femalevocalist, female, dancepop, r&b
- indierock, indiepop, lofi, folk, altcountry, alternativerock, indiefolk, experimental, favorites, postrock
- postpunk, newwave, 80s, punk, indierock, british, punkrock, goth, grunge, noiserock
- newrave, electropop, electro, synthpop, indiepop, indietronica, electronic, electronica, indierock, dance
- blackmetal, deathmetal, thrashmetal, metal, brutaldeathmetal, norwegian, technicaldeathmetal, heavymetal, swedish, melodicdeathmetal
- grunge, hardrock, seattle, funkrock, alternativerock, 90s, postgrunge, punk, funk, metal
- triphop, downtempo, ambient, idm, chillout, electronica, lounge, electronic, nujazz, chill
- britpop, british, indierock, britrock, garagerock, alternativerock, postpunkrevival, postpunk, indiepop, britishrock
- poppunk, emo, punkrock, punk, poprock, notemo, screamo, powerpop, alternativerock, emocore
- lofi, dreampop, shoegaze, experimental, indierock, psychedelic, indiepop, postrock, freakfolk, ambient
- britpop, acoustic, poprock, pianorock, singersongwriter, acousticrock, mellow, british, indierock, softrock
- melodicdeathmetal, deathmetal, folkmetal, finnish, blackmetal, finnishmetal, metal, swedish, powermetal, swedishmetal
- hiphop, rap, undergroundhiphop, gangstarap, alternativerap, eastcoastrap, eastcoast, rnb, alternativehiphop, undergroundrap
- postgrunge, hardrock, alternativemetal, numetal, emo, metal, metalcore, alternativerock, grunge, screamo

Fig. 8.4 Most important tags for the topics found by LDA

look reveals that topic 25 encompasses foremost hard rock and heavy metal bands (among others, Iron Maiden, AC/DC, and Megadeth), whereas topic 42 includes well-known female pop performers (e.g., Lady Gaga, Rihanna, and Britney Spears). This impression is confirmed by looking at Fig. 8.4, which shows the most important tags gathered from *Last.fm* for all artists and aggregated for each topic. Each colored bullet represents a topic, and topics are sorted in decreasing order of importance. Topic colors match the ones used in Fig. 8.3. In addition to general observations such as pop artists being represented in topic 42 and metal bands in 25, we can also make out more fine-grained differences between seemingly very similar topics. For instance, the yellow cluster in the right of Fig. 8.3 represents black metal and (brutal) death metal bands. Also being metal bands, Eluveitie, Korpiklaani, Ensiferum, and similar, which are situated towards the top left and colored in blue, can be described as folk metal, viking metal, or melodic death metal. Although these differences may seem subtle for the listeners not into the metal genre, they bear intriguing additional information for those who are interested.

Finally, we perform a quantitative analysis by computing the mean average cosine distance between the top 50 artists in each topic (d_{topic}) and the mean average cosine distance between all pairs of artists irrespective of the topic (d_{all}). As it turns out, calculating these two values on the topic vectors yields $d_{topic} = 0.341$ and $d_{all} = 0.949$, thus a ratio $d_{all}/d_{topic} = 2.785$, which indicates that artist distances within topics are significantly lower than average overall artist distances. Based on this finding, it seems reasonable to use topic models not only for artist clustering but also for recommendation purposes. As already said, one way to do so is to compute a similarity matrix from the probability distribution vectors of artists and exploit these similarities to select unknown artists similar to the target user's known and liked ones. Another even simpler way is to directly recommend artists belonging to the same cluster. In addition, since LDA is a generative model, items can directly be sampled from the model for a given user, according to the probabilities of the user liking each topic.

8.3.3 Music Recommendation

Latent Dirichlet allocation has successfully been applied for the task of music recommendation. Since LDA is a generative model, learned from listening data, it is straightforward to use it for playlist generation and recommendation. For instance, Zheleva et al. [537] use listening histories extracted from usage logs of the *Zune Social* platform[8] covering 14 weeks. They propose two probabilistic graphical models based on LDA: a *taste model* and a *session model*. The former models the process of music listening as a distribution over a set of latent taste clusters. These clusters are represented as distribution over songs. Each song in a user's playlist can thus be generated by selecting one cluster and picking a song based on the cluster's distribution. This model is a direct adaptation of the LDA model. The second model Zheleva et al. propose, i.e., the session model, incorporates information on the users' listening sessions, where a session is defined based on a threshold for the length of the pause between listening events. This means a new session is identified each time the user does not listen to music for a certain amount of time. The authors presume that a listening session is characterized by the mood of the listener. Thus, clusters are regarded as conveying latent moods, and each user is represented as a mixture over different moods. Each cluster is again a mixture of songs. Evaluation of the two models is carried out on a playlist generation task. Perplexity is shown to be lower for the session model than for the taste model, which means that the predictive power of the former outperforms that of the latter. However, recommendation quality is

[8]http:/www.zune.net.

evaluated on the level of genres, instead of artists or songs. A prediction is thus a distribution over genres, which represents a very coarse level of prediction.

Another LDA-based music recommendation approach targeting serendipitous recommendations in particular (cf. Sect. 7.6) is proposed by Zhang et al. [536]. The authors introduce *artist-based LDA*, which in contrast to Zheleva et al.'s models treats artists as documents and users as words. Consequently, topics can be thought of as user communities, instead of music genres or styles, and artists are represented as distribution over such communities of listeners. Artist similarities are computed via cosine similarity between the respective artists' topic vectors [cf. Eq. (6.8)]. The authors further propose a community-aware diversification technique that rewards artists with particularly diverse user communities. They compute the diversity of an artist a_i as the entropy of a_i's topic vector, which encodes a probability distribution over the user communities. The underlying assumption is that artists producing very specific styles of music, e.g., "brutal technical death metal" such as Visceral Bleeding, have a rather narrow community base, whereas bands like The Rolling Stones attract a much wider audience. In addition, Zhang et al. introduce a declustering method that first builds an artist network, in which two artist nodes are connected by weighted edges whose weights correspond to the similarity found by LDA. The weights are then thresholded in order to compute the clustering coefficient for each node, which indicates how tightly connected the node and its neighbors are within the network. Assuming that artists with a high clustering coefficient represent rather mainstream and boring recommendations, the declustering emphasizes artists with a low coefficient as they are more likely to lie at the borders of similar artist clusters and for this reason are presumed to be more interesting. Zhang et al. combine the three approaches artist-based LDA, diversification, and declustering by weighted rank aggregation of the approaches' individual recommendations. The authors call this hybrid system "Full Auralist." They further propose a set of performance measures to gauge the aspects of diversity, novelty, and serendipity in music recommendation (cf. Sect. 7.6). For evaluation, they use a data set proposed by Celma [62], comprising listening events by 360,000 users. The authors treat listening events as unary preference indications, i.e., they neglect how often a user listens to a certain item, just retaining whether or not she listens to it. In a conducted user study, the hybrid Full Auralist approach produced more serendipitous recommendations than the basic artist-based LDA, though at the cost of accuracy.

8.4 Learning from Explicit and Implicit User Feedback

As discussed in Sect. 7.4, feedback on music items is sometimes given explicitly in the form of ratings. These ratings can be given on different levels or scales, ranging from "continuous" slider values (e.g., between 1 and 100) to Likert-style discrete steps (e.g., 1–5 or 10 stars, possibly in half steps) to binary ratings (e.g., "thumbs up/down") to unary ratings ("Like"); cf. [461].

When aiming at recommending items, it is often more desirable to directly predict how a user would rate (or would relate to) an item and then present only those items to the user that have the highest predicted value. To perform this type of prediction, one must have access to a (large and active) community and its interaction traces in order to apply collaborative filtering (CF) techniques. Thus, these techniques are often to be found in online music platforms such as *Last.fm* or *Amazon*.

Collected explicit ratings are usually represented in a $f \times n$ user–item matrix R, where $r_{u,i} > 0$ indicates that user u has given a rating to item i. Higher values indicate stronger preference. Given implicit feedback data, e.g., listening data, such a matrix can be constructed in a similar fashion. For instance, a value of $r_{u,i} > 0$ could indicate that u has listened to artist i at least once, bought the CD, or has that artist in her collection (cf. Sect. 7.2), $r_{u,i} < 0$ that u dislikes i (e.g., u has skipped track i while listening), and $r_{u,i} = 0$ that there is no information available (or neutral opinion). Note that without additional evidence, a number of assumptions must usually be made in order to create such a matrix from implicit data (see Sect. 7.1). For reasons of simplicity, we will refer to all types of entries as ratings in the following.

The goal of the approaches discussed in this section is to "complete" the user-item matrix, i.e., to populate all cells for which no feedback is available. Based on the completed matrix, the unrated (i.e., unseen) items most likely to be of interest to the user are recommended. Broadly, there are two types of rating prediction approaches, namely, memory-based and model-based collaborative filtering.

8.4.1 Memory-Based Collaborative Filtering

Memory-based (or neighborhood-based) collaborative filtering approaches operate directly on the full rating matrix (which they keep in the memory, thus the name). Although this requirement usually makes them not very fast and resource-demanding, they are still widespread due to their simplicity, explainability, and effectiveness [100]. In their most traditional form, *user-based CF*, a predicted rating $r'_{u,i}$ for a user u and an item i is calculated by finding K_u, the set of k most similar users to u according to their rating preferences (nearest neighbors) and combining their ratings for item i [236]:

$$r'_{u,i} = \frac{\sum_{g \in K_u} sim(u,g)(r_{g,i} - \bar{r}_g)}{\sum_{g \in K_u} sim(u,g)} + \bar{r}_u \tag{8.15}$$

where $r_{j,i}$ denotes the rating of user j given to item i, $\bar{r}_u$ the average rating of user u, and $sim(u,j)$ a weighting factor that corresponds to the similarity to the neighboring user.

While user-based CF identifies neighboring users by examining the rows of the user×item matrix, *item-based CF*, on the other hand, operates on its columns to find similar items [396]. Predictions $r'_{u,i}$ are then made analogously to Eq. (8.15) by finding the set of the k most similar items to i that have been rated by u and combining the ratings that u has given to these. Since typically the number of rated items per user is small compared to the total number of items, item-to-item similarities can be pre-calculated and cached. In many real-world recommender systems, item-based CF is thus chosen over user-based CF for performance reasons [274].

For determining nearest neighbors (NNs), i.e., the k most similar vectors, in both user-based and item-based CF, *cosine similarity* or a "cosine-like measure" [328] is utilized. This includes *Pearson correlation* (typically chosen in user-based CF) and *adjusted cosine similarity* (typically chosen in item-based CF).

Pearson correlation has already been introduced in Eq. (3.18). Equation (8.16) shows the formula again using the task-related symbols and specifics.

$$sim_{pearson}(u, v) = \frac{\sum_{p \in P}(r_{u,p} - \bar{r}_u)(r_{v,p} - \bar{r}_v)}{\sqrt{\sum_{p \in P}(r_{u,p} - \bar{r}_u)^2}\sqrt{\sum_{p \in P}(r_{v,p} - \bar{r}_v)^2}} \tag{8.16}$$

Here, u and v represent users, P the set of items rated by both u and v and $\bar{r}_u = \frac{1}{|P|}\sum_{p \in P} r_{u,p}$ the mean rating of user u. The Pearson correlation coefficient ranges from -1 to $+1$ and requires variance in the user ratings; else it is undefined. Subtraction of the mean rating accounts for rating biases of the users, i.e., whether there are tendencies to generally give high or low ratings. Calculating the correlation only on the set of co-rated items P is necessary due to the sparsity of the data, i.e., the fact that the vast majority of entries in the rating matrix are zero. On the other hand, with few rating overlaps between u and v, i.e., a small $|P|$, the calculated values tend to become unreliable.

Adjusted cosine similarity is similar to the Pearson correlation coefficient, as can be seen in Eq. (8.17).

$$sim_{adjcos}(i, j) = \frac{\sum_{b \in B}(r_{b,i} - \bar{r}_b)(r_{b,j} - \bar{r}_b)}{\sqrt{\sum_{b \in B}(r_{b,i} - \bar{r}_b)^2}\sqrt{\sum_{b \in B}(r_{b,j} - \bar{r}_b)^2}} \tag{8.17}$$

Here, i and j represent items and B the set of users that have rated both i and j. Despite their similarity, note that adjusted cosine similarity is not just a Pearson correlation on R^T, as also here the user average is subtracted from the ratings (not the item average). Like Pearson correlation, adjusted cosine values range from -1 to $+1$. For selecting NNs, typically only neighbors with $sim > 0$, i.e., positively correlated entities, are considered [171].

As can be seen from the above formulation, in contrast to the methods described in Chaps. 5 and 6, CF does not require any additional meta-data describing the music items. Due to the nature of rating and feedback matrices, similarities can be calculated without the need to associate occurrences of meta-data with actual items. In addition to rating prediction, the methods described can also be used to obtain item similarity. For example, Slaney and White [451] analyze 1.5 million user ratings by 380,000 users from the *Yahoo! Music* service and obtain music piece similarity by comparing normalized rating vectors over items and computing respective cosine similarities. However, these simple approaches are very sensitive to factors such as popularity biases and data sparsity. Especially for items with very few ratings, recommendations performed in a fashion as outlined are not very reliable.

As a final note on memory-based CF, we have already discussed and illustrated in Sect. 3.4 that computing item similarity typically benefits from distance space normalization and hub removal. Therefore, it does not come as a surprise that these approaches have also shown useful in memory-based collaborative filtering; cf. [227].

8.4.2 Model-Based Collaborative Filtering

To avoid expensive calculations directly on the rating matrix and to create more robust representations, model-based methods build upon a latent factor representation obtained by learning a factorization model of the rating matrix, e.g., [184, 239]. The assumption behind this is that the observed ratings are the result of a number of latent factors of user characteristics (e.g., gender, age, preference or taste, "personality") and item characteristics (e.g., genre, mood, instrumentation, era, male vs. female singer). The goal is to obtain a representation that, ideally, reflects such properties in order to predict ratings. Similar to LSI, which uncovers latent factors in a term–document matrix (cf. Sect. 6.1.2) the purpose of matrix factorization is to estimate hidden parameters from user ratings. These parameters should explain the observed ratings by characterizing both users and items in an ℓ-dimensional space of factors. Note that these computationally derived factors basically describe variance in the rating data and are not necessarily human interpretable.

The matrix factorization model described in the following is based on truncated singular value decomposition (truncated SVD; cf. Sect. 6.1.2). In the form discussed here, the rating matrix R can be decomposed into two matrices W and H such that $R = WH$, where W relates to users and H to items.[9] In the latent space, both users and items are represented as ℓ-dimensional column vectors. Each user u is represented as a vector $w_u \in \mathbb{R}^\ell$ and each item i as a vector $h_i \in \mathbb{R}^\ell$. $W = \left[w_1 \dots w_f\right]^T$ and $H = [h_1 \dots h_n]$ are thus $f \times \ell$ and $\ell \times n$ matrices, respectively. Entries in w_u and h_i measure the relatedness to the corresponding latent factor. These entries can be negative. The number of latent factors is chosen such that $\ell \ll n, m$ (values $20 < \ell < 200$ are typical), which yields a significant decrease in memory consumption in comparison to memory-based CF. Also rating prediction is much

[9]From the formulation of SVD in Sect. 6.1.2 that decomposes R into three matrices U, Σ, and V, such that $R = U\Sigma V^T$, W and H can be calculated as $W = U_\ell \Sigma_\ell^{1/2}$ and $H = \Sigma_\ell^{1/2} V_\ell^T$ [397].

more efficient once the model has been trained as this is now simply achieved by calculating the inner product $r'_{u,i} = w_u^T h_i$. More generally, after learning the factor model from the given ratings, the complete prediction matrix $R' = WH$ contains all rating predictions for all user–item pairs. It needs to be pointed out that, in order to learn the factors, it is essential that only available rating information is used, i.e., entries $r > 0$, rather than an SVD approach that factorizes the complete matrix, treating missing values as numeric zeros.

Mathematically, the matrices W and H are found by minimizing the reconstruction error (equaling the squared Frobenius norm) as shown in Eq. (8.18).

$$err = \|R - WH\|_F^2 = \sum_{(u,i)\,|\,r_{u,i}>0} (r_{u,i} - w_u^T h_i)^2 \tag{8.18}$$

Minimizing this error is equivalent to finding the SVD of R.

However, experiments conducted on rating data from various domains showed that the pure user–item interaction $w_u^T h_i$ cannot explain the full rating value. As we have seen for the memory-based approaches, it is important to account for biases, such as a skeptical user's bias of giving generally lower ratings or an item's bias of receiving high ratings because it is popular. Biases should therefore also be embedded in the model and accounted for in the factorization approach. To reflect biases, a rating can thus be estimated as a linear combination of the global average rating μ, the user bias c_u, the item bias d_i, and the interaction term, as shown in Eq. (8.19).

$$r'_{u,i} = \mu + c_u + d_i + w_u^T h_i \tag{8.19}$$

In addition to the latent factor matrices W and H, these new parameters need to be modeled and can be simultaneously learned from the given data. However, optimizing a model with that many parameters also bears the danger of "overfitting." This means that the learned model will produce predictions that are optimal for the training data, but for unseen data, this might lead to bad predictions. Overfitting the training data harms therefore generalization, which is the primary objective of learning. The concrete problem is that fitting of the model is not bound to any value constraints, and, as a result, optimized values can become extreme. As a hypothetical example, imagine that some jazz pieces are disliked by a user in the training data and so the model not only learns to predict values close to 0 for jazz but actually values $\ll 0$ that are far off the used rating scale, to be on the "safe side," so to speak. Since predictions are anyways clipped to the range of ratings, this does not harm the overall performance during evaluation. As a consequence, no jazz track will ever be predicted with a positive rating, even if one is comparatively predicted higher, which would agree with the taste of the user. The remedy to this situation is to penalize the magnitude of feature values by introducing regularization terms (Tikhonov regularization). Including the biases introduced above as well as the regularization term λ, we obtain a new, regularized squared error function to minimize as seen in Eq. (8.20).

Algorithm 1 A simple variant of gradient descent in pseudocode, cf. [237, 354, 509]

procedure LEARNFACTORS
init:
  $w_u, h_i \leftarrow rand$
training:
  loop over training epochs or until overall error increases
    for all $(u, i) \mid r_{u,i} > 0$ **do**:
      calculate prediction error: $err_{u,i} \leftarrow r_{u,i} - (\mu + c_u + d_i + w_u^T h_i)$
      for all latent factors $k = 1 \ldots \ell$ **do**:
        /* *update parameters in direction of gradient*
          proportional to learning rate γ *(e.g.,* $\gamma = 0.001$*)* */
        $w_u \leftarrow w_u + \gamma(err_{u,i} \cdot h_i - \lambda \cdot w_u)$
        $h_i \leftarrow h_i + \gamma(err_{u,i} \cdot w_u - \lambda \cdot h_i)$
        $c_u \leftarrow c_u + \gamma(err_{u,i} - \lambda \cdot c_u)$
        $d_i \leftarrow d_i + \gamma(err_{u,i} - \lambda \cdot d_i)$
    optional: decrease learning rate γ
predict:
  calc $r'_{u,i} = \mu + c_u + d_i + w_u^T h_i \quad \forall (u, i) \mid r_{u,i} = 0$
  clip predictions to range

$$\min_{w*,h*,c*,d*} \sum_{(u,i) \mid r_{u,i}>0} (r_{u,i} - \mu - c_u - d_i - w_u^T h_i)^2 + \lambda(\|w_u\|^2 + \|h_i\|^2 + c_u^2 + d_i^2) \tag{8.20}$$

The value for the regularization term λ is either determined by cross-validation or set manually, e.g., to 0.02 [509] or 0.05 [354].

To solve the regularized model shown in Eq. (8.20), a simple *gradient descent* technique can be applied. The gradient descent algorithm iterates through all examples for which a rating is given and calculates the error between the modeled rating $\mu + c_u + d_i + w_u^T h_i$ and the real rating and modifies the parameters towards predicting the real rating. The advantages of gradient descent are its easy implementation and fast running time. A variant of gradient descent in pseudocode can be seen in Algorithm 1.

Koren [237] proposes *SVD++*, an extension of the outlined approach that, on top of explicit rating data, can also include additional information on implicit feedback by the user. This implicit feedback can, for instance, stem from the user's browsing history or from listening events and is expressed in binary form. To this end, additional latent factors are included that are optimized in conjunction with the latent factors modeling the explicit ratings. Another work also takes into account the temporal scope of feedback [238].

Hu et al. [193] adapt the SVD approach to deal with implicit feedback given as the times an item has been consumed (*weighted matrix factorization*). This can also contain fractions of integer values, e.g., a movie has been watched 2.5 times or a song has been listened to 12.3 times. Consequently, $r_{u,i} \in \mathbb{R}^+ \cup \{0\}$; therefore, all entries in the user–item matrix can be considered populated. Thus, in contrast to the SVD, the method by Hu et al. also exploits entries that equal zero. Furthermore, the implicit information is split into a *preference* and a *confidence* value to account for the fact that there is uncertainty about the meaning of an entry equaling zero (e.g., unawareness vs. dislike). While preference is defined as $p_{u,i} = sgn(r_{u,i})$ and indicates whether we can assume that a user likes an item, confidence is defined as $c_{u,i} = 1 + \alpha r_{u,i}$ with α set to, e.g., 40 and directly correlates with the amount an item has been consumed. In order to learn the latent model from the full rating matrix, the cost function $\min_{w*,h*} \sum_{u,i} c_{u,i}(p_{u,i} - w_u^T h_i)^2$ (regularization terms omitted) is minimized. Similar to memory-based CF, this method allows to explain recommendations made by linking latent factor variables to well-known anchor items.

For the task of modeling large-scale music listening data, Johnson [200] suggests *logistic matrix factorization* (logistic MF), a probabilistic approach to derive latent factor matrices from implicit feedback. In the probabilistic formulation of the problem, the probability of a user–item interaction $l_{u,i}$ is distributed according to a logistic function as shown in Eq. (8.21).

$$p(l_{u,i}|c_u, d_i, w_u, h_i) = \frac{\exp(c_u + d_i + w_u^T h_i)}{1 + \exp(c_u + d_i + w_u^T h_i)} \tag{8.21}$$

For regularization, priors on the latent factors are set to zero-centered spherical Gaussians. Assuming independence of all $r_{u,i} \in R$, latent factors are optimized by maximizing the log posterior of the likelihood of the observations $r_{u,i}$ using a gradient ascent algorithm.

For comparison with two other methods, namely, weighted matrix factorization and a popularity-based baseline that always returns the most popular artists, a large-scale data set is constructed from *Spotify* listening data. This (non-disclosed) data set contains listening data of the 10,000 most popular artists for randomly chosen 50,000 users, resulting in a matrix with 86 million nonzero entries. With only 83 % sparsity, this data set is very atypical for recommender tasks. Each entry $r_{u,i}$ is again a nonnegative real value and corresponds to the amount of listens, where a listening event is registered if it exceeds 30 s of streaming. The author also hints at the possibility of an alternative transformation to aggregate listening events, such that explicitly selected events are weighted higher than events originating from automatic streams. Furthermore, more recent events can be given higher weights to account for changes in the user's taste over time.

Logistic MF is shown to outperform both weighted MF and the popularity baseline in a recall-oriented evaluation scenario. The popularity baseline proves to be a hard contender as the majority of users usually listen to the same subset of artists which already yields high agreement. Johnson also makes use of the learned latent item vector space for similarity calculations for nearest neighbor retrieval and exemplifies the higher quality in the rankings.

8.5 Multimodal Combination

After focusing on methods to exploit individual sources of user interaction traces, we discuss approaches that combine user context, feedback, or interaction data with music content, music context, or both. In particular, we address two areas. First, we deepen the discussion on hybrid recommender systems, i.e., approaches that incorporate content and context information in order to improve item recommendation and rating prediction. Second, we look into metric learning, i.e., adaptive similarity measures that optimize combinations of descriptors in order to match human feedback.

8.5.1 *Hybrid Recommender Systems*

The collaborative data and filtering approaches mentioned so far have the advantage of making recommendations that are close to "what the user wants" as their underlying data stems directly from interactions between users and music items [450]. However, they have several drawbacks due to their dependency on the availability of (a substantial amount of) user data. A combination with other types of data, particularly content-based information, should mitigate these drawbacks and therefore provide a number of benefits; cf. [50, 51, 62, 108]:

- *Reducing cold-start problems*: With new items entering the system, audio content analysis and comparison to all existing items can be performed instantly. Thus, when no preference data is available, a hybrid system could resort to audio similarity for recommendation. While this can help remove the impact of item cold start, for dealing with user cold start, i.e., the sparseness of user profiles immediately after entering the system, other strategies need to be deployed.
- *Avoiding popularity biases*: The availability of user preference data and content meta-data may be concentrated on popular items. Content-based information, on the other hand, is available uniformly. Objective content descriptors can lower the impact of such biases and improve accessibility of music in the long tail [62].
- *Promoting diversity and novelty*: Popularity biases can result in a limited range of recommended items (cf. Sect. 7.6), whereas audio-based approaches are agnostic to whether music is a hit or from the long tail. Therefore, new and lesser known items are more likely to be recommended when both sources are exploited.
- *Integrating interaction data with musical knowledge*: As music is a highly multifaceted domain, methods should reflect different aspects of music perception. As collaborative data and music descriptors, in particular those extracted from the audio signal, exhibit complementary features, combination is expected to improve recommendation quality.

With all these advantages in mind, the challenge is to combine sources in a manner that avoids the individual drawbacks rather than propagating them. For instance, a simple feature concatenation or unsupervised linear combination can easily preserve the data sparsity problems of preference-based approaches and therefore does not yield any improvements; cf. [444]. In the following section, we review selected methods for hybrid recommendation that benefit from the combination of data sources. We start with methods that incorporate additional information, such as item relations, content similarity, collaborative tags, or user context, into matrix factorization approaches and continue with examples of other methods that can be used to predict ratings and recommend items based on information from the discussed sources. In particular, we review approaches that employ classification or regression techniques as well as probabilistic and graph-based models.

8.5.1.1 Multimodal Extensions to Matrix Factorization

Baltrunas et al. [22] include information on the listening context into matrix factorization, namely, environmental and user-related factors (**context-aware collaborative filtering** [2]). Their goal is to recommend music to the user when driving a car. For incorporating environmental (such as *traffic* and *weather*) and user-related factors (such as *mood* and *sleepiness*) into rating prediction, they extend matrix factorization by introducing one additional parameter for each pairwise combination of contextual condition and musical genre to the model. The parameters of the model are then learned using stochastic gradient descent. It is shown that mean absolute error (MAE) between the predicted and the actual ratings decreases when incorporating contextual factors. MAE is formally defined in Eq. (8.22), where $r'_{u,i}$ and $r_{u,i}$ are the predicted and actual ratings, respectively, and T contains all user–item pairs in the test set, i.e., for which predictions are effected and can be compared to the actual ratings.

$$MAE = \frac{1}{|T|} \cdot \sum_{(u,i)\in T} \left| r'_{u,i} - r_{u,i} \right| \tag{8.22}$$

However, using this strategy, improvements in prediction accuracy come at the price of increased cold-start issues as specific ratings for music need to be given for every contextual condition.

The incorporation of additional music-related information into matrix factorization methods is a strategy to reduce cold-start problems while still substantially improving their prediction accuracy in the music domain. This could, for instance, be witnessed in the *KDD-Cup 2011* competition [113], where the goal was to predict music ratings based on the *Yahoo! Music* data set. In addition to ratings information, relations between the items were available, given as a multilevel taxonomy information containing associations between genre, artist, album, and track. Dror et al. [112] show that factorization models for music recommendation can be easily extended to incorporate such additional **relational information** while also modeling **temporal dynamics** in listening behavior and temporal dynamics in item histories. Incorporating relations helps in overcoming problems with sparsity, as missing values can be implicitly compensated for by using information from other levels. The usage of listening timestamps allows the modeling of both temporal changes in the user's taste in music, including effects such as short-lived interest in current chart hits, as well as information on listening sessions, i.e., sequences of pieces listened to within the same context. Their matrix factorization model incorporates these additional aspects using a rich bias model. This model contains explicit terms to model the potential biases induced by users, items, sessions, temporal dynamics, and genres. In addition to the bias model, the user–item interaction is also further extended by including taxonomy information to the item factors and session information to the user factors. Depending on the type of item, different formulations are used. Additionally, for each learned parameter, different learning rates and regularization factors are assigned. This results in a model with over 20 meta-parameters. For optimization of these, the Nelder–Mead simplex search algorithm is used [331]. Nelder–Mead does not guarantee convergence to the global minimum; however, it is frequently applied to tasks where complex models with a high number of variables need to be optimized. From the experiments conducted, the authors can conclude that the incorporation of temporal dynamics at the user factor level is a crucial component for the superiority of the approach. Yang et al. [523] further investigate the temporal context of music consumption. In contrast to modeling a gradual change in general taste over time, they focus on a smaller and more consistent time scale that reflects changes that are strongly influenced by the listening context and occurring events ("local preferences").

To address data sparsity in collaborative music tagging, Nanopoulos et al. [329] incorporate content-based audio similarity information. As tags are given to music items by individual users, a three-way dependency needs to be modeled. This is achieved using three-order **tensor factorization** to capture cubic correlations between users, tags, and music items. For dealing with the sparsity in the cubic model, a tag propagation step based on audio similarity is performed. For audio similarity, Nanopoulos et al. resort to single Gaussian MFCC models that are compared using the KL divergence. Latent space projection of this audio-augmented model is performed by *higher order singular value decomposition* (HOSVD) [251]. This can be seen as an extension to the classic two-dimensional semantic space model of music items and tags proposed by Levy and Sandler [263]; cf. Sect. 6.1.3. Experiments lead to two conclusions. First, the usage of a tensor model to explicitly deal with the three-dimensional nature of tagging data is superior to models that perform early aggregation and thus lose this information. Second, tag propagation using audio similarity yield better results than (sparse) social tagging alone.

Van den Oord et al. [491] learn to **predict latent item factor vectors from audio features**. First, the weighted matrix factorization algorithm by Hu et al. [193] is used to learn latent factor representations of users and songs from usage data in the *Million Song Dataset* [36]; cf. Sect. 8.4.2. Second, log-compressed Mel-spectrograms of randomly sampled 3-second windows from the songs are presented to a *convolutional neural network* (CNN) [176]. In contrast to a bag of frames approach, this feature representation method preserves temporal relations in music to some extent. The crucial point of this step is that the latent factor item vectors obtained from the weighted matrix factorization step represent the learning targets during training of the network. This allows to derive latent factor vectors for items directly from an audio representation.

It is shown that this latent factor modeling of audio optimized for latent factor information on usage outperforms traditional MFCC-based vector quantization methods using linear regression or a multilayer perceptron for latent factor prediction, as well as the metric learning to rank method by McFee et al. [303] (see Sect. 8.5.2). For retrieval, similarities between music pieces can be calculated in this latent space. The examples given also indicate that retrieved pieces are more diverse than pieces retrieved with weighted MF. The authors also point out, however, that the semantic gap between audio features and user preference remains large, despite all improvement in accuracy.

Donaldson [108] proposes a hybrid recommender model built upon item similarity derived from playlist co-occurrences and acoustic features. First, latent characteristics of the item-to-item similarity graph's edge matrix are uncovered by using eigenvalue decomposition. Using the first ℓ dimensions corresponding to the first ℓ eigenvectors and eigenvalues, a low-dimensional representation of items is obtained. For each item, this is combined with acoustic feature vectors relating to timbre, rhythm, and pitch. For combination, vectors are z-score normalized to permit statistically meaningful comparability of the different value ranges stemming from different spaces. As a query, a playlist with at least two entries is required in order to substantiate and **disambiguate the user's music-seeking intentions**. This is achieved by identifying significant correlations in the individual dimensions throughout the pieces in the query. For retrieval, these dimensions receive more weight and thus have a larger influence at selecting candidates for recommendation.

8.5.1.2 Classification and Regression

Pursuing a machine learning strategy, classification and regression techniques can be applied to learn user categories from audio and other sources. For instance, Stenzel and Kamps [465] learn a mapping from low-level audio features such as spectral centroid as well as tempo and chroma features to collaborative data profiles using support vector machines. The predicted collaborative vectors are then used to calculate similarity. To directly predict binary preference (i.e., "like" vs. "dislike"), Moh et al. [321] use MFCCs and low-level audio features to train a classifier. They also experiment with support vector machines, as well as with a probabilistic Gaussian model. Beyond binary classification, Reed and Lee [382] propose ordinal regression for rating prediction. The audio features they utilize for this task describe the temporal evolution of MFCCs within each track.

As previously discussed in Sect. 6.4.1, a straightforward approach to incorporating multiple sources is to learn individual predictors and combine their outputs, potentially by learning an output combination model, i.e., a meta-classifier. This technique is known as **ensemble learning** or, particularly in the context of retrieval, as *late fusion*. Following this direction, Tiemann and Pauws [475] implement an item-based CF recommender as well as a content-based recommender that integrates timbre, tempo, genre, mood, and release-year features. Both recommenders predict ratings as weighted combinations of the most similar items' ratings. As meta-classifier, a simple nearest neighbor approach is applied that finds the closest example from the training phase in the two-dimensional space defined by the output values of the two individual recommenders.

8.5.1.3 Probabilistic Combination

Another strategy for multimodal combination is a probabilistic formulation. Li et al. [265] propose a probabilistic model in which music tracks are pre-classified into groups by means of both audio content (timbral, temporal, and tonal features) and user ratings. Predictions are made for users considering the Gaussian distribution of user ratings given the probability that a user belongs to a group.

Yoshii et al. [529] propose a hybrid probabilistic model, in which each music track is represented as a vector of weights of timbres (a "bags-of-timbres"), i.e., as a GMM over MFCCs. Each Gaussian corresponds to a single timbre. The Gaussian components are chosen universally across tracks, being predefined on a certain music collection (cf. [70, 117], Sect. 3.6). Both, ratings and "bags-of-timbres," are mapped with latent variables which conceptually correspond to genres. In such a way, music preferences of a particular listener can be represented in terms of proportions of the genres. The mapping is performed via a three-way aspect model (a Bayes network), expressing that a user stochastically chooses a genre according to preference, and then the genre stochastically "generates" pieces and timbres.

The *Just-for-me* system by Cheng and Shen [75] is an extension to the three-way aspect model by Yoshii et al. While the three-way aspect model only captures long-term music preference, *Just-for-me* also considers the user's short-term music needs. Based on the assumption that users prefer different types of music in different places, short-term needs are accounted for by using the additional information sources of **location and music popularity**. Thus, within the system, the user's location is monitored, music content analysis is performed, and global music popularity trends are inferred from microblogs. In their model, an observed listening event is represented as a quadruple $< u, l, s, w >$ denoting that user u, at location l, listens to track s containing the (discretized) audio content word w. Similar to [529], observations are generated by u randomly selecting a latent topic z according to u's interest distribution. In turn, z generates l, s, and w based on their probabilistic distributions over z. The joint probabilistic distribution of u, l, s, and w can be expressed as shown in Eq. (8.23).

$$P(u, l, s, w) = \sum_{z} P(z)P(u|z)P(l|z)P(s|z)P(w|z) \tag{8.23}$$

By marginalizing s and w, we can obtain the probability of user u choosing a track s at location l, as shown in Eq. (8.24).

$$P(s|u,l) = \frac{P(u,l,s)}{P(u,l)} = \frac{\sum_{w \in W_s} P(u,l,s,w)}{\sum_s \sum_{w \in W_s} P(u,l,s,w)} \tag{8.24}$$

where W_s denotes the audio words present in track s. This allows locating the tracks that best fit the music needs of user u at location l. In terms of popularity, users generally have a higher chance to consume current popular music, but their preferences on different contents may vary, also depending on the location. To reflect this in the model, popularity is associated with the number of observations $n(u,l,s,w) = n(u,l,s) \cdot n(s,w) \cdot pop(s)$, where $n(u,l,s)$ is the number of times u has listened to s at l, $n(s,w)$ the frequency of audio word w in s and $pop(s)$ the popularity score of s estimated via microblogs; cf. Sect. 9.4. By including popularity, a track containing many audio words that also appear in popular tracks has a higher chance to be liked by the user. To estimate the model parameters $P(z)$, $P(l|z)$, $P(u|z)$, $P(s|z)$, and $P(w|z)$ while assuming independence of observation, the log-likelihood as shown in Eq. (8.25) is maximized using a variant of the EM algorithm; cf. Sect. 3.2.4.

$$\mathscr{L} = \sum_{u,l,s,w} n(u,l,s,w) \log(P(u,l,s,w)) \tag{8.25}$$

An experimental study shows that popularity information is a relevant dimension in music recommendation and that its inclusion is advantageous compared to standard context-aware collaborative filtering methods.

8.5.1.4 Graph-Based Combinations

Several approaches follow a *graph-based interpretation* of musical relations to integrate different sources. In the resulting models, the vertices correspond to the songs, and the edge weights correspond to the degree of similarity. Shao et al. [444] build such a model upon a hybrid similarity measure that automatically re-weights a variety of audio descriptors in order to optimally reflect user preference. On the resulting song graph, rating prediction is treated as an iterative **propagation of ratings** from rated data to unrated data.

Multiple dimensions of similarity can be expressed simultaneously using a **hypergraph**—a generalization of a graph in which "hyperedges" can connect arbitrary subsets of vertices. Bu et al. [49] compute a hybrid distance from a hypergraph which contains MFCC-based similarities between tracks, user similarities according to collaborative filtering of listening behavior from *Last.fm*, and similarities on the graph of *Last.fm* users, groups, tags, tracks, albums, and artists, i.e., all possible interactions that can be crawled from *Last.fm*. The proposed approach is compared with user-based collaborative filtering, a content-based timbral approach, and their hybrid combination, on a listening behavior data set. Again, the performance of a timbral approach fell behind the ones working with collaborative filtering, while incorporation of all types of information showed the best results.

McFee and Lanckriet [308] build a hypergraph on a wide range of music descriptors to model and, subsequently, generate playlists by performing **random walks** on the hypergraph. Hypergraph edges are defined to reflect subsets of songs that are similar in some regard. The different modes of similarity are derived from the *Million Song Dataset* [36] and include:

- *Collaborative filtering similarity*: connects all songs via an edge that are assigned to the same cluster after k-means clustering for $k = \{16, 64, 256\}$ on a low-rank approximation of the user–song matrix.
- *Low-level acoustic similarity*: connects all songs assigned to the same cluster after k-means clustering for $k = \{16, 64, 256\}$ on audio features.
- *Musical era*: connects songs from the same year or same decade.
- *Familiarity*: connects songs with the same level of popularity (expressed in the categories low, medium, and high).
- *Lyrics*: connect songs assigned to the same topic derived via LDA; cf. Sect. 8.3.1.
- *Social tags*: connect songs assigned to the same *Last.fm* tag.
- *Pairwise feature conjunctions*: create a category for any pairwise intersection of the described features and connects songs that match both.
- *Uniform shuffle*: an edge connecting all songs in case no other transition is possible.

The weights w of the hypergraph are optimized using a large set of playlists S, i.e., a succession of music pieces. The assumption made for the model is that the selection of a song only depends on the previous song and the edge weights between them. This makes the chosen model a first-order Markov process and the likelihood of a playlist a product of the likelihood of the first piece in the playlist and the transition probabilities. The model is trained by finding the maximum a posteriori estimate of the weights: $w \leftarrow \text{argmax}_w \log P(w|S)$.

The corpus of training data comes from the *AotM-2011* data set [308], a collection of over 100,000 unique playlists crawled from *Art of the Mix*; cf. Sect. 7.3. In addition to playlist information, this data set also contains a timestamp and a categorical label, such as *romantic* or *reggae*, for each playlist. Experiments on a global hypergraph with weights learned from all playlists and on category-specific hypergraphs trained only on the corresponding subsets of playlists show that performance can be improved when treating specific categories individually ("playlist dialects"). In terms of features, again, social tags have the most significant impact on the overall model; however, audio features are more relevant for specific categories such as *hip-hop*, *jazz*, and *blues*, whereas lyrics features receive stronger weights for categories like *folk* and *narrative*.

The category labels of the *AotM-2011* data set exhibit further interesting aspects. While most labels refer to genre categories, some refer to a user's activity (*road trip*, *sleep*), emotional state (*depression*), or social situation (*breakup*). The results indicate that the influence of different aspects of musical content can vary dramatically, depending on contextual factors.

8.5.2 *Unified Metric Learning*

In this section, we want to focus on the idea of learning optimized and unified similarity measures from a variety of feature representations and similarity functions and continue the discussion of methods in Sect. 6.4.1 by putting emphasis on user data. Metric learning is a topic that is of interest far beyond the domain of music, e.g., [94, 169, 246, 431, 510]. The goal of the particular metric learning approaches discussed in the following is to find a transformation that combines different music-related sources such that a set of given user judgments on music similarity is approximated. Optionally, user interaction data is sometimes also found to be used as additional features fed into the system. We divide the discussed approaches into methods that learn unified similarity metrics with the goal of optimizing for a distance function and methods with the goal of learning a metric for ranking. For the latter, usually no assumptions regarding transitivity or symmetry need to be made for the target space.

8.5.2.1 Learning Similarity Metrics

Sotiropoulos et al. [460] present a content-based music retrieval system that learns from relevance feedback (see also Sect. 8.7) in order to find the best matching music similarity measure for each user. Multiple feature subsets are used to model the given similarity ratings given by each user. After training, the feature subset and model that approximate a user's feedback data best are selected. Here, the unification of different similarity measures consists of selecting the most appropriate model for each user. In subsequent evaluation tasks, subjects' responses confirmed the improved approximation of perceived similarity.

Stober and Nürnberger [466, 468] focus on **linear combinations** of different facets of similarity. A similarity (or distance) facet represents a specific aspect of computational music similarity and is computed from single features or a combination of features. The combined distance model is a weighted sum of m facet distances $\delta_{f_1} \dots \delta_{f_m}$ and can be easily understood and even directly manipulated by the user, allowing him or her to remain in control of the overall system. The weights for individual facets can be learned from **relative distance constraints** in the form of $d(s, a) < d(s, b)$, requiring object a being closer to seed object s than object b. Optimization consists of finding a weight mapping that violates as few constraints as possible, either by making adaptations to violating weights based on gradient descent [19] or by translating the constraints—and therefore the metric learning task—into a binary classification problem [74], allowing the use of support vector machines and other classifiers for optimization. The idea is to transform each

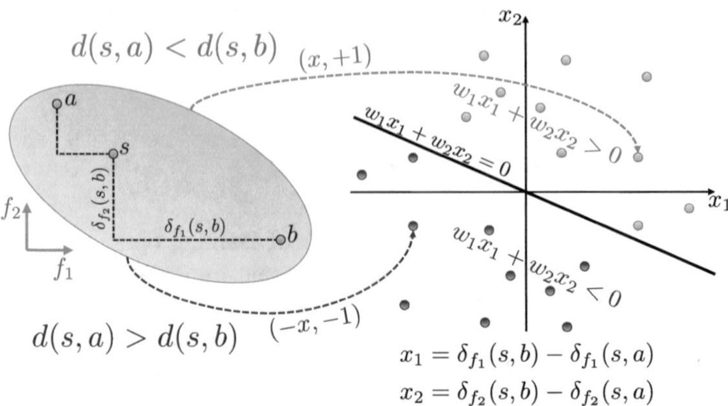

Fig. 8.5 Transforming a relative distance constraint into two training instances for the corresponding binary classification problem, cf. [74] (reprinted with permission from [419])

relative distance constraint into $\sum w_i(\delta_{f_i}(s,b) - \delta_{f_i}(s,a)) = \sum w_i x_i = w^T x > 0$, where x_i is the distance difference with respect to facet f_i [468]. From this, two training examples are generated: $(x, +1)$ as the positive example representing the satisfied constraint and $(-x, -1)$ as the negative example representing its violation. This process is visualized in Fig. 8.5.

Comparative similarity assessments, namely, those collected by Ellis et al. [121] in the *aset400* corpus, are used by McFee and Lanckriet [307] for integrating heterogeneous data into a single, unified, multimodal similarity space using a **multiple kernel learning** technique. Extending their work [305] discussed in Sect. 6.4.1, they demonstrate the applicability of their technique on a music similarity task on the artist level by including five data sources representing different aspects of an artist. In contrast to [305], they exclude chroma information and instead include collaborative filtering to complement artist timbre, auto-tags, social tags, and biographical text data. The included data describes an artist as the corresponding row in a user–artist listening matrix created from *Last.fm* listening histories [62]. For each type of data, a kernel maps the artist representation into an individual nonlinear space using partial order embedding. Then, an individual transformation is learned for each type of data and optimized with respect to the comparative similarity assessments. The vectors resulting from each individual transformation are then combined, resulting in a joint Euclidean space representation.

Comparing the unified similarity space with individual similarity spaces (and partial combinations) against the human-annotated ground truth shows that the multiple kernel learning technique outperforms an unweighted combination of individual kernels and therefore corresponds best to the perceived similarity of users. It can also be seen that the timbre similarity performs poorly (potentially since it is originally targeting the song level rather than the artist level) and that social tags contribute the most valuable information.

Slaney et al. [452] evaluate six strategies to learn a Mahalanobis distance metric to embed songs into a Euclidean metric space. Mahalanobis metrics are generic variants of the Euclidean metric and allow for a linear transformation of the feature space. Slaney et al.'s models are trained on data extracted from music blogs (audio content, meta-data, and blog context). One of their goals is to compare the similarity of music within a blog to album and artist similarity. Each of the six compared approaches rotates and scales the feature space with a linear transform. Wolff and Weyde [520] also learn an optimized similarity space from comparative similarity assessments via Mahalanobis distance metrics. To this end, they use data (audio features and similarity assessments) from the *MagnaTagATune* data set collected via the *TagATune* game; cf. Sect. 5.3.3 and compare an SVM approach [431] with the metric learning to rank approach [306] (see Sect. 8.5.2.2). Chen et al. [72] learn an embedding of songs and tags into a Euclidean space from playlists modeled as Markov chains. The distances in this space (song to song, song to tag, and tag to tag) are given a probabilistic interpretation, allowing to be used for playlist generation (even of unseen songs based on the tag model) and natural language querying (even of untagged songs).

8.5.2.2 Learning to Rank

Methods that learn to rank are very similar to the methods discussed so far. However, the primary goal of these approaches is not to model similarity relations and distances in an embedded space (although that can be a result as well) but to find a unified projection that results in optimized rankings given a query. Thus, otherwise important aspects of a learned metric such as transitivity or symmetry are negligible. The objective that these methods optimize for is therefore typically a rank-based criterion.

Lu and Tseng [286] combine three rankings from three recommenders in order to obtain a final, **personalized ranking**. The first produces a ranking according to content similarity based on features extracted from the symbolic musical score. The second ranker implements user-based CF over a data set of user surveys. The third approach ranks items based on emotions (according to manual expert annotations). In the combination step, a personalization component is introduced. This component re-weighs the individual rankings according to user feedback gathered in an initial survey in which users specified preference assessments (like/dislike) and the underlying reasons (such as preference by tonality, rhythm, etc.) for a sample of tracks. More precisely, the user's preferred music pieces as given in the survey are identified in the rankings produced by the individual approaches. Their positions relate to the weight that an individual method should be given, i.e., a ranker's weight is bigger if the sum of the ranks of preferred pieces is smaller. These weights can be continuously updated and thus account for changes in preference, as more user feedback is gathered in the process of using the system. A final score for re-ranking the pieces is simply obtained by a linear weighted combination of the individual ranks.

McFee et al. [303, 306] optimize a content-based similarity metric for query-by-example tasks by learning from a sample of collaborative data. Starting from a codebook representation of delta-MFCCs that represents songs as a histogram over the derived codewords, ranking candidates given a query q is based on Euclidean similarities in the vector space, i.e., the ordering of music pieces a is obtained by sorting according to increasing distance $\|q - a\|$ between q and a. In order to optimize the distance function according to rankings based on user similarity, a Mahalanobis distance matrix W is learned such that $\|q - a\|_W = \sqrt{(q-a)^T W(q-a)}$. To learn W, the **metric learning to rank** (MLR) algorithm is presented. MLR builds upon *structural support vector machines* [199] to adapt W according to a ranking loss function. This is in contrast to methods for metric learning for embedded similarity spaces, where loss functions are defined at the level of pairwise distances rather than rankings. Optimization of the model is again performed using a gradient descent solver (for details, the reader is referred to [306]).

Through MLR, the above-described distance yields optimal rankings of the training data given each contained song as query. Knowledge of optimal ranking comes from item similarity as obtained from collaborative filtering data, precisely from the Jaccard index of the user sets given by the item vectors, which normalizes the number of overlapping users by the number of users contained in either of the user sets of the two item vectors under comparison. More formally, the Jaccard index is defined in Eq. (8.26), where S_i and S_j are two sets of elements, in this case given by the users in the user–item matrix for the two items to compare.

$$J(S_i, S_j) = \frac{|S_i \cap S_j|}{|S_i \cup S_j|} \tag{8.26}$$

After MLR optimization of W, it is possible to find similar items even for novel and unpopular queries based on audio content while maintaining high recommendation accuracy resulting from feedback data. An extension to MLR called *robust MLR* is presented by Lim et al. [273]. Robust MLR adds regularization for sparsity in both input and output spaces and thus detects and suppresses noisy features, resulting in improvements in both low- and high-noise settings.

8.6 Summary

In this chapter, we discussed how data on interaction traces between users and music items can be used to compute similarities and in turn build music recommender systems. We first reviewed approaches that infer similarity from co-occurrence counts, presuming that music items that more frequently co-occur, for instance, in microblogs, playlists, or peer-to-peer networks, are more similar than items with lower co-occurrence frequency. These methods work quite well, but performance comes at the cost of typically hard-to-acquire and noisy data.

We proceeded with an analysis of graph- or network-based techniques that infer similarity from properties of connections between users or between users and music items, by exploiting playlists, social networks, or peer-to-peer network data. In this case, estimating similarity is again frequently performed via co-occurrence analysis or variants thereof. In case of data sources that provide a sequential ordering of items, e.g., playlists, co-occurrence counts are sometimes modified by a factor that weights co-occurrences between two items by their distance within a playlist. Unfortunately, the performance of this category of approaches has not been fully investigated yet. While they show some merit, information about user connections is commonly hard to acquire on a large scale (e.g., from *Facebook*). The same holds for playlist data, where large repositories are rare or data is hard to gather on a substantial scale.

Another set of approaches that exploit co-occurrence information are methods based on topic models, in particular LDA. In this case, latent factors, topics, or concepts are derived from listening sessions or listening histories. By building a probabilistic model, LDA provides all advantages of operating within Bayesian probability theory. After a topic model has been learned, each music item can be represented by a vector of probabilities over topics, i.e., probabilities that the item belongs to each of the topics learned. Similarities can subsequently be computed by applying a similarity function to pairs of these topic vectors. We illustrated the application of LDA for an artist clustering task, using a set of almost half a billion listening events as input. Building a 50-topic model, we showed that the resulting clustering of artists pretty much corresponds to a separation between different styles, in which sometimes even fine-grained distinctions were observable, e.g., a cluster of pop music with female vocals or a folk metal cluster. We further evidenced that LDA or extensions thereof have been successfully used for music recommendation. While topic models represent a powerful way to infer and encode latent factors from listening data, building these models typically requires a large amount of data and is time-consuming.

The task of rating prediction is a central one in recommender systems. Based on existing ratings of items by users, which can be effected on various scales (from binary liking or disliking indications to discrete 5- or 10-point Likert scales), the goal is to predict how a user would rate an unseen item. The items with highest predicted ratings are then recommended. This approach is flexible in that it can also estimate which items a user will not like. The standard approach to recommender systems is collaborative filtering, which comes in the flavors memory based and model based. The former compute similarities directly on the rating matrix—either between users or items—and predict the (weighted) average rating of the most similar users or items for the target user–item pair. Model-based approaches, in contrast, factorize the rating matrix and represent both items and users in the same latent factor space that is supposed to capture aspects of user and item characteristics concealed in the rating data. In comparison to memory-based CF, model-based CF is more efficient at the prediction stage and less memory-consuming once the factor model has been learned. However, the computational complexity of decomposing the rating matrix into user and item components (matrix factorization) is not negligible. Since model-based CF is nevertheless the de facto standard in current recommender systems, we detailed different matrix factorization techniques, in particular, SVD and several extensions that can account for the temporal scope of ratings or for implicit feedback (e.g., browsing behavior or skipping a song). We further reviewed how logistic matrix factorization can be used in a music artist recommendation task.

In the last part of the chapter, we discussed hybrid recommender systems that integrate content and context information to improve rating prediction and recommendation. In particular, we looked into matrix factorization approaches that consider item relations, content similarity, collaborative tags, user context, or combinations thereof. We concluded with a discussion of adaptive similarity measures, reviewing metric learning and learning to rank strategies. In our case, the goal of metric learning is to create a metric that integrates similarities from various music-related sources in a way that a set of given user judgments (e.g., relative similarity assessments between a seed song and two target songs) is approximated. Learning to rank, in contrast, does not intend to create a full metric, but rather an optimal (personalized) ranking for a given query, based on user feedback information and rankings or similarities derived from different music-related data sources.

8.7 Further Reading

The importance of the user in music retrieval and recommendation is evident and has been highlighted frequently in the literature. A general, early approach to incorporating user feedback information (relevance feedback) into a retrieval process is presented by Rocchio [385]. Exploiting relevance feedback is done in an iterative process in which the user is presented with a ranked list of results most relevant to the query according to the system. After examination of the list, the user marks those pieces which are relevant in his or her opinion. Using the vector space model for retrieval, this information is then used by moving the query vector towards the relevant results and away from the irrelevant; cf. [218]. Such relevance assessment could also be collected from implicit feedback, e.g., in a music retrieval scenario such as the one presented in Sect. 6.2 by measuring the time a user is listening to or looking at the returned tracks.

A more extensive discussion of the role of the user than provided in this chapter is given, for instance, in some parts of the edited book *Multimodal Music Processing* [327], which provides chapters on user-aware music retrieval [419] and on the use of MIR techniques in music education [104]. In addition, a survey article on recent developments and applications in MIR [422] provides a discussion of computational user modeling, user-adapted music similarity, music discovery based on user preferences, and user studies for evaluating MIR approaches. Taking both a computer scientist's and a psychologist's point of view, we discuss in [420] why the listeners themselves have largely been neglected in music retrieval so far and we reflect on why the breakthrough in this field has not been achieved yet. We further compare the role of the users in MIR with their role in related fields, including (movie) recommender systems and text information retrieval. In addition, we review works that incorporate user-centric aspects at least to a limited extent and conclude the article by ideas on how user-centric MIR algorithms could be evaluated.

Apart from modeling the individual user, there are manifold sources on collaborative methods beyond the ones discussed in this chapter, with those from the domain of recommendation being most relevant. The field of recommender systems is treated in much detail in the second edition of the *Recommender Systems Handbook* [383], one of the landmark books on the topic and a highly recommended read. The second edition comes with a chapter on music recommender systems [424], next to comprehensive treatments of collaborative filtering; content-based method; context-aware recommendation; location-based and social recommender systems; personality-based recommendation; cross-domain recommendation; aspects of privacy, novelty, and diversity in recommender systems; group recommender systems; and industrial perspectives, to illustrate a few topics covered in the book.

For readers who want to gain a deeper understanding of topic modeling and LDA, an easy-to-access work is the survey article by Blei [39]. More technical treatments of the topic are provided in [40, 41] as well as in [533], which further discusses the related technique of probabilistic latent semantic indexing (PLSI).

Part IV
Current and Future Applications of MIR

Part IV presents current applications of MIR, identifies challenges of current technology, and discusses possible future directions of the field. In Chap. 9, a comprehensive overview of applications of MIR including automatic playlist generation, visualizations, browsing and retrieval interfaces, active listening systems, and popularity estimation that are built upon the methods described in the preceding chapters is given. The book concludes with a discussion of major challenges MIR is facing, an outlook to the possible future of MIR, and the directions and applications to come (Chap. 10).

Chapter 9
Applications

In this chapter, we present a selection of applications of the music search and retrieval techniques discussed in the previous chapters. In particular, we address several use cases, which we organize into those that require user interaction and those that do not, i.e., assume passive users. As for the former, we first discuss music information systems that make use of information extraction techniques to acquire knowledge about music items and present it to the user in a structured way (Sect. 9.1). Subsequently, we review applications for exploration and browsing in music collections via intelligent user interfaces, in particular map-based interfaces (Sect. 9.2). Use cases that do not require direct user interaction include automatic playlist generation (Sect. 9.3) and creating music charts from contextual social data on music (Sect. 9.4).

9.1 Music Information Systems

A classic way for accessing music collections or information on music is via a (typically web-based) information system. Examples of music information systems are *allmusic*, *Yahoo! Music*, *Last.fm*, and also (if restricted to the music domain) *Wikipedia*. Informally defined, a music information system serves as a kind of multimedia encyclopedia that has entries for different musical entities on different levels of granularity as well as links between these entity descriptions. In practice, typical categories included in such a system are discographies, biographical information, and lineup for bands, as well as track listings, cover artwork, and other meta-data for albums. Furthermore, recommendations such as similar artists or albums are frequently included. The presented information may originate from editors (as is the case for *allmusic* and *Yahoo! Music*, for instance) or from a community (as in the case of *Last.fm* and *Wikipedia*).

P. Knees, M. Schedl, *Music Similarity and Retrieval*, The Information Retrieval Series 36, DOI 10.1007/978-3-662-49722-7_9

Alternatively, web mining techniques can be used to automatically extract music-related pieces of information from unstructured or semi-structured data sources via natural language processing and machine learning. This process is sometimes referred to as music information extraction and is a small but important subfield of MIR. Based on the mined information, a music information system can then be built [417]. An overview of work addressing the automatic data acquisition of music-related information, in particular, members and instrumentation of bands, country of origin of artists, and album cover images, is given in the following paragraphs.

9.1.1 Band Members and Their Roles

Automatic prediction of the members of a band and their respective roles, i.e., the instruments they play, can be achieved by crawling web pages about the band under consideration [405]. From the set of crawled web pages, *n*-grams are extracted and several filtering steps (e.g., with respect to capitalization and common speech terms) are performed in order to construct a set of potential band members. A rule-based approach is then applied to each candidate member and its surrounding text. The frequency of patterns such as *[member] plays the [instrument]* is used to compute a confidence score and eventually predict the *(member, instrument)* pairs with highest confidence.

Extending work by Krenmair [243], we have investigated two further approaches to band member detection from web pages [217]. We use a part-of-speech tagger [48] (cf. Sect. 5.4), a gazetteer annotator to identify keywords related to genres, instruments, and roles, among others, and finally perform a transducing step on named entities, annotations, and lexical meta-data. This final step yields a set of rules that can be applied for information extraction; cf. [405]. We further investigate support vector machines [494] (cf. Sect. 4.1) to predict whether each token in the corpus of music-related web pages is a band member or not. To this end, we construct feature vectors including orthographic properties, part-of-speech information, and gazetteer-based entity information.

9.1.2 Artist's or Band's Country of Origin

Identifying an artist's or a band's country of origin provides valuable clues about their background and musical context. For instance, a performer's geographic, political, and cultural context or a songwriter's lyrics might be strongly related to their origin.

Govaerts and Duval [158] search for occurrences of country names in biographies on *Wikipedia* and *Last.fm*, as well as in properties such as *origin*, *nationality*, *birthplace*, and *residence* on *Freebase*.[1] The authors then apply simple heuristics to predict the most probable country of origin for the artist or band under consideration. An example of such a heuristic is predicting the country that most frequently occurs in an artist's biography. Another one favors early occurrences of country names in the text. Govaerts and Duval show that combining the results of different data sources and heuristics yields the best results.

Other approaches distill country of origin information from web pages identified by a search engine [413]. One approach is a heuristic that compares the page count estimates returned by *Google* for queries of the form *"artist/band" "country"* and simply predicts the country with highest page count value for a given artist or band. Another approach takes into account the actual content of the web pages. To this end, up to 100 top-ranked web pages for each artist are downloaded and term frequency–inverse document frequency (TF·IDF) weights are computed (cf. Sect. 6.1.1). The country of origin for a given artist or band is eventually predicted as the country with highest TF·IDF score using as query the artist name. A third approach relies on text distance between country name and key terms such as "born" or "founded." For an artist or band a under consideration, this approach predicts as country of origin c the country whose name occurs closest to any of the key terms in any web page retrieved for c. It is shown that the approach based on TF·IDF weighting outperforms the other two methods.

9.1.3 Album Cover Artwork

An important aspect in music information systems is the display of the correct album cover. There are approaches to automatically retrieve and determine the image of an album cover given only the album and performer name [410, 417]. First, search engine results are used to crawl web pages of artists and albums under consideration. Subsequently, both the text and the HTML tags of the crawled web pages are indexed at the word level. The distances at the level of words and at the level of characters between artist/album names and `<img>` tags are computed thereafter. Using Eq. (9.1), where $p(\cdot)$ refers to the position of artist name a, album name b, and the image tag i in the web page and τ is a threshold, a set of candidate images of cover artwork is constructed by fetching the corresponding images.

$$|p(a) - p(i)| + |p(b) - p(i)| \leq \tau \tag{9.1}$$

The resulting set still contains a lot of irrelevant images. To resolve this, simple content-based filtering techniques are employed. First, non-square images are discarded, using filtering by width/height ratios. Also images showing scanned compact discs are identified by a circle detection technique and consequently

[1] http://www.freebase.com.

removed. The images with minimal distance according to Eq. (9.1) are eventually chosen as album covers.

9.1.4 Data Representation

As for the structured organization and representation of music-related information, a few scientific works deal with the use of semantic web technologies and representation of music and listener entities via knowledge graphs. The most popular ontology in this context is certainly the *Music Ontology*,[2] proposed by Raimond et al. [373, 374]. It was later extended, among others, by Gängler who added personalization components, such as a playback ontology, recommendation ontology, and cognitive characteristics ontology [139]. An example for a music playlist created within the playback ontology expressed in the resource description framework (RDF) is shown in Fig. 9.1. Raimond further assesses methods to interlink existing open database available on the web [375], for instance, *MusicBrainz*, *DBpedia*,[3] *Jamendo*,[4] and *Magnatune*.[5]

9.2 User Interfaces to Music Collections

This section reviews different user interfaces for music collections. In general, such systems use various kinds of information on music to automatically structure a given repository and aid the user in exploring the contents. We provide an overview of the large number of available map-based interfaces and briefly discuss other intelligent music interfaces.

9.2.1 Map-Based Interfaces

The idea of map-based music interfaces is to organize musical entities in a two-dimensional layout to display the global composition of collections and intuitively visualize similarity by relating it to closeness on the map. Furthermore, orientation on a map is a concept familiar to people and therefore a particularly good choice to be incorporated into interfaces for novel purposes. This kind of structuring allows

[2]http://www.musicontology.com.

[3]http://www.dbpedia.org.

[4]https://www.jamendo.com.

[5]http://www.magnatune.com.

```
@prefix xsd:    <http://www.w3.org/2001/XMLSchema#> .
@prefix dc:     <http://purl.org/dc/elements/1.1/> .
@prefix olo:    <http://purl.org/ontology/olo/core#> .
@prefix pbo:    <http://purl.org/ontology/pbo/core#> .
@prefix ao:     <http://purl.org/ontology/ao/core#> .
@prefix mo:     <http://purl.org/ontology/mo/> .
@prefix ex:     <http://example.org/> .
@prefix sim:    <http://purl.org/ontology/similarity/> .

ex:FunkyPlaylist a pbo:Playlist ;
   dc:title "Funky Playlist"^^xsd:string ;
   dc:description "A playlist full of funky legends"^^xsd:string ;
   dc:creator <http://foaf.me/zazi#me> ;
   olo:length 2 ;
   sim:association ex:ZazisAssociation ;
   sim:association ex:BobsAssociationInUse ;
   pbo:playlist_slot [
      a pbo:PlaylistSlot ;
      olo:index 1 ;
      pbo:playlist_item ex:SexMachine
   ] ;
   pbo:playlist_slot [
      a pbo:PlaylistSlot ;
      olo:index 2 ;
      pbo:playlist_item ex:GoodFoot
   ] .

ex:SexMachine a mo:Track ;
   dc:title "Sex Machine"^^xsd:string ;
   dc:creator <http://dbpedia.org/resource/James_Brown> .

ex:GoodFoot a mo:Track ;
   dc:title "Good Foot"^^xsd:string .

ex:ZazisAssociation a sim:Association ;
   dc:creator <http://foaf.me/zazi#me> ;
   ao:genre "Funk"^^xsd:string ;
   ao:mood "party"^^xsd:string ;
   ao:occasion "my birthday party 2008"^^xsd:string .

ex:BobsAssociation a sim:Association ;
   dc:creator <http://foaf.me/zazi#me> ;
   ao:genre ex:Funk ;
   ao:mood "happy"^^xsd:string ;
   ao:occasion "good feeling music"^^xsd:string .

ex:BobsAssociationInUse a ao:LikeableAssociation ;
   ao:included_association ex:BobsAssociation ;
   ao:likeminded <http://moustaki.org/foaf.rdf#moustaki> .

ex:Funk a mo:Genre .
```

Fig. 9.1 A sample representation of a music playlist in RDF/Turtle format, according to http://smiy.sourceforge.net/pbo/spec/playbackontology.html#sec-example, "Creative Commons Attribution 1.0 Generic" license

then for browsing music collections by examining different regions on the map as well as for implicit recommendation by exploring areas surrounding known pieces.

Most systems that create a map for music organization and exploration rely on a self-organizing map (SOM) or on multidimensional scaling (MDS) techniques, which are introduced in the following before presenting interfaces that rely on them.

Both are unsupervised learning techniques, used in particular for clustering. Their objective is to map items given in a high-dimensional input space onto positions in a low-dimensional (typically, two-dimensional) output space while preserving the original pairwise distances between items as faithfully as possible.

9.2.1.1 Self-organizing Map

The self-organizing map (SOM) [235] is a powerful and popular neural network technique. It is a nonlinear clustering method, thus allows one to preserve nonlinear relationships in the high-dimensional data space when mapping them to the low-dimensional visualization or output space.

The SOM layout consists of an ordered set of interconnected map units (or neurons), arranged in a two-dimensional array, which forms either a rectangular or hexagonal grid. Each of the u map units is assigned a model vector $\mathbf{m_i}$ of the same dimensionality d as the data items in the input space. The set of all model vectors $m_1 \ldots m_u$ of a SOM is referred to as its codebook. An illustration of the two principal layouts is given in Fig. 9.2.

The input to a SOM is a data set $\mathbf{X}$, which is an $n \times d$ matrix composed of n feature vectors of dimensionality d. In our case, these feature vectors represent the music items under consideration and are likely the result of applying feature extraction techniques presented in Parts I and II.

Before training the SOM, i.e., fitting the model vectors to the data set, the model vectors are initialized. This can be accomplished by creating *random* model vectors in the same range as the input data set or by mapping the high-dimensional input vectors onto a two-dimensional space by employing *linear* projection techniques, such as principal components analysis (PCA) [187]. Doing so, the space spanned by the two greatest eigenvectors is mapped to the two-dimensional SOM grid, and the model vectors are initialized accordingly. While random initialization is faster than linear initialization, two equally parametrized training runs performed on the same data set can yield different trained SOMs.

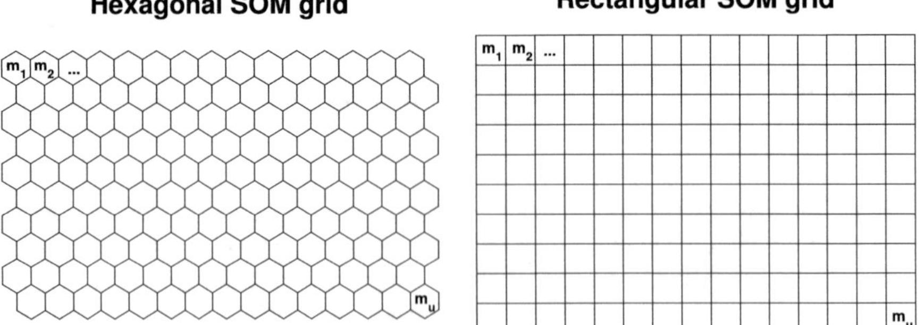

Fig. 9.2 Hexagonal and rectangular SOM grid layout with u map units

SOMs can either be trained sequentially (online learning) or using the batch map algorithm [497]. Both are iterative methods that loop until some convergence criterion is fulfilled, for instance, no significant change in some error function, e.g., mean quantization error (MQE),[6] can be seen anymore. In **sequential training** [233], a data item $\mathbf{x}_j$ from $\mathbf{X}$ is randomly chosen in each iteration. Subsequently, the best matching unit (BMU) for $\mathbf{x}_j$ is identified. This is the map unit whose model vector exhibits the smallest distance to $\mathbf{x}_j$, formally defined in Eq. (9.2), where d is a distance function, typically Euclidean distance.

$$bmu(\mathbf{x}_j) = \underset{i}{\operatorname{argmin}} \ d(\mathbf{m}_i, \mathbf{x}_j) \tag{9.2}$$

After having identified the BMU of $\mathbf{x}_j$, all model vectors of the SOM are updated to better represent data item $\mathbf{x}_j$. The amount of adaptation of each model vector depends on two factors: (1) the distance of the model vector to $bmu(\mathbf{x}_j)$ and (2) the learning rate. The former ensures that model vectors close to $bmu(\mathbf{x}_j)$ receive a bigger amount of adaptation than model vectors far away from the BMU on the SOM grid. This in turn results in neighboring map units representing similar data items. The learning rate $\alpha(t)$ is used to gradually decrease the amount of adaptation over time, where t is the iteration count. The adaptation function for the model vectors is given in Eq. (9.3), where $h(\mathbf{x}_j, \mathbf{m}_i, t)$ models the neighborhood of the model vector on the two-dimensional SOM grid, as defined in Eq. (9.4). More precisely, $y_{bmu(\mathbf{x}_j)}$ and y_i represent the two-dimensional locations in the output space of map units $bmu(\mathbf{x}_j)$ and y_i, respectively; d is again a distance function, usually Euclidean distance, and $\delta(t)$ is referred to as neighborhood kernel radius, a time-dependent parameter that ensures a decreasing size of the neighborhood during training. This enables the formation of large clusters at early training stages and supports fine-tuning towards the end.

$$\mathbf{m}_i(t+1) = \mathbf{m}_i(t) + \alpha(t) \cdot h(\mathbf{x}_j, \mathbf{m}_i, t) \cdot [\mathbf{x} - \mathbf{m}_i(t)] \tag{9.3}$$

$$h(\mathbf{x}_j, \mathbf{m}_i, t) = \exp\left(-\frac{d(y_{bmu(\mathbf{x}_j)}, y_i)}{\delta(t)^2}\right) \tag{9.4}$$

The **batch map** [234] is also an iterative algorithm. Instead of presenting a single data item to the SOM in each iteration, however, the whole data set is taken into account. Each iteration includes two steps, which are performed again until convergence. First, the BMU of each data item is determined. Second, the model vectors are updated based on the assignment between data items and BMUs, according to Eq. (9.5), in which n denotes the number of data items. Hence, the new model vector $\mathbf{m}_i(t+1)$ is computed as a normalized linear combination of the data items, where the coefficients or weights are given by the value of the neighborhood

[6]The MQE assesses the average difference between the model vectors and the data items of the respective map units.

kernel function $h(\mathbf{x}_j, \mathbf{m}_i, t)$. The farther away (in the output space) the model vector to update $\mathbf{m}_i$ is from the BMU of $\mathbf{x}_j$, the less influence $\mathbf{x}_j$ exerts on the new value of the model vector $\mathbf{m}_i(t+1)$.

$$\mathbf{m}_i(t+1) = \frac{\sum_{j=1}^{n} h(\mathbf{x}_j, \mathbf{m}_i, t) \cdot \mathbf{x}_j}{\sum_{j=1}^{n} h(\mathbf{x}_j, \mathbf{m}_i, t)} \tag{9.5}$$

The batch map does not require a learning rate parameter since the entire data set is presented to the SOM in each iteration. It can further be regarded as a generalization of k-means clustering [288]. In fact, if the neighborhood kernel takes the simple form shown in Eq. (9.6), the batch map yields the same clustering as k-means.

$$h(\mathbf{x}_j, \mathbf{m}_i, t) = \begin{cases} 1 & \text{if } bmu(\mathbf{x}_j, t) = i \\ 0 & \text{otherwise} \end{cases} \tag{9.6}$$

Comparing sequential training to batch map, the former easily adapts to data sets that dynamically vary in size. The latter, on the other hand, is much faster and retraining on the same data yields the same SOM, up to rotation and mirroring.

After having trained a SOM, there are several ways to illustrate the resulting map. An overview of the most important ones is given in [497]. As it is used in several approaches to visualize clusters of music, we here discuss the **smoothed data histogram** (SDH) [347]. The idea is fairly simple. Each data item votes for a fixed number s of map units which best represent it. Voting is effected in a way that the best matching unit receives s points, the second best matching unit receives $s-1$ points, and so on, up to the sth best matching unit which receives only 1 point. All other units do not receive any points by the data item. Each data item votes, and the votes are accumulated for each map unit, resulting in a voting matrix. This matrix contains large values for densely populated areas on the map and small values for sparse areas. The voting matrix is then normalized to the range $[0,\ 1]$, and its values are mapped to a color space for visualization. An SDH visualization of a 4×6 SOM trained on web features extracted for the toy music data set is depicted in Fig. 9.3. The SOM grid is also shown for convenience. As you can see, the clustering is far from perfect, but classical music is clustered well in the top right. However, pop artists are spread across the map, which is the result of their term features not being very discriminative from other genres.

9.2.1.2 SOM-Based Interfaces for Music Collections

In their seminal work, Rauber and Frühwirth propose the use of a SOM to form clusters of similar songs and to project them onto a two-dimensional plane [380]. Pampalk extends this approach to create the *Islands of Music* interface [342], which is later frequently adopted to organize music collections. For Pampalk's *Islands*

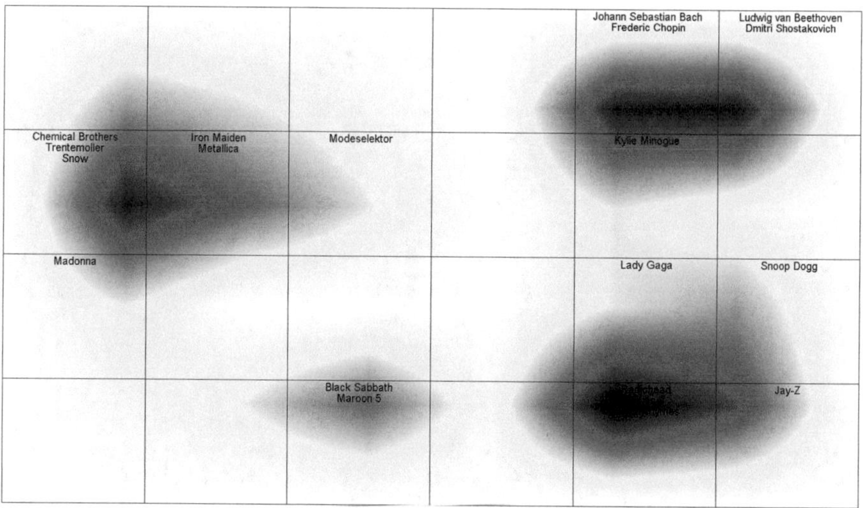

Fig. 9.3 A smoothed data histogram (SDH) visualization of a self-organizing map (SOM) trained on the toy music data set

of Music, a SOM is trained on rhythm descriptors, more precisely, fluctuation patterns (cf. Sect. 3.3.1). It is visualized via an SDH, in which a color model inspired by geographical maps is applied. An illustration of the result can be seen in Fig. 9.4. Using this approach, clusters of similarly sounding music (according to the extracted audio features) become visible. The red and yellow squares offer further functionality. While the red ones allow the landscape to be refined by revealing a more detailed SOM for the map unit at hand, the yellow ones provide a link to the directory in which the displayed songs reside. Several extensions to this *Islands of Music* approach have been proposed, e.g., the usage of aligned SOMs [348] to enable a seamless shift of focus between different aspects of similarity, for instance, between a SOM visualization created only on rhythm features and one created only on timbre features, or a hierarchical component to cope with large music collections [400]. Neumayer et al. use SOMs for browsing in collections and intuitively generate playlists on portable devices [332]. Leitich and Topf propose mapping songs to a sphere instead of a plane by means of a GeoSOM [521] to create their *Globe of Music* interface [261]. Other approaches use SOM derivatives, for instance, the one proposed by Mörchen et al. [323]. Their *MusicMiner* browsing interface employs an emergent self-organizing map and a U-map (or U-matrix) visualization technique [488] that color-codes similarities between neighboring map units. Vembu and Baumann use textual information mined from *Amazon* reviews to structure music collections via a SOM by incorporating a fixed list of musically related terms to describe similar artists [496].

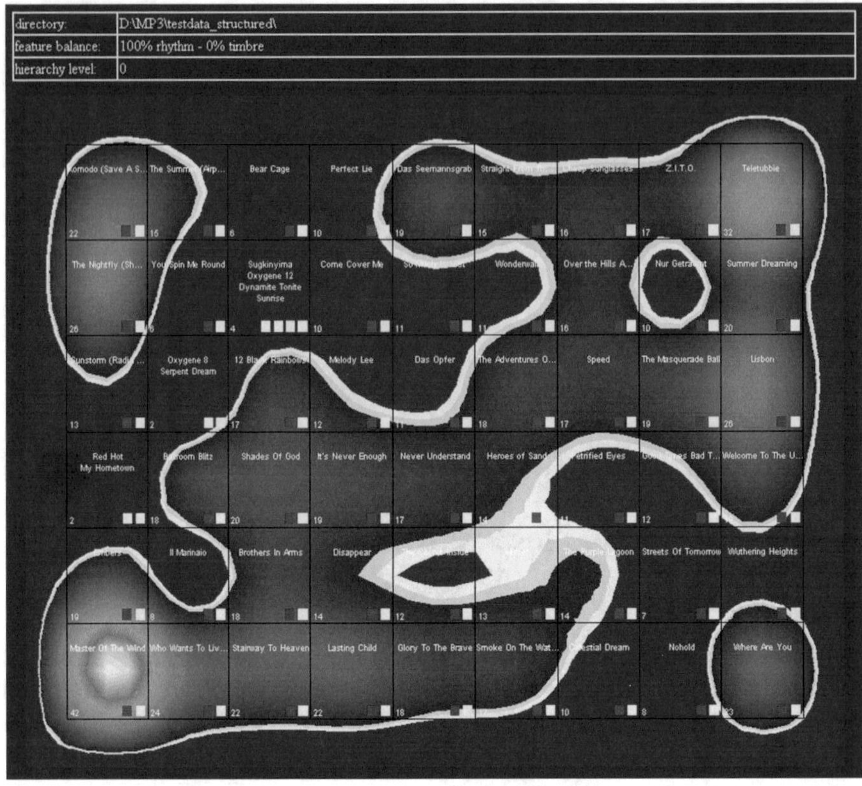

Fig. 9.4 A variant of the *Islands of Music* browsing interface [400]

For deepening impression and immersion when exploring music collections, three-dimensional browsing interfaces employing a SOM are also available. Such extensions enable the user to move through a virtual terrain to discover new music, much like in a computer game. The first approach in this direction is *nepTune*,[7] which employs the *Islands of Music* metaphor of islands representing clusters of similar music [222]. A version for mobile devices is proposed by Huber et al. [194]; cf. Fig. 9.5. A shortcoming of approaches that use a single SOM to visualize an entire music collection of real-world size is the information overload caused by visual clutter. To alleviate this issue, the *deepTune* interface [414] employs a growing hierarchical self-organizing map (GHSOM) [103] that automatically structures the data set into hierarchically linked individual SOMs. In *deepTune*, prototypical music tracks are determined by integrating audio similarity with popularity information mined from the web. This is performed for each map unit in the GHSOM and allows the user to explore largely unknown music collections using

[7] http://www.cp.jku.at/projects/neptune.

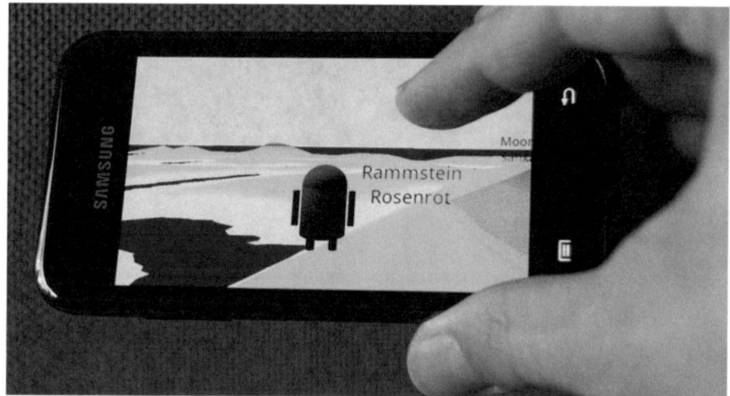

Fig. 9.5 The *nepTune* music browsing interface for mobile devices; cf. Fig. 1.3

the prototypes as anchor points. Screenshots of the *deepTune* interface are shown in Fig. 9.6. The topmost screenshot shows a visualization of the whole collection on the highest hierarchy level, by a 16×7 SOM. If the user then decides to focus on the cluster whose representative song is "Many a Mile to Freedom" by Traffic, he is shown the 4×5 SOM depicted in the middle. If he again decides to dig deeper into the visualization, he is shown the 4×4 SOM plotted at the bottom of Fig. 9.6.

More recent alternatives of three-dimensional, SOM-based interfaces include [287], where Lübbers and Jarke present a browser that also forms landscapes but which uses an inverse height map in comparison to the *Islands of Music* interfaces. This means that agglomerations of songs are visualized as valleys, while clusters are separated by mountains rather than by the sea. Unlike the previously discussed interfaces, Lübbers and Jarke's allows the user to adapt the landscape by building or removing mountains, which triggers an adaptation of the underlying similarity measure.

9.2.1.3 Multidimensional Scaling

Unlike the SOM which maps data items into a predefined number of clusters, i.e., map units, the multidimensional scaling (MDS) [88] approach projects data items from the input space to arbitrary positions in the visualization space, in an attempt to preserve the distances between data items. MDS can be regarded as an optimization problem, in which the rationale is to minimize the pairwise differences between distances in the input and in the output space. To this end, MDS algorithms utilize an error function defined on the item distances in the input space p_{ij} and in the output space d_{ij}, i.e., p_{ij} and p_{ij} are the distances between data items i and j in the input and output space, respectively. There are a variety of error functions. One of the most common is proposed by Sammon in [393] and is show in Eq. (9.7), where n is the number of data items. In contrast to simpler functions, such as

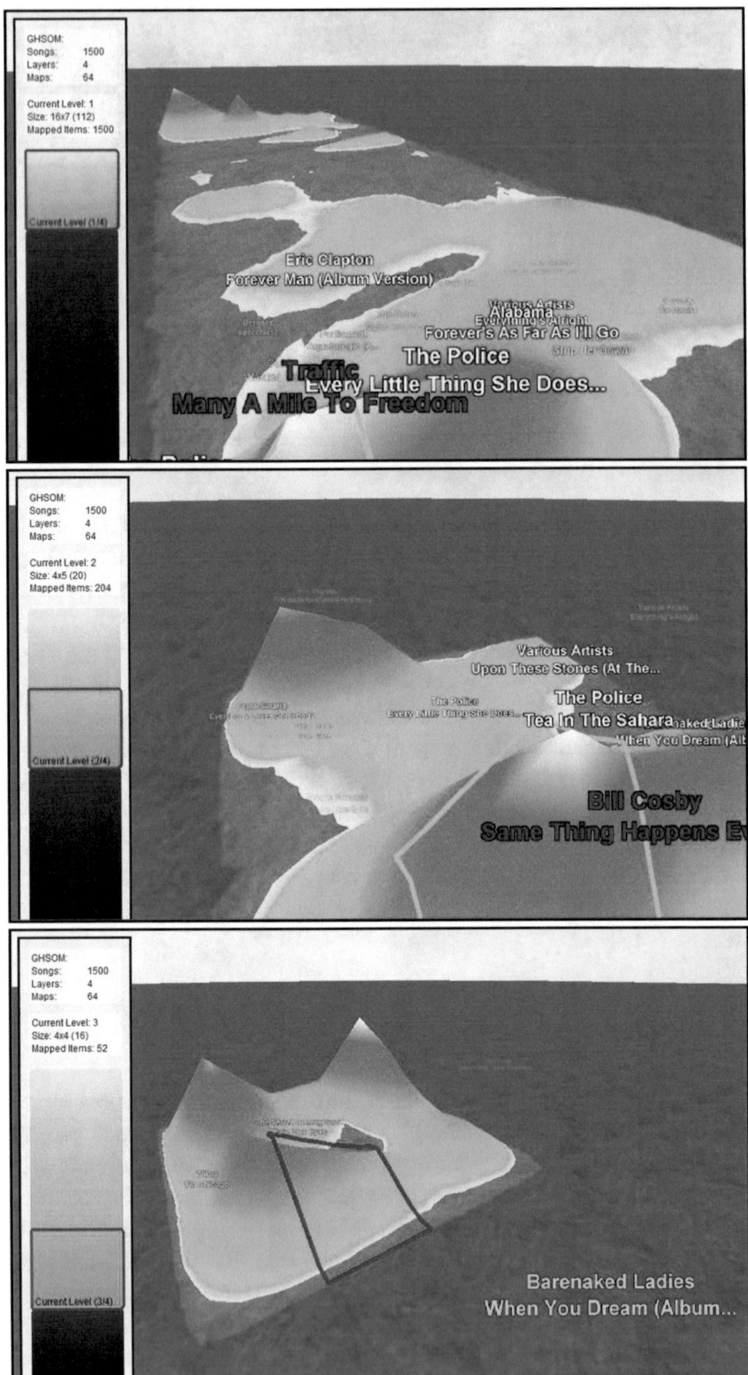

Fig. 9.6 The hierarchical *deepTune* music browsing interface. A view on the music collection is shown at different hierarchy levels, from highest (*top*) to lowest (*bottom*)

$\epsilon_{ij} = |p_{ij} - d_{ij}|$, the formulation by Sammon emphasizes the preservation of smaller distances during projection. This is due to the term p_{ij} in the denominator, which particularly penalizes large errors made for small values of p_{ij}, i.e., small distances in the input space.

$$\epsilon = \sum_{i=1}^{n-1} \sum_{j=i+1}^{n} \frac{(p_{ij} - d_{ij})^2}{p_{ij}} \tag{9.7}$$

The MDS approach by Sammon, nowadays also referred to as Sammon's mapping, first positions the data items randomly in the visualization space. However, Sammon also suggests the use of PCA to obtain an initial linear projection. Subsequent to this initial placement, two steps are performed iteratively, either for a fixed number of iterations or until convergence of ϵ: (1) the section of two data items i and j and (2) the adjustment of their positions in the output space in a way that reduces ϵ. Step (1) can again be performed by randomly selecting two data items. Step (2), however, should be addressed in a more informed way. Sammon proposes a method based on gradient descent, i.e., the computation of the partial derivatives of the error function, which gives the direction of steepest descent and modifies the data points accordingly. More formally, Eq. (9.8) shows how the output space configuration, i.e., the position of all items in the visualization space, is obtained in each iteration $t + 1$; $\mathbf{y}_i$ is a two-dimensional vector containing item i's coordinates on the map, $\Delta_i(t)$ is the gradient computed according to Eq. (9.9), and MF is what Sammon calls the "magic factor" which he empirically determines as $0.3 \leq MF \leq 0.4$. It can be interpreted as a learning rate. More details and all derivatives can be found in [393].

$$\mathbf{y}_i(t+1) = \mathbf{y}_i(t) - MF \cdot \Delta_i(t) \tag{9.8}$$

$$\Delta_i(t) = \frac{\frac{\partial \epsilon(t)}{\partial \mathbf{y}_i(t)}}{\left| \frac{\partial^2 \epsilon(t)}{\partial \mathbf{y}_i(t)^2} \right|} \tag{9.9}$$

9.2.1.4 MDS-Based Interfaces for Music Collections

Two-dimensional music browsing interfaces employing MDS include *SongSurfer*, presented in [59], which incorporates a FastMap in the visualization. In the *MusicGalaxy* interface [467] by Stober and Nürnberger, a pivot-based MDS is applied. *MusicGalaxy* combines multiple sources of similarity information, giving the user control over their influence. Furthermore, when exploring the map, a magnification of similar tracks supports browsing by compensating for data projection deficiencies. In the *Search Inside the Music* interface [250], Lamere and Eck utilize an MDS approach for similarity projection into a three-dimensional space. Their interface provides different views that arrange images of album covers according to the output of the MDS, either in a cloud, a grid, or a spiral. A similarity-graph-based interface for portable devices is presented by Vignoli et al. in [498]. Lillie applies principal components analysis (PCA), a linear variant of MDS, to project multidimensional music descriptors to a plane in the *MusicBox* framework [272]. In addition to browsing, *MusicBox* allows users to create playlists by selecting individual tracks or by drawing a path in the visualization space, as shown in the top and bottom of Fig. 9.7, respectively. Circles with the same color are from the same genre. Seyerlehner proposes in [436] the *MusicBrowser* interface, which applies an MDS algorithm for audio similarities in order to create a map of music; cf. Fig. 9.8. Modeling each data item by a Gaussian centered at its position in the visualization space and aggregating the values given by all Gaussians for each point on the map, Seyerlehner obtains an estimation of the density distribution. He then normalizes these density values and maps them to a color scale resembling the one used in the *Islands of Music*.

9.2.1.5 Map-Based Interfaces for Music Collections Based on Other Techniques

Other map-based interfaces enable additional interaction methods by exploiting specific devices, foremost tabletop displays. A tabletop display is a horizontal screen that usually allows multi-touch input. Examples of such interfaces for music composition are the *reacTable* [201] and the *Bricktable* [181]. For instance, Hitchner et al. use a tabletop display to facilitate browsing and rearrangement of a music collection structured via a SOM [180]. A similar approach that makes use of a *Bricktable* multi-touch interface for structuring electronic music for DJs is presented by Diakopoulos in [101]. The *SongExplorer* interface by Julià and Jordà structures large collections and allows for interaction using a *reacTable*-like interface [202]. Baur et al. exploit lyrics-based descriptors to organize collections in the *SongWords* tabletop application [29]. The *MUSICtable* interface presented by Stavness et al. integrates different approaches to map creation (including manual construction) and is intended to enforce social interaction when creating playlists [464].

To elaborate map-based interfaces for music collections, the underlying maps do not have to be artificially created. For instance, music collections (or music-related data) can be made available by distributing them over real geographical maps. To this end, Celma and Nunes use data from *Wikipedia* for placing artists or bands on their corresponding location on a world map, according to their place of birth or foundation [64]. Govaerts and Duval extract geographical information from biographies mined from the web and utilize this information to visualize, among other items, radio station playlists [158].

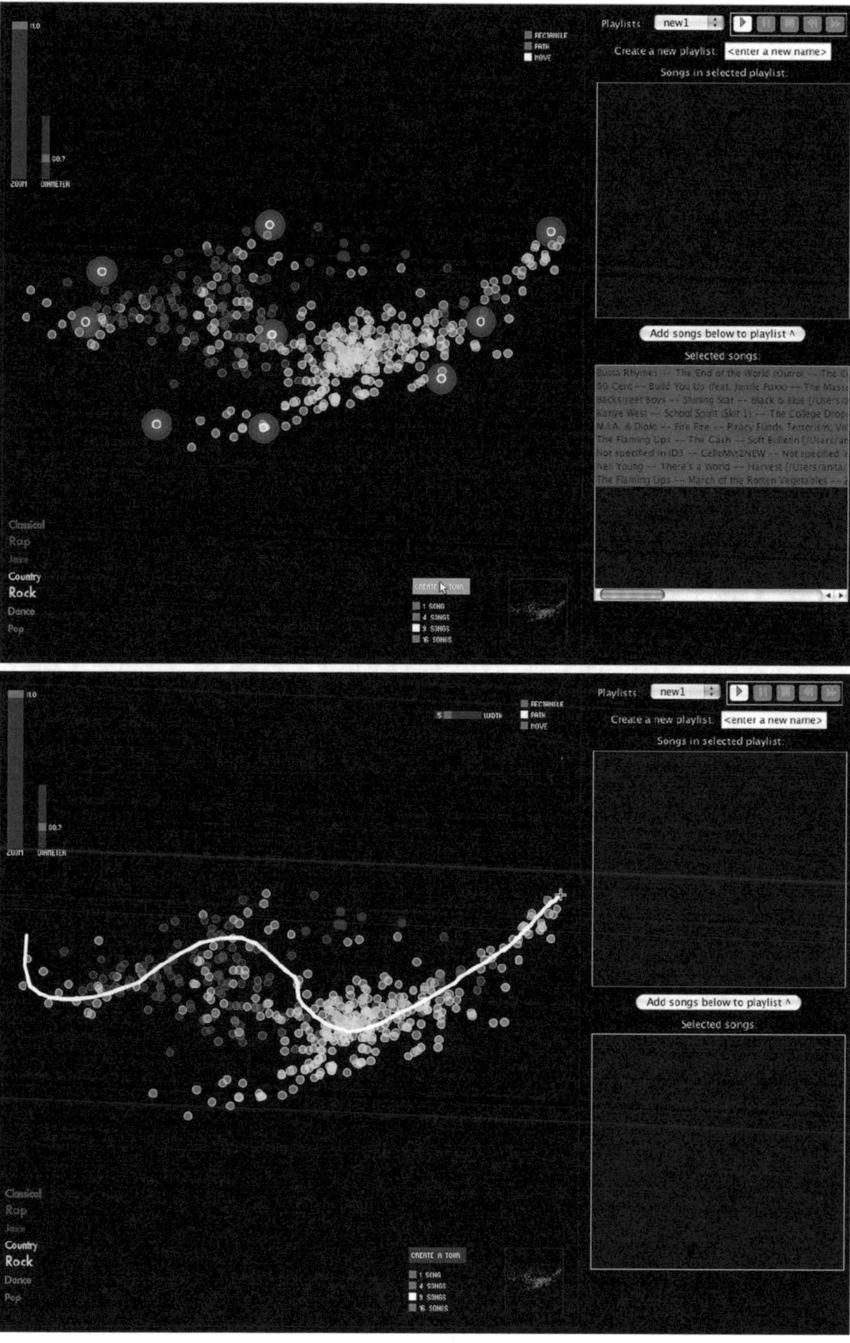

Fig. 9.7 The *MusicBox* browsing interface showing a few selected songs (*top*) and a user-drawn path to create a playlist (*bottom*) [272] (reprinted with permission)

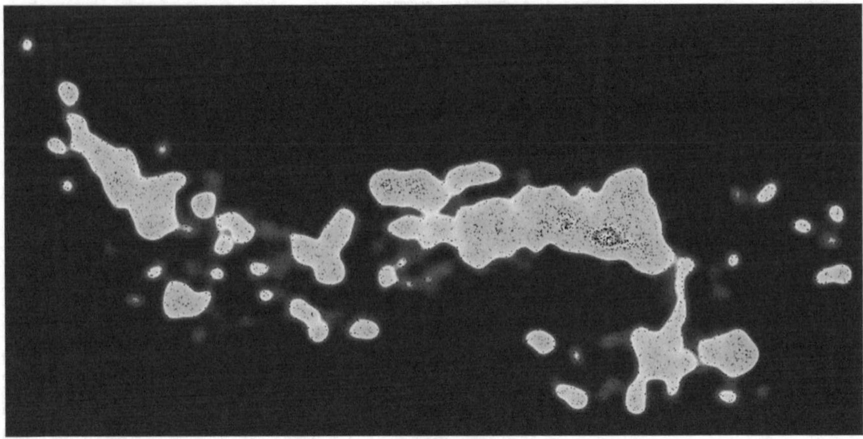

Fig. 9.8 A music collection visualized in the *MusicBrowser* interface that employs multidimensional scaling (MDS)

9.2.2 Other Intelligent Interfaces

In addition to map-based interfaces, which are the focus of scientific work on the task of music browsing techniques, there are a variety of other applications that do not make direct use of clustering or data projection techniques. Among those with a commercial background are the widely known web radios and music recommendation systems, such as *Spotify*, *Last.fm*, or *Deezer*.[8] Lesser known interfaces include *Liveplasma*,[9] which is a search engine for music discovery that, based on a seed artist, shows a network in which similar artists are interconnected. Each node represents an artist, and its size corresponds to the artist's popularity. The creators of such commercial applications obviously do not reveal technical or algorithmic details of their underlying approaches. Furthermore, there are a few applications that are based on information by experts, typically built by music enthusiasts. Examples are *Ishkur's Guide to Electronic Music*[10] and the *Map of Metal*[11] that allow the user to learn about the respective genres and styles while exploring corresponding music.

[8]http://www.deezer.com.

[9]http://www.liveplasma.com.

[10]http://techno.org/electronic-music-guide.

[11]http://mapofmetal.com.

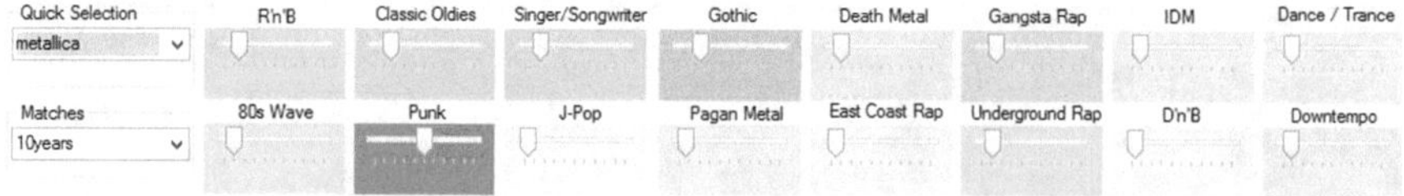

Fig. 9.9 The *Artist Browser* interface to search music via semantic concepts

Intelligent music interfaces with a scientific background include the following works: The *Musicream*[12] interface for discovering new pieces and easily generating playlists is presented by Goto and Goto [154]; cf. Fig. 1.4. From streams of music pieces (represented as discs), the user can simply pick out a piece to listen to or collect similar pieces by dragging a seed song into one of the streams. The different streams describe different moods. The frequency of released discs can be regulated for each mood separately by "faucets." Furthermore, the system invites users to experiment with playlists as all modifications can be undone easily by a so-called time-machine function. The intuitive drag-and-drop interface also facilitates the combination of playlists.

Pohle et al. [365] employ NMF (cf. Sect. 6.1.3) on web term profiles of artists, which yields a weighted affinity of each artist to each resulting semantic concept. In their proposed *Artist Browser* music search application (cf. Fig. 9.9) the influence of each concept on the search results can be adjusted manually to find the best fitting artists and to display related web content, given by *Last.fm*. Note that concepts are manually annotated in Fig. 9.9.

Pampalk and Goto propose the *MusicSun*[13] interface [345] that fosters exploration of music collections by different types of descriptive terms and item similarities; cf. Fig. 9.10. Terms for each artist are extracted from album reviews found via search engine requests. Indexing these reviews using different vocabularies that contain, for instance, instruments, genres, moods, and countries, the authors determine a set of most descriptive terms via TF·IDF weighting (cf. Sect. 6.1.1) and visualize them as "sun rays" (7) in the user interface. In addition, similarities are computed on audio content features, on the TF·IDF profiles of the artists, and on term weights between the term of the currently selected ray and all artists' TF·IDF representations. Initially, the user selects a number of seed artists (6) via meta-data search (1), for which the "sun" is created based on the seeds' most descriptive terms (7). When the user then selects a term, the system provides a list of recommendations (9) that best fit the seed artists and the selected term.

[12]https://staff.aist.go.jp/m.goto/Musicream.

[13]https://staff.aist.go.jp/m.goto/MusicSun.

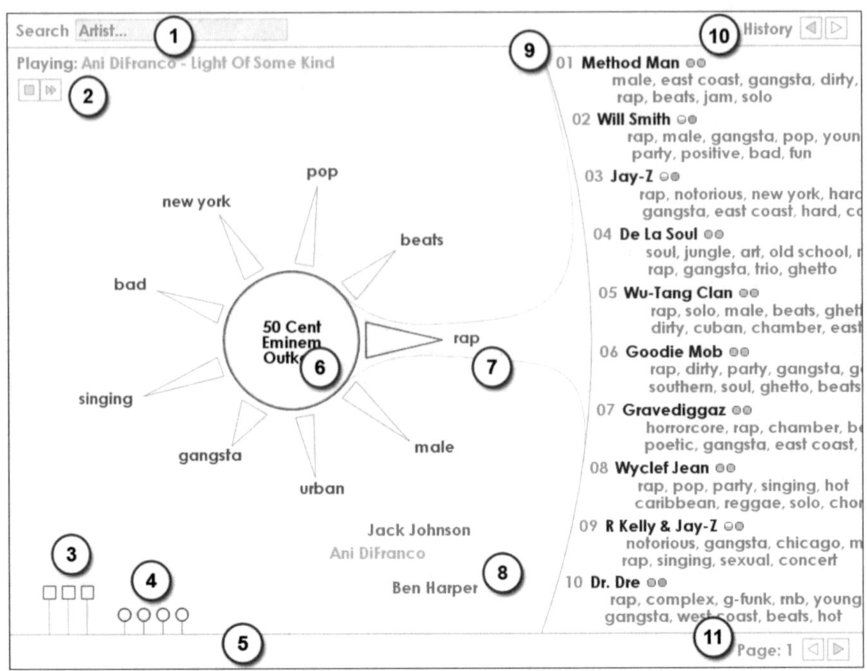

Fig. 9.10 The *MusicSun* interface to browse music collections via meta-data and different kinds of similarity [345] (reprinted with permission)

She can also adjust the weight of each type of similarity used to determine the recommendations (3). For the web-based similarity component, the user can decide on which combination of the four vocabularies the similarities are computed (4). A dedicated area on the interface (8) can be used to drag artists to, for later use. Audio playback components (2), mouse over help (5), undo functionality (10), and moving to the previous and next page of recommended artists (11) complement the *MusicSun* interface.

The *Three-Dimensional Co-Occurrence Browser* can be used to explore collections of artist-related web pages via multimedia content by selecting descriptive terms [403, 411]. An example is given in Fig. 9.11, in which three *Sunburst* visualizations [463], representing video, image, and audio content, respectively, are stacked. The figure illustrates the web pages of the band Bad Religion. The user has selected all web pages on which the name of the band and the terms "punk" and "guitar" co-occur. The height of the individual arcs corresponds to the amount of multimedia content available for the given user selection.

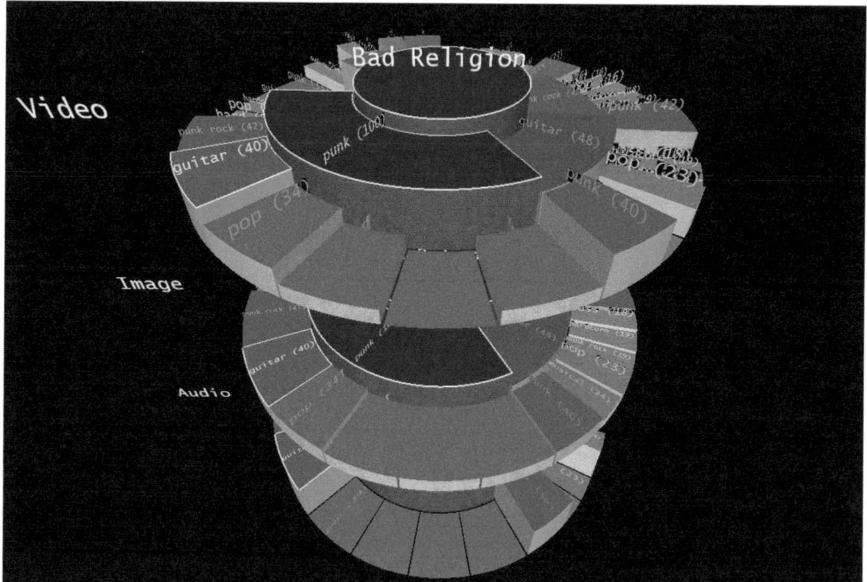

Fig. 9.11 The *Three-Dimensional Co-Occurrence Browser* to explore music artists via web pages, multimedia content, and descriptive terms

The *Songrium*[14] interface by Hamasaki et al. [163] is a collection of web applications designed to enrich the music listening experience. It offers various ways to browse and explore music; cf. Fig. 9.12. The *Music Star Map* (A) visualizes original songs in a graph using audio-based similarity for placement, while derivative works (e.g., cover songs) can be explored in the *Planet View* (B), via a solar system-like structure. Any particular area can be expanded for viewing by double clicking and then scrolled to by dragging. Furthermore, music can be explored by following directed edges between songs, which may be annotated by users (E). *Songrium* also embeds a video player for *Niconico*,[15] an online video sharing platform with overlay commenting functions. In addition, a playback interface for trial listening, called *SmartMusicKIOSK* (cf. Fig. 1.5), is included.

9.3 Automatic Playlist Generation

The automatic generation of a music playlist, sometimes referred to as "automatic DJing," is an interesting application that heavily draws on MIR techniques. There are two principle approaches to address the problem: (1) static playlist generation,

[14]http://songrium.jp.

[15]http://www.nicovideo.jp.

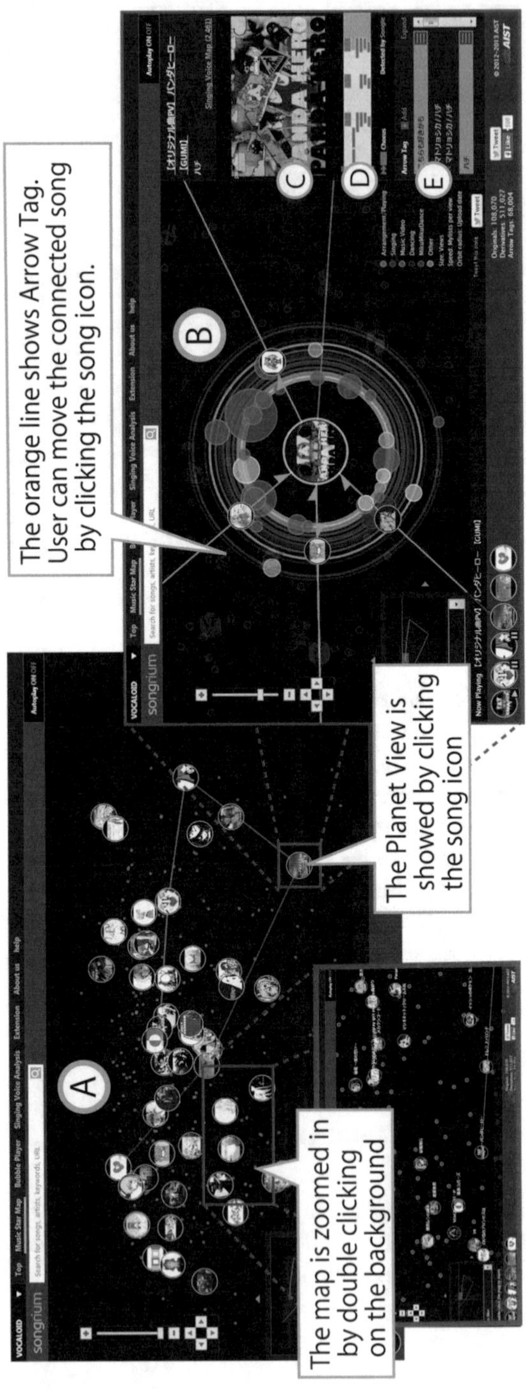

Fig. 9.12 The *Songrium* interface for exploring music collections [163] (reprinted with permission)

which is typically based on similarity computation between audio or web features, and (2) dynamic playlist modification or adaptation, which takes into account either explicit user feedback or contextual user aspects. Furthermore, there are quite a few studies that investigate how humans manually create playlists, e.g., work by Cunningham et al. [91], and how handcrafted playlists can be modeled, for instance, via random walks as proposed by McFee and Lanckriet [308] or via latent topic models as proposed by Zheleva et al. [537]. We focus here, in contrast, on computational approaches to automatically create playlists.

Early research on automatic playlist generation typically proposes creating static playlists or at most incorporates explicit user feedback. Given a music collection and pairwise similarities between its items, it is already possible to compute a simple playlist based on a **single seed song**. For instance, Logan uses similarities computed from Mel frequency cepstral coefficients (MFCC) representations for a collection of 8000 songs to assemble a playlist of songs most similar to the seed [282]. She measures the quality of the resulting playlists based on whether the songs are by the same artist, from the same album, or from the same genre as the seed. However, this kind of meta-data-based evaluation has some drawbacks; cf. [418, 536]. In particular, the listener might not be interested in being recommended songs by the same artist, or even by the same genre, over and over again.

Another category of approaches allow the definition of **start and end songs**, sometimes also the **length of the playlist**. Such research includes Alghoniemy and Tewfik's approach that models songs as a set of binary attributes and treats automatic playlist generation as a constraint satisfaction problem [4]. Pachet and Aucouturier also model the task as a constraint satisfaction problem using constraints defined on meta-data, such as duration, beats per minute, instrumentation, or perceived energy [13]. They employ an adaptation of local search [77], which models the user-defined constraints and the meta-data of each song via a cost function. Starting with a random sequence of tracks, the playlist is iteratively refined with regard to the cost function, rather than performing an exhaustive search. Flexer et al. employ the single Gaussian model explained in Sect. 3.2.5 to compute content-based distances between songs in the given collection [128]. Playlists are then created by (1) identifying the songs with highest distance to the start song and those with highest distance to the end song; (2) discarding songs that occur in both groups; (3) for each remaining song i that is either close to the start song s or the end song e of the playlist, computing the ratio $r(i) = d(i,s)/d(i,e)$; (4) determining the ideal distance ratios between consecutive playlist items based on the desired length of the playlist and the distance between start and end song; and (5) selecting and positioning in the playlist songs that best match these ideal distance ratios.

Automatic playlist generation can also be addressed by **organizing the whole music collection as a long playlist**, which enables the listener to start listening at each song in the collection. Aiming to create playlists of arbitrary length, in which consecutive songs sound as similar as possible, Pohle et al. propose the *Traveller's Sound Player*, which is an interface for accessing music on mobile music players [363]. It computes song similarities from extracted MFCC features and creates a placement for all songs on a circular path, using a traveling salesman

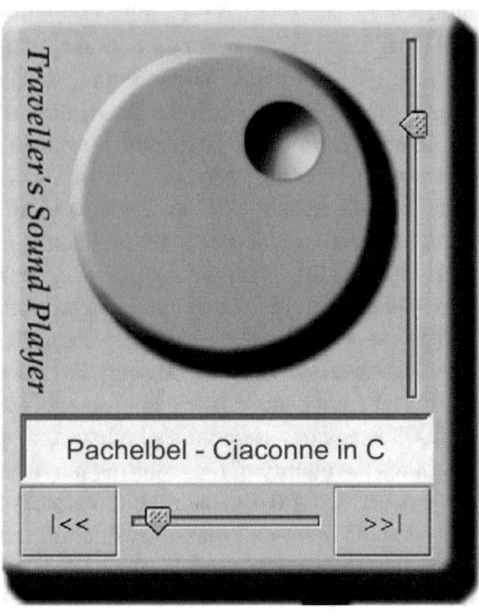

Fig. 9.13 The *Traveller's Sound Player* interface including automatic playlist generation

problem (TSP) algorithm. This arrangement permits the user to quickly locate music of a particular style by simply turning a wheel, much like searching for radio stations on a radio; cf. Fig. 9.13. The authors further propose an approach employing web mining techniques to speed up the creation of playlists. To this end, they extract term features from artist-related web pages, train a SOM from these TF·IDF features, and restrict the audio content-based similarity search to songs by artists mapped to nearby regions on the SOM [221]. Pohle et al. follow two evaluation strategies. First, they compute the entropy of the genre distribution of playlists with different lengths. Second, a qualitative evaluation of automatically generated playlists shows that similarity between consecutive tracks is an important requirement for a good playlist. On the other hand, playlists that reveal too much similarity between consecutive tracks may easily bore the listener.

Schnitzer et al. make use of *Last.fm* tags for pre-clustering and playlist labeling [429]. In the resulting *Intelligent iPod*[16] interface (cf. Fig. 1.2), the click wheel on *Apple*'s *iPod* can be used to navigate linearly through the entire music collection. While doing so, the descriptive labels of the current selection are displayed. Another variant of the *Traveller's Sound Player* is the *MusicRainbow*[17] interface by Pampalk and Goto [344]. In this interface, artist similarity is computed from audio features (a combination of spectral and rhythm features) extracted from songs and aggregated for each artist. Again a TSP algorithm is subsequently employed to create the playlists. Artists are placed on a "circular rainbow," where colored arcs reflect the

[16]http://www.cp.jku.at/projects/intelligent-ipod.

[17]https://staff.aist.go.jp/m.goto/MusicRainbow.

genre of each artist. Furthermore, the interface is enriched with descriptive terms gathered from artist-related web pages.

Another approach to playlist generation is taken by Chen et al., who model the sequential aspects of playlists via Markov chains and learn to embed the occurring songs as points in a latent multidimensional Euclidean space [72]; cf. Sect. 8.5.2.1. The resulting generative model is used for playlist prediction by finding paths that connect points. Although the authors only aim at generating new playlists, the learned projection could also serve as a space for Euclidean similarity calculation between songs.

Applications that feature automatic playlist adaptation based on **explicit user feedback** include [350], in which Pampalk et al. analyze song skipping behavior of users and exploit such information to adapt their playlists. While the authors base their playlist generation approach on similarities derived from audio features, skipping behavior is incorporated in a variety of ways. First, they propose the updating of the playlist by adding songs similar to the last accepted (i.e., not skipped) song by the user. Their second approach recommends songs that are most similar to all previously accepted songs. Pampalk et al.'s third approach considers for each candidate song (i.e., song neither played nor skipped yet) the distance to the most similar accepted song d_a and to the most similar skipped song d_s. The songs with smallest ratio d_a/d_s are added to the playlist. The authors evaluate their approaches on a collection of 2522 tracks from 22 genres. They show that the third ratio-based approach by far outperforms the others, when quality is judged by the total number of skips.

Automatically adapting playlists based on changes in the **user context** is addressed by Breitschopf et al. in their mobile *Android* application *Mobile Music Genius*.[18] To create initial static playlists, the authors propose to linearly combine term weights inferred from *Last.fm* tags on the level of artists, albums, and tracks to describe each song [47, 421]. Cosine similarities between the resulting term vectors are computed, and nearest neighbor search is performed to create a playlist based on a user-defined seed song; cf. Fig. 9.14. The user can further set the length of the playlist and whether she wants to have the seed song and other tracks by the seed artist included. An option to randomize the songs in the playlist can increase the level of perceived diversity. Dynamically updating the playlist is then realized by continuously monitoring a wide variety of contextual user factors (e.g., time, location, weather, running tasks, lighting conditions, and noise level) and computing the changes of these factors over time. If these changes exceed a threshold, a classifier trained on previously encountered relationships between user context and music preferences is triggered to suggest songs based on the new user context. These songs are then added to the playlist after the current track.

A few context-aware applications of automatic playlist adaptation target specific scenarios, a frequently-addressed one being to seamlessly update a playlist while doing sports. This is a scenario in which avoiding user interaction is very important

[18] http://www.cp.jku.at/projects/MMG.

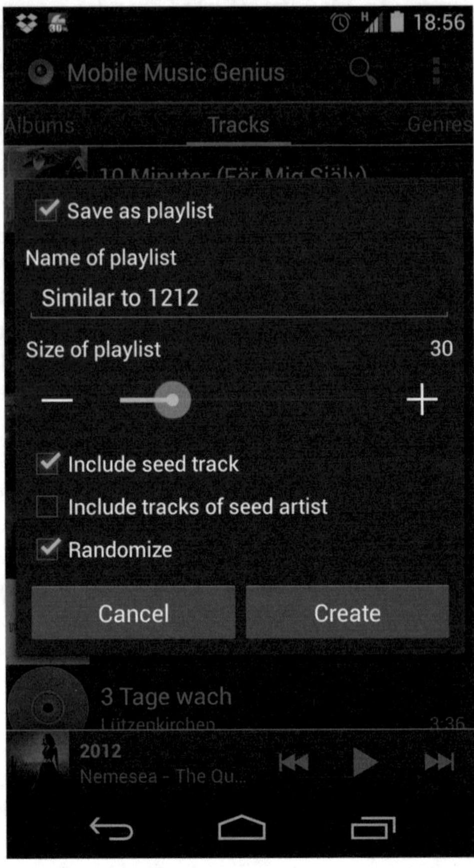

Fig. 9.14 The *Mobile Music Genius* interface that dynamically updates playlists. Here, the process of automatically creating a playlist is depicted

to keep user distraction at a minimum while exercising. Typically, such players try to match the music with the pace of the runner, in terms of bpm, to achieve a certain training outcome. For instance, Biehl et al. propose the *DJogger* interface for pace regulation [37]. It selects subsequent songs depending on the current performance of the user. If they are underperforming, the system selects a faster song; if their pace exceeds the healthy range, the next song selected will be a slower one. The authors further suggest as an alternative to manipulate songs while they are being played, i.e., to dynamically adjust the bpm according to the user's pace. A similar system, called *D-Jogger*, is proposed by Moens et al. in [320]. The focus of this work is to analyze if and how listeners synchronize their pace to the music played. The authors exploit a variety of physical sensors that are investigated while participants are using the player in a treadmill. The player continuously monitors the user's pace and chooses a song whose tempo is a better fit for the current pace if bpm and step rate diverge by more than 10 %. Moens et al.'s main findings are that most users synchronize their pace to the beat, irrespective of the difference between bpm and pace, and that users tend to feel more motivated when their pace is in sync with the tempo of the music. In related work, Liu et al. target music therapy by considering the user's heartbeat instead of his or her pace in order to adjust songs in a playlist [279]. They aim at regulating the user's heartbeat in order to keep it within a healthy range. To this end, the authors use a Markov decision process to select songs from the preferred genre of the user but adjusted to his or her current heart rate (lower bpm if the user's heart rate is too high, higher bpm if it is too low, and similar bpm if it is in a healthy range). Experiments show that music can be used to increase the heart rate to a normal level.

Sometimes, automatic playlist generation is built on top of map-based music browsing interfaces (cf. Sect. 9.2.1). For instance, the *MusicBox* application [272], the *Play SOM*[19] and the *Map of Mozart*[20] interfaces [301, 332], and the *nepTune* browser [222] enable the listener to create playlists by drawing a curve on the map, as shown in Fig. 9.7. The playlist is then generated by applying a pathfinding algorithm, e.g., the A* algorithm [165]. A more detailed treatment of the topic of automatic playlist generation can be found in [424].

9.4 Music Popularity Estimation

As a last application of MIR techniques, we discuss the topic of accurately measuring the popularity of an artist or a song in a certain region of the world. Of course, the music industry is highly interested in such information, but they are not the only ones. Music aficionados are likely interested in which music is currently "hot" in different parts of the world. Furthermore, popularity information can serve serendipitous music retrieval algorithms, e.g., [62, 418]. Ultimately, performers are also interested to know where in the world their music is particularly popular or unpopular.

Popularity of artists, albums, or performances can be estimated by a variety of predictors. Traditionally, record sales were used to gauge popularity via music charts, such as the *Billboard Hot 100*,[21] released weekly for the United States of America by the *Billboard Magazine*.[22] Since the advent of online music distribution and music streaming services, more and more companies specialize in social-media-based services to approximate music popularity and provide other analysis tools. Examples include *Band Metrics*,[23] *BigChampagne Media Measurement*,[24] and *Next Big Sound*,[25] which claims to analyze "social, sales, and marketing signals to help

[19]http://www.ifs.tuwien.ac.at/mir/playsom.html.

[20]http://www.ifs.tuwien.ac.at/mir/mozart/index_en.html.

[21]http://www.billboard.com/charts/hot-100.

[22]http://www.billboard.com.

[23]http://www.bandmetrics.com.

[24]http://www.bigchampagne.com.

[25]https://www.nextbigsound.com.

you make smarter, braver decisions." Although the details of their algorithms are a business secret, it can reasonably be assumed that these commercial service providers harvest multiple data sources to create their popularity estimates. Also music streaming services sometimes offer charts based on the listening data of their users, for instance, *Spotify Charts*[26] and *Last.fm* via their API.[27] *Echonest* even offers a public API function[28] to retrieve a ranking based on the so-called "hotttnesss" of an artist. This ranking is based on editorial, social, and mainstream aspects. Scientific work on the topic of music popularity estimation either exploits content-based audio features or music context information. The former have also been used to predict whether a song will become a hit or not, for instance, by Pachet and Roy [339]. However, audio features proved incapable of accurately predicting popularity. On the other hand, contextual data sources can be used to some extent to model, and even predict, artist and song popularity. The sources mentioned in the literature have already been introduced in Chaps. 5 and 7. In particular, corresponding indicators include microblogging activity, play counts on web radio stations or music streaming services, and occurrences on web pages and in folders shared on peer-to-peer (P2P) networks.

In the following, we first provide an overview of work that makes use of these data sources to predict the popularity of an artist or a song. Subsequently, we discuss properties of the data sources and highlight aspects such as particular biases, availability, noisiness, and time dependence.

9.4.1 Popularity Estimation from Contextual Data Sources

The popularity of a music item can be defined on various levels of granularity, for instance, for a particular user, a user group, a country's population, a cultural region, or even on a global scale. Music charts are typically created on the level of countries (e.g., *Billboard Hot 100*) or of communities (e.g., *Spotify Charts*), the latter frequently being formed by users of a certain web platform.

Focusing on web data sources to derive artist popularity on the level of countries, we investigate in [412] various popularity estimators: page counts of web pages given by a search engine's index, artist occurrences in geolocated microblogs, metadata of files shared within the *Gnutella* P2P network, and play count data from *Last.fm*. Page count information is acquired by querying the web search engines *Google* and *Exalead*[29] for *"artist" "country" music*. To model artist popularity specific to each country, we dampen artists that are popular everywhere in the world by introducing a factor that resembles the inverse document frequency

[26] http://charts.spotify.com.

[27] http://www.last.fm/api/show/chart.getTopTracks.

[28] http://developer.echonest.com/docs/v4/artist.html#hotttnesss.

[29] http://www.exalead.com/search.

(cf. Sect. 6.1.1). The ranking is eventually calculated as shown in Eq. (9.10), where $pc(c, a)$ is the page count estimate returned for the conjunctive query for artist a and country c, $|C|$ refers to the total number of countries for which data is available, and $df(a)$ is the number of countries in whose context artist a appears at least once, i.e., the number of countries with $pc(c, a) > 0$.

$$r(c, a) = pc(c, a) \cdot \log_2 \left(1 + \frac{|C|}{df(a)}\right) \tag{9.10}$$

Geolocated microblogs are retrieved from *Twitter*, using the hashtag `#nowplaying` to distill microblogs that relate to music listening events. We categorize these tweets according to the country in which they were posted. The number of occurrences of artist a in the tweets of country c is then used as factor $pc(c, a)$ in Eq. (9.10), which again gives the final artist popularity ranking. As another data source, we exploit meta-data of shared folders in the P2P network *Gnutella*. To this end, we identify music files and extract artist and song information. Mapping the users' IP addresses to countries enables us again to define a popularity ranking for each country with active users. We further identify, for each country, the 400 most active listeners on *Last.fm* and subsequently aggregate their artist play counts, also for their respective country. This way, we again obtain a popularity ranking of artists for each country. Comparing the four approaches, we find that the popularity approximations of music artists correlate only weakly between different data sources. A remarkable exception is a higher correlation found between shared folders in P2P networks and page count estimates. This can be explained by the fact that these data sources accumulate data rather than reflect current trends, which is the case for charts that are based on record sales or postings of *Twitter* users (cf. Sect. 9.4.2).

Koenigstein et al. also analyze in [232] shared folders in the *Gnutella* network, employing a two-stage process. First, a crawler component discovers the highly dynamic network topology. Second, a browser queries the active nodes, which correspond to users, for meta-data of files in their shared folders. The crawler treats the network as a graph and performs breadth-first exploration. Discovered active nodes are queued in a list that is processed by the browser. Shared digital content is associated with artists by matching the artist names of interest against ID3 tags of shared music files. Occasionally ID3 tags are missing or misspelled. Artist names are therefore also matched against the file names. Creating popularity charts for a specific country necessitates the determination of the geographical location of the users. This identification process is based on IP addresses. First, a list of all unique IP addresses in the data set is created. IP addresses are then geolocated using the commercial *IP2Location*[30] service. Next, each IP address is attached a country code, a city name, and coordinates given by latitude and longitude. The geographical information obtained in this way pinpoints fans and enables tracking of spatial diffusion of artist popularity over time.

[30]http://www.ip2location.com.

In another work, Koenigstein and Shavitt analyze search queries issued in the *Gnutella* network [231]. They again infer user locations from IP addresses and are thus able to compare charts created from the query terms with official music charts, in particular the *Billboard Hot 100* charts. The authors find a remarkable time-shifted correlation between searches for artists on P2P networks and chart rankings. In particular, many artists that enter the charts have frequently already been sought for 1–2 weeks.

Another approach to infer popularity from social media data is presented in [159], where Grace et al. compute popularity rankings from user comments posted in the social network *MySpace*. To this end, the authors apply various annotators to crawled *MySpace* artist pages in order to spot, for example, names of artists, albums, and tracks, sentiments, and spam. Subsequently, a data hypercube (OLAP cube) is used to represent structured and unstructured data and to project the data to a popularity dimension. A conducted user study evidences that the ranking generated by Grace et al.'s approach is, on average, preferred to the ranking of the *Billboard* charts.

Recently, *Twitter* has become a frequently researched source for estimating the popularity of different kinds of items, for instance, *YouTube* videos [530] or news articles [23]. To look into this task for the music domain, one can focus on tweets including hashtags that typically indicate music listening events, for instance, `#nowplaying` or `#itunes` [167]. In order to map such tweets to the mentioned artists and songs, a cascade of pattern matching approaches is required.

From this activity, two data sets have emerged: *MusicMicro* [402] and *Million Musical Tweets Dataset* [168]. These data sets differ in size, but both offer information on listening activities inferred from microblogs, along with temporal and spatial annotations for each tweet. Based on these data sets, an analysis tool to explore the music listening behavior in the microblogosphere is presented [167]. Figure 9.15 shows a screenshot of the corresponding *Music Tweet Map*[31] interface that visualizes listening events around the world. The display can be focused on by a location, by a time range, and by a social similarity measure (where similarity is inferred from co-occurrences in playlists), among others. Color is used to distinguish different genres.

Accumulating the listening events per artist or per song over time yields detailed listening statistics and time-dependent popularity estimates. An example is given in Fig. 9.16, which depicts the popularity of songs by Madonna over the time span of November 2011 to August 2012. Taking a closer look at the figure, an interesting observation can be made. While the share of all-time hits such as *Like a Prayer* or *La Isla Bonita* remains quite constant over time, songs with a sudden and steep increase in listening activity clearly indicate new record releases. For instance, *Girl Gone Wild* from the album MDNA started its rise in the end of February 2012. However, the album was not released before March 23. Such a prerelease observation is in line with Koenigstein and Shavitt's investigation of P2P data [231] and may be used as

[31] http://www.cp.jku.at/projects/MusicTweetMap.

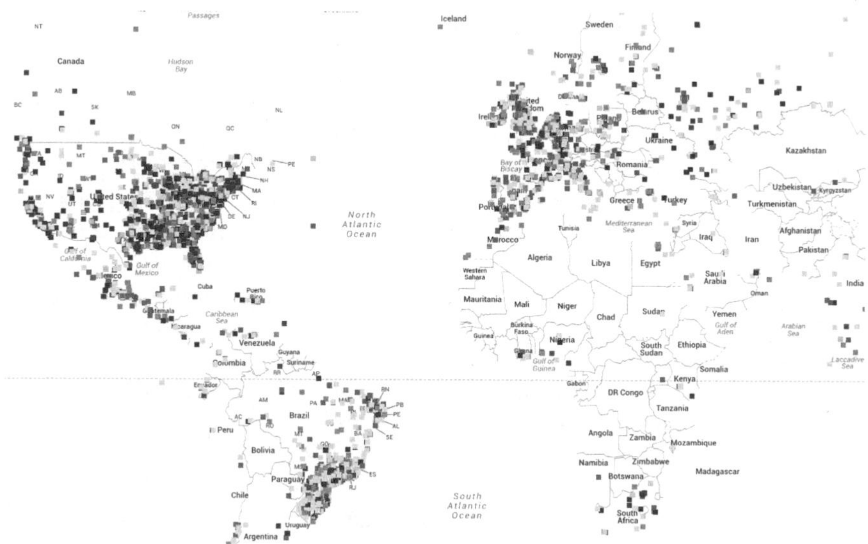

Fig. 9.15 Global visualization of music listening events, as shown in the *Music Tweet Map*

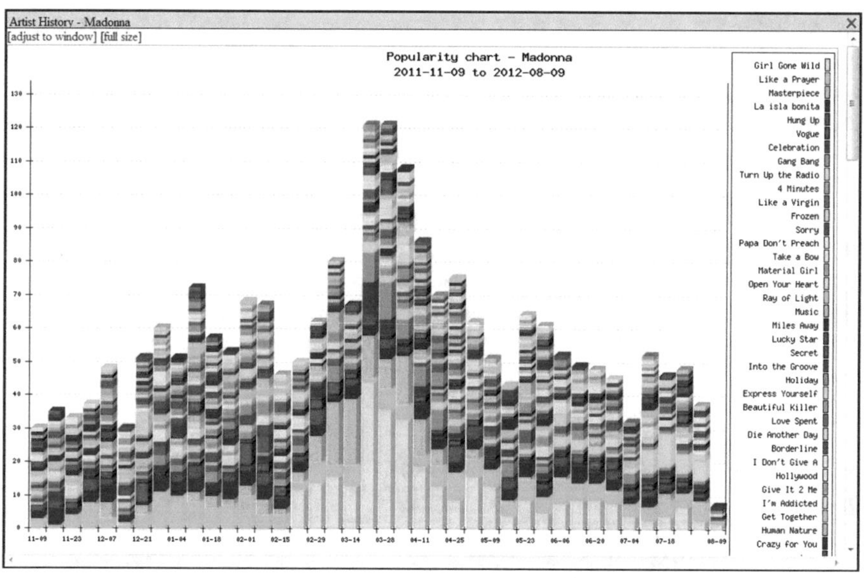

Fig. 9.16 Popularity of songs by Madonna on *Twitter*. The steady large *green bar* near the base is her consistently popular song Like A Prayer, while the *light blue* and the *orange bars* are Girls Gone Wild and Masterpiece, respectively, by her March 2012 release MDNA

a predictor of popularity. However, one has to keep in mind that this kind of social data might be noisy, biased, manipulated, or otherwise distorted.

9.4.2 Comparison of Data Sources

We show in [412] that popularity estimates obtained from the different inhomogeneous data sources correlate rather weakly. Different data sources thus seem to capture different aspects of popularity, which indicates that estimating popularity is a challenging and multifaceted task. Table 9.1 compares major music context data sources and traditional music charts, according to several criteria relevant to the task of popularity estimation.

All approaches are prone to a certain *bias*. For instance, the music taste of *Last.fm* or *Twitter* users may be very different from that of the general public. Exploiting such data is hence prone to a *community bias*. Traditional charts, on the other hand, are commonly biased towards the record sales figures the music industry uses as proxy. Although nowadays they incorporate also digital online sales, view counts of music videos on *YouTube* or microblogging activity is not accounted for. Web page counts reflect the popularity of music items for which web pages exist. As there are zillions of web pages on every possible topic available, web data is likely biased to a lesser extent than other sources. However, coverage of certain music items still differs between different regions of the world.

Another aspect to consider is *availability* of data. While page count estimates are available for all countries in the world, the approaches based on *Twitter* and on peer-to-peer data suffer from an unbalanced coverage. Traditional music charts also vary strongly between countries in terms of availability. We analyze coverage of different countries in [412] and find that *Last.fm* and web pages perform best with 100 % of all countries covered. Coverage is 63.5 % for *Twitter* and only 35.3 % for P2P data.[32]

In terms of *noisiness*, traditional charts perform best as the business figures used to gauge popularity are most robust against noise. In contrast, page count estimates are easily distorted by ambiguous item names (e.g., countries, songs, or artists). *Last.fm* data suffers from hacking and vandalism, as shown by Lamere and Celma [249], as well as from unintentional input of wrong information and misspellings.

Table 9.1 A comparison of various contextual indicators to predict music popularity

Source	Bias	Availability	Noisiness	Temporal focus
Web page counts	Web users	High	High	Accumulating
Twitter	Community	Country-dependent	Medium	Both
Peer-to-peer	Community	Country-dependent	Low–medium	Accumulating
Last.fm	Community	High	Medium–high	Both
Traditional charts	Music industry	Country-dependent	Low	Instantaneous

[32]These results are based on a list of 240 countries retrieved from *Last.fm*.

With respect to the *temporal focus*, data sources can be categorized into those that reflect an instantaneous popularity and those that encode overall, time-independent popularity information. The latter type of sources accumulates popularity information over time, while the former only considers a temporal snapshot, e.g., a time window of a few weeks. For instance, the corpus of web pages indexed by major search engines is certainly an accumulating data source. In contrast, traditional charts only consider record sales or plays within a certain time window. The time dependence of listening data from *Twitter*, *Last.fm*, and other social media sources depends on the availability of respective data access functionality, such as API functions. While these data sources accumulate user information over time, such information might not be accessible publicly or easily.

9.5 Summary

In this chapter, we discussed a few applications of MIR techniques. Of high commercial interest are music information systems, examples of which include *allmusic*, *Last.fm*, and *Discogs*. We demonstrated how techniques from music information extraction can be used to build or enhance such information systems. To this end, we reviewed approaches to identify certain pieces of music information in web pages, including members and instrumentation of bands, their country of origin, and images of album cover artwork. Most of these approaches rely on rule-based pattern extraction, classification, and heuristics. However, with the prevalence of platforms supporting collaborative music annotation and information gathering, web mining techniques for music information extraction are experiencing a decline.

Another important category of applications we discussed are user interfaces to music collections. Their common rationale is to provide listeners with intuitive and powerful ways to navigate in music repositories, in particular for browsing purposes. They frequently incorporate music features (e.g., extracted from audio or web resources) and employ clustering techniques such as self-organizing maps (SOMs) or multidimensional scaling (MDS) to create visualizations in which a given music collection is automatically organized as groups of similar sounding pieces. While these interfaces often provide elaborate ways to access music collections, we have not seen their commercial breakthrough yet. In our opinion, the most important reason for this is that current systems are not capable of dealing with catalogs of tens of millions of tracks, which are common in today's era of music streaming services.

In contrast, an important industrial application of MIR techniques is automatic playlist generation or serial recommendation. We categorized approaches into ones that create static playlists based either on a single seed song or on given start and end songs and ones that dynamically update or extend playlists based on user feedback or changes in the user's context. In commercial systems, the output of such automatic playlist generators is sometimes also referred to as "personalized radio stations."

Finally, we discussed the topic of music popularity estimation, which aims at measuring the attention an artist or song is receiving, frequently on a global or country-specific scale. On the one hand, approaches drawing from web and social media data have shown some merit to gauge popularity by building models based on the content of music-related web pages, peer-to-peer networks, microblogs, or social networks. In contrast, forecasting whether a song, album, or artist will become popular in the future, a task sometimes referred to as "hit song science" [339] or music popularity prediction, seems much harder as there are too many factors that influence popularity.

Chapter 10
Grand Challenges and Outlook

In this book, we provided an introduction to important techniques and interesting applications of music similarity and retrieval. Although music information retrieval is a relatively young research field, it has managed to establish itself as a highly multidisciplinary area. The constant evolution of the field during the past decade has been driven both by academia and industry. It has led to mature techniques and appealing applications, ranging from industry-strength music fingerprinting and identification algorithms to creative user interfaces for browsing music collections and interacting with music; from techniques to automatically determine tempo and melody of a song from the audio to personalized music recommender systems.

Despite the great potential current MIR techniques have shown, there are still exciting challenges to be faced. In the following sections, we first discuss such major challenges, related to research methodology, data collection and usage, and user centric aspects, in addition to general challenges. We conclude this book by identifying future directions of research in music similarity and retrieval and eventually provide some of our own visions for future MIR applications.

10.1 Major Challenges

MIR as a field is facing several challenges. Revisiting existing discussions [111, 152, 435] and adding our own perception of limitations in current MIR research, we structure the corresponding challenges into challenges that can be addressed by novel methods, challenges that relate to the availability and creation of data sets, challenges which arise when following user-centric approaches, and general challenges that are pertinent to the field as a whole.

P. Knees, M. Schedl, *Music Similarity and Retrieval*, The Information Retrieval Series 36, DOI 10.1007/978-3-662-49722-7_10

10.1.1 Methodological Challenges

Task orientation: MIR research is often targeted at improving algorithms, particularly with respect to performance on existing data sets (cf. Sect. 10.1.2). However, these efforts are only valuable when they actually serve a user performing a task (including the passive act of consuming music). In fact, the requirements of MIR systems, the definition of similarity (see below), the goals of retrieval, and—not least—the mode of evaluation all depend on the targeted task. This is strongly intertwined with a need for identifying the users involved in these tasks (cf. Sect. 10.1.3), i.e., the target groups, stakeholders, and beneficiaries of MIR technology. For instance, the needs of off-line music listeners [90] are different to those of users of commercial music services [259], as well as of musicologists [196] and music composers and creators [6, 228]. Hence, MIR, being the heterogeneous field it is, needs to better articulate, who is targeted by and benefitting from individual research progress in which task.

Music similarity measurement: To develop an understanding of how the low-level and mid-level content and context features discussed in this book relate to the human perception of music is a crucial point for constructing multifaceted similarity measures that reflect human perception of music similarity, which is a frequently communicated requirement for improved music retrieval and recommendation tasks. However, human judgment of music similarity varies highly, as shown, for instance, by Novello et al. [334]. This subjectivity can even occur among people with the same cultural background but different music taste, music education, or age. To identify the relevant musical facets, corresponding features to extract them from content and context sources, and in turn to elaborate similarity measures, we thus need to investigate whether generally valid similarity relations exist or music perception is too individual.

Scalability and robustness: When it comes to creating MIR techniques and applications that scale to tens of millions of songs, there is frequently a mismatch between the requirements of industry and academia. While researchers from academia typically aim at showing that their approaches substantially improve over existing works, industry demands methods that are robust and that scale to music collections of commercial size. Standardized and publicly available collections, such as the Million Song Dataset [36],[1] allow researchers to test their methods on larger data sets. However, when it comes to tasks involving music similarity computation, which has a quadratic runtime complexity, there is still a gap between a collection of one million items and a commercial collection of tens of millions, such as the music catalogs offered by *iTunes*[2] or *7digital*.[3]

[1]http://labrosa.ee.columbia.edu/millionsong/.

[2]http://www.apple.com/pr/library/2012/09/12Apple-Unveils-New-iTunes.html.

[3]http://www.7digital.com/about.

Cultural and multilingual aspects: While MIR techniques have already made exciting new services and applications possible, they are highly centered on Western music, in particular on pop and rock music, as evidenced by Serra [434] and others. Methods to automatically determine the key in famous Rock songs or gather the lyrics of well-known pop songs perform remarkably well. In contrast, identifying melody or rhythm in Carnatic music [230] or computing web-based similarity for songs with lyrics in languages other than English or for non-anglophone artists requires the development of new methods or at least considerable adaptation of existing ones [220, 416]. Making MIR techniques work well, independent of the kind of music as well as the cultural background of the composer, performer, and audience, is a major goal for these reasons. Considering cultural factors when creating MIR systems is, on the other hand, also important from the user's point of view, as the access to, consumption, and perception of music differs substantially between cultures. The same is true for general music taste [168, 402].

10.1.2 Data-Related Challenges

Data sets and data availability: High-quality data sets that include a wide variety of music and music-related information are a necessity for conducting substantial experiments to validate music similarity, retrieval, and recommendation approaches, among others. In contrast to classical (text) information retrieval, where standardized corpora have existed for decades,[4] MIR suffers from a lack of such data sets. While there are quite a few data sets that were specifically developed for a particular task, e.g., key detection, melody extraction, or beat tracking, these are typically the result of individual efforts that did not undergo quality control. Moreover, unlike data sets used in text or image retrieval, large-scale music collections are protected by intellectual property rights; sharing them is impossible. The use of royalty-free alternatives, such as *Jamendo*, on the other hand, severely restricts the applicability of many music context and user context approaches due to data sparsity, since the music offered by such services is usually not well known. The big challenge here is thus to convince record labels to make their music available for research purposes.

Collaborative creation of resources: Adding another point to the previous discussion on data availability, the creation of verified high-quality data sets is very time consuming, much more so than for data sets of other media types. The main reason is that annotation efforts are considerably higher in music. To give an example, creating a ground truth for algorithms that align the score of a symphony with an audio recording of its performance requires the precise manual location of individual notes in the audio and connecting them to the notes in the digitized score, which is typically given as MIDI file. Annotating onsets, beats, or key is no easier of a task. For this reason, these annotation tasks should

[4] See, for instance, the corpora used in the Text REtrieval Conference (TREC): http://trec.nist.gov/.

be performed in a collaborative effort, both to ensure a rapid availability (e.g., by crowdsourcing the task) and a certain quality of resources (via annotators' agreements on annotations). User-generated annotations could further complement expert-generated annotations, provided that a comparable quality can be ensured. In a similar fashion, not only data sets but also code to perform MIR tasks should be made publicly available under an open source license. Doing so will not only facilitate reproducibility of results but also provide researchers novel to MIR with a starting point for implementing their ideas.

Noisiness: Both content-based approaches and contextual methods suffer from different kinds of noise in the data they rely upon. The former have to deal with noise in the audio signal, such as bad recording quality or artifacts introduced by compression. The quality of the latter is strongly influenced by factors such as typos in tags, unrelated web pages or microblogs, technical and legal restrictions of service providers (e.g., limits for the amount of available or retrievable data imposed by *Google* or *Twitter*), and distorted reflection of the average population when gathering data from specific community-based sources, such as music taste of *Last.fm* users or microbloggers, which differs from the taste reflected by record sales [404].

10.1.3 User-Centric Challenges

Personalization and context-awareness: In order to build truly user-centric music retrieval and recommendation systems, we first need to learn much more about the mutual influence of music listened to and the user's mental and emotional state. As the latter is affected by a wide variety of factors, identifying corresponding user characteristics is vital. Examples range from simple aspects, such as demographics or musical education and training, to highly complex models of personality. Other important aspects include the temporal, spatial, social, environmental, and cultural context of the user [271, 420]. If we are able to distill relationships between these user characteristics and musical aspects, individually by user, fulfilling their individual information or entertainment needs by providing the right technical solutions will greatly improve user satisfaction with MIR technology and enable new music experiences. For example, a relevant solution could be the automatic creation or adaptation of a playlist that fits the user's current mood or activity, when the user is in a relaxed state and not willing to actively interact with his music collection.

Multimodal integration: Content-based and context-based MIR research already yielded methods to extract or gather various kinds of semantic music descriptors, such as tempo, melody, key, lyrics, or background information about the performer. While these methods have limitations (cf. Sect. 10.1.1), they still provide a wide range of knowledge, each piece of which is interesting to at least a certain type of user. However, some other multimedia data sources, such as music videos or album cover artwork, have not been thoroughly exploited so far. On the other hand, it is still unclear how the diverse multifaceted descriptors that are already available can be integrated to build multimodal systems that support access to music in a way that is most beneficial for the user.

Semantic gap and user modeling: Even though recent MIR research has made good progress in bridging the semantic gap, the discrepancy between computational models of music features and high-level concepts such as similarity is still evident. In addition, another semantic gap was pointed out by some researchers, namely, the semantic gap between people. For instance, musical factors deemed important by musicologists are likely to differ strongly from aspects important to end users. These factors and concepts are further different between people with different cultural backgrounds or other characteristics of personality. We need to thoroughly investigate these differences, but also similarities, to determine (1) concepts important to a wide range of people, if not all, and (2) aspects that are dependent on certain user characteristics. If we succeed, tailoring computational models to the actual music perception and the specific requirements of the users will help to further narrow the semantic gap. To this end, existing user models need to be revised and novel user modeling strategies need to be elaborated.

10.1.4 General Challenges

Evaluation: Assessing the performance of MIR algorithms and systems has been carried out to a large extent on artificial data sets or using experimental settings that do not adequately reflect the actual requirements of the users. Examples are data sets that are biased or not representative of the task to be solved or a setup that assesses the quality of a music similarity algorithm via a genre proxy (cf. Sect. 1.5.1), instead of asking humans for their judgment of the algorithm's results. Therefore, it is crucial to conduct quantitative and qualitative experiments that (1) are tailored to the task to be evaluated, (2) involve end users for whom the task is important, (3) employ a standardized methodology, and (4) make use of real-world data sets.

Certainly, there are tasks for which evaluation can be carried out without end user involvement, because a solid and objective ground truth is available, e.g., chord detection or instrument identification. When it comes to tasks that involve users in an active or passive way, however, well-designed user studies need to be carried out. For instance, the assessment of techniques for emotion regulation with music or for serendipitous music recommendation is not feasible without user-centric evaluation strategies. Elaborating such techniques can and should be supported by borrowing from other disciplines, in particular psychology.

Multidisciplinarity: MIR is a highly multidisciplinary field, bringing together researchers from computer science, signal processing, musicology, social sciences, and even law. However, collaborations of people from different fields are always challenging for several reasons, among others, different validation methodologies and attitudes towards conducting experiments (e.g., quantitative versus qualitative, artificial data versus user studies), or a strong focus on particular minor details (e.g., musicological details a non-musicologist does not care about or even does not recognize). Further strengthening interdisciplinary communication and exchange, sharing resources and methodologies, or simply making researchers from different disciplines, but with a mutual interest in MIR, aware of others' research problems will certainly raise mutual understanding and be beneficial for all communities involved.

Openness to other fields: Even within the broad field of computer science, there are subfields that share many techniques applied in MIR, even though the respective communities are pretty closed and separated from the MIR community. Related subfields include multimedia, recommender systems, data mining, machine learning, and of course (text) information retrieval. Remarkably little overlap between the MIR community and the text information retrieval community can be made out. From conversations with major figures in text IR, we learned that the limited interest in MIR research among a broader IR audience can be explained by at least two reasons. First, the inhibition threshold to understand the most important content-based techniques is quite high for someone without a solid signal processing background. Second, the shortage of standardized data sets and widespread habit of validating novel approaches on small sets created ad hoc is seen as a big shortcoming by text IR researchers, who are used for decades to have access to well-established and standardized corpora. The underlying criticism of limited reproducibility is certainly a very valid one and has already been discussed previously. If MIR wants to be recognized as a serious research field, its researchers must proactively seek connection to other related communities and communicate that their research is not only fun but also scientifically solid.

10.2 Future Directions

To conclude this book, we would like to present some visions for MIR applications, which we would like to contemplate in the future. Given the fast pace at which the field is evolving, we are sure that some of these will soon become a reality.

Predicting music trends: From an industry perspective, one highly requested tool is a reliable predictor of whether or not a music piece, an album, or an up-and-coming band will become successful. Even though "hit song science is not yet a science" [339], given the wealth of data sources from which listening events can be extracted and the fast increase of corresponding data, it should soon be possible to model the evolution of popularity and eventually use the respective models to make predictions. However, such a popularity predictor may severely limit the diversity of music and foster a fast converge towards production and consumption of mainstream music.

Musical companion: Music has a strong impact on many aspects of our daily lives. It is used to regulate our emotions, support our intellectual development, and even as treatment for some diseases. Integrating current and future MIR technologies into a single system that accompanies us in our lives and provides personalized support in every situation is what we foresee as a major application of MIR [422]. Such a universal music companion will create entirely new experiences in both music creation and music consumption scenarios.

To give two use cases, such a universal music companion could, for instance, support pianists. A fraction of a second after the pianist started to play an arbitrary piece, it automatically figures out which piece she is performing, via music identification techniques [10]. It then loads and displays the score of the piece, rendering unnecessary the need to go through the stack of sheet music in order to find the right score manually. During the pianist's play, the system further performs online audio-to-score alignment (score following) to interactively highlight the position in the visualized score while she is playing. It is also capable of investigating her expressive style and automatically constructing a personalized virtual orchestral accompaniment via music synthesis and performance rendering techniques. This gives the pianist the impression of being part of a real performance, playing together with a world-famous orchestra of her choice.

As a second use case, consider a very personal and emotional moment in your life, be it falling in love or grieving for a beloved friend. The music companion will be able to connect music to your memories, experiences, and emotional states. It will also know about a general as well as your individual perception of emotional content in music via music emotion recognition techniques. Sophisticated emotion regulation methods, drawing from both psychology and MIR, will in turn enable the companion to provide the right music to support joyful moments but also to more easily overcome depressing episodes or just a stressful day.

User-aware and multimodal music access: Musical information comes in various modalities, such as score sheets, recordings of performances, audio streams, images of bands or album covers, video clips, and lyrics. How the wealth of musical and music-related material contributes to a holistic model of music perception and how the various sources influence music taste is, however, still an open question. After having addressed this challenge successfully, we will have a comprehensive knowledge about the multifaceted relationship between the different music-related data sources and the user's musical preferences. Given the current trends of sensor-packed smart devices, the "always-on" mentality, and the inclination for leaving digital traces, future systems will also know details about the user's current intrinsic and extrinsic characteristics, such as personality, state of mind, or social contacts.

The resulting comprehensive user profiles will include relational information about the individual and the music, derived among others by machine learning techniques. This makes imaginable systems that anticipate the user's musical preference, anywhere and anytime, and correspondingly deliver the most appropriate music for each place and time. The success of music streaming services, which have made most music in the (Western) world available to everyone, sometimes overwhelms the user by their sheer amount of music offered. For this reason, the described systems are likely to become highly requested by a large audience.

At the same time, maintaining control over their music collections and how to access it is important to many users. User-aware music players will therefore also learn about the preference of their users when it comes to access collections.

Depending on the current situation and user characteristics, as mentioned above, systems will offer the most suited modalities for music access, drawing from a wide range of options, for instance, keyword search, query by example, browsing, faceted search, or recommendation of serendipitous songs. By doing so, they will provide the best possible experience for the user.

Appendix A
Description of the Toy Music Data Set

Writing about music is like dancing about architecture.

—Martin Mull

Here we want to give background information on the toy music data set that serves as example throughout the book (see Table 1.3). Unfortunately, we cannot distribute the audio data that we used to create the examples due to legal reasons, although it would be favorable to provide a CD or a download link to allow for convenient reproducibility of the examples.

For each of the selected genres, artists, and music pieces, we provide a collection of short quotes from *Wikipedia* and other sources to allow for a quick reception of the type and specifics of the music data set. Also, this highlights the type of data that often forms the basis of music-context approaches, particularly those described in Part II, and therefore represents a "semantic" view on the used audio recordings.

A.1 Electronic Music

"Electronic music is music that employs electronic musical instruments and electronic music technology in its production (...) Today electronic music includes many varieties and ranges from experimental art music to popular forms such as electronic dance music."[1]

The Chemical Brothers

> The Chemical Brothers are a British electronic music duo composed of Tom Rowlands and Ed Simons originating in Manchester in 1991. (...) They were pioneers at bringing the big beat genre to the forefront of pop culture.[2]

[1] http://en.wikipedia.org/w/index.php?title=Electronic_music&oldid=603643931.

[2] http://en.wikipedia.org/w/index.php?title=The_Chemical_Brothers&oldid=604156729.

P. Knees, M. Schedl, *Music Similarity and Retrieval*, The Information Retrieval Series 36, DOI 10.1007/978-3-662-49722-7

> Big beat tends to feature distorted, compressed breakbeats at tempos between 120 to 140 beats per minute, acid house-style synthesizer lines, and heavy loops from 60s and 70s funk, jazz, rock, and pop songs. They are often punctuated with punk-style vocals and driven by intense, distorted basslines with conventional pop and techno song structures. Big beat tracks have a sound that includes crescendos, builds, drops, extended drum rolls and dramatic sound effects, such as explosions, air horns, or sirens. As with several other dance genres at the time, the use of effects such as filters, phasing, and flanging was commonplace.[3]

Their track **Burst Generator** can be found on their sixth album *We Are the Night* and features an atypical sound, described as "where synths explode into coruscating My Bloody Valentine waves of sound."[4] At the time, this led to questions if this should be seen as "experiments or signal a new direction (…), perhaps swaying from their genre defining big beat albums of the past."[2]

Modeselektor

"Modeselektor is an electronic music band formed in Berlin, featuring Gernot Bronsert and Sebastian Szary. The group draws heavily from IDM, glitch, electro house and hip hop."[5] Their track **Blue Clouds** opens their album *Monkeytown* and is described as "celestial techno"[6] that "coasts to life on hollow drums."[7] *Pitchfork* describes it as "a slightly more placid take on (Flying Lotus') Cosmogramma highlight '… And the World Laughs With You', swapping out its frantic, future-junglist breaks for a less-cluttered rhythm and letting the synths simmer instead of boil."[8]

Radiohead

"Radiohead are an English rock band from Abingdon, Oxfordshire, formed in 1985."[9] Their song **Idioteque** "is (…) featured as the eighth track from their 2000 album *Kid A*."[10] "*Kid A* (…) marked a dramatic evolution in Radiohead's musical style, as the group incorporated experimental electronic music, krautrock and jazz influences. (…) *Kid A* featured a minimalist and textured style with more

[3] http://en.wikipedia.org/w/index.php?title=Big_beat&oldid=603932978.

[4] http://www.bbc.co.uk/music/reviews/d3jw.

[5] http://en.wikipedia.org/w/index.php?title=Modeselektor&oldid=602696899.

[6] http://www.bbc.co.uk/music/reviews/g3fd.

[7] http://www.residentadvisor.net/review-view.aspx?id=9789.

[8] http://pitchfork.com/reviews/albums/15867-modeselektor-monkeytown/.

[9] http://en.wikipedia.org/w/index.php?title=Radiohead&oldid=602454542.

[10] http://en.wikipedia.org/w/index.php?title=Idioteque&oldid=597794029.

diverse instrumentation including the ondes Martenot, programmed electronic beats, strings, and jazz horns."[9]

"'Idioteque' contains two credited samples of experimental 1970s computer music" and "has been described by others as an 'apocalyptic' song, with possible references to natural disaster, war and technological breakdown."[10]

Trentemøller

"Anders Trentemøller is a Copenhagen based musician and producer who started in the late 1990's with different Indie Rock projects, before he turned to electronic music."[11] His work is attributed to the genre labels alternative, indietronic, minimal, electronica, techno, chillout, and ambient.[11]

We chose his track **Moan (feat. Ane Trolle—Trentemøller Remix Radio Edit)** in his album *The Trentemøller Chronicles* as an example. According to *Resident Advisor*, "'Moan' proves the perfect bridge between shoegazing and barnstorming, when a wailing synth line cuts through Ane Trolle's sweet voice and turns serene smiles to grimaces."[12] It is the remix of a track from his album *The Last Resort* which *Pitchfork* describes as "a mood piece. (...) This is sultry, sexy music equally suited to chillout areas, wine bars, and bedrooms; it may well be this decade's equivalent to Kruder & Dorfmeister's downtempo classic, The K&D Sessions."[13]

A.2 Classical Music

> Classical music is art music produced or rooted in the traditions of Western music (both liturgical and secular). It encompasses a broad period from roughly the 11th century to the present day. (...) The term 'classical music' did not appear until the early 19th century, in an attempt to distinctly 'canonize' the period from Johann Sebastian Bach to Beethoven as a golden age. (...) The instruments (...) consist of the instruments found in an orchestra or in a concert band, together with several other solo instruments (such as the piano, harpsichord, and organ). The symphony orchestra is the most widely known medium for classical music and includes members of the string, woodwind, brass, and percussion families of instruments.[14]

[11] http://en.wikipedia.org/w/index.php?title=Trentem%C3%B8ller&oldid=604196393.

[12] http://www.residentadvisor.net/review-view.aspx?id=4878.

[13] http://pitchfork.com/reviews/albums/9500-the-last-resort/.

[14] http://en.wikipedia.org/w/index.php?title=Classical_music&oldid=604168025.

Ludwig van Beethoven

"Ludwig van Beethoven was a German composer and pianist. A crucial figure in the transition between the Classical and Romantic eras in Western art music, he remains one of the most famous and influential of all composers."[15] "The **Piano Sonata No. 14 in C-sharp minor 'Quasi una fantasia', Op. 27, No. 2**, popularly known as the **Moonlight Sonata**, is (...) one of Beethoven's most popular compositions for the piano."[16] The rendition used was performed by Korean pianist William Youn (Stomp Music).

Johann Sebastian Bach

"Johann Sebastian Bach was a German composer and musician of the Baroque period. He enriched established German styles through his skill in counterpoint, harmonic and motivic organisation, and the adaptation of rhythms, forms, and textures from abroad, particularly from Italy and France."[17]

> The **Well-Tempered Clavier (German: Das Wohltemperierte Klavier)**, BWV 846-893, is a collection of solo keyboard music. (...) Musically, the structural regularities of the Well-Tempered Clavier encompass an extraordinarily wide range of styles, more so than most pieces in the literature. The Preludes are formally free, although many individual numbers exhibit typical Baroque melodic forms, often coupled to an extended free coda (e.g. Book I preludes in C minor, D major, and B-flat major). The Preludes are notable also for their odd or irregular numbers of measures, both as to phrases and as to the entire length of a given Prelude.[18]

Prelude In C Major (BWV 846) is "a simple progression of arpeggiated chords. The technical simplicity (...) has made it one of the most commonly studied piano pieces for students completing their introductory training."[18] The rendition used was performed by Maurizio Pollini and published on the album *The Well-tempered Clavier Part 1, BWV 846-869* (Deutsche Grammophon).[19]

---

[15] http://en.wikipedia.org/w/index.php?title=Ludwig_van_Beethoven&oldid=601999837.

[16] http://en.wikipedia.org/w/index.php?title=Piano_Sonata_No._14_(Beethoven)&oldid=601348728.

[17] http://en.wikipedia.org/w/index.php?title=Johann_Sebastian_Bach&oldid=604035698.

[18] http://en.wikipedia.org/w/index.php?title=The_Well-Tempered_Clavier&oldid=603462752.

[19] http://www.deutschegrammophon.com/en/cat/4778078.

Frédéric Chopin

"Frédéric François Chopin (...) was a Romantic-era Polish composer."[20] "The Nocturnes, Op. 9 are a set of three nocturnes written by Frédéric Chopin between 1830 and 1832."[21] **Nocturne in B-flat minor, Op. 9, No. 1** has

> a rhythmic freedom that came to characterise Chopin's later work. The left hand has an unbroken sequence of eighth notes in simple arpeggios throughout the entire piece, while the right hand moves with freedom in patterns of seven, eleven, twenty, and twenty-two notes. The opening section moves into a contrasting middle section, which flows back to the opening material in a transitional passage where the melody floats above seventeen consecutive bars of D-flat major chords. The reprise of the first section grows out of this and the nocturne concludes peacefully with a Picardy third.[21]

The rendition used was performed by Peter Schmalfuss and published on the album *Chopin Nocturne Collection, Opus No. 9, 15, 27 & 32* (Stradivari Classics).

Dmitri Shostakovich

"Dmitri Dmitriyevich Shostakovich was a Soviet Russian composer and pianist and a prominent figure of 20th-century music. (...) Sharp contrasts and elements of the grotesque characterize much of his music."[22]

"**24 Preludes and Fugues, Op. 87** by Dmitri Shostakovich is a set of 24 pieces (that is, 24 prelude-fugue pairs) for solo piano, one in each of the major and minor keys of the chromatic scale." **Prelude and Fugue No. 2 in A Minor** "is a toccata mostly for one voice, with semiquavers running through in the style of a perpetuum mobile. It is followed by a three-part fugue with a characteristic theme of sevenths and acciaccaturas."[23] The rendition used was performed by Konstantin Scherbakov and published on the album *SHOSTAKOVICH: 24 Preludes and Fugues, Op. 87* (Naxos, Catalogue No: 8.554745-46).[24]

A.3 Heavy Metal

"Heavy metal (or simply metal) is a genre of rock music that developed in the late 1960s and early 1970s, largely in the United Kingdom and the United States. With roots in blues rock and psychedelic rock, the bands that created heavy metal

[20]http://en.wikipedia.org/w/index.php?title=Fr%C3%A9d%C3%A9ric_Chopin&oldid=604299659.

[21]http://en.wikipedia.org/w/index.php?title=Nocturnes,_Op._9_(Chopin)&oldid=604030036.

[22]http://en.wikipedia.org/w/index.php?title=Dmitri_Shostakovich&oldid=603768409.

[23]http://en.wikipedia.org/w/index.php?title=24_Preludes_and_Fugues_(Shostakovich)&oldid=592452009.

[24]http://www.naxos.com/catalogue/item.asp?item_code=8.554745-46.

developed a thick, massive sound, characterized by highly amplified distortion, extended guitar solos, emphatic beats, and overall loudness. Heavy metal lyrics and performance styles are often associated with masculinity, aggression and machismo."[25]

Black Sabbath

"Black Sabbath are an English rock band (…). Originally formed in 1968 as a heavy blues rock band (…), the band began incorporating occult themes with horror-inspired lyrics and tuned-down guitars. (…) Although Black Sabbath have gone through many line-ups and stylistic changes, their original sound focused on ominous lyrics and doomy music, often making use of the musical tritone, also called the 'devil's interval'."[26]

The song "**Paranoid** is (…) featured on their second album *Paranoid.* (…) The song focuses on a paranoid man and the theme of paranoia, with the driving guitar and bass creating a nervous energy."[27]

Iron Maiden

"Iron Maiden are an English heavy metal band formed in Leyton, east London, in 1975."[28] They are considered "pioneers of the New Wave of British Heavy Metal,"[28] a style that "toned down the blues influences of earlier acts, incorporated elements of punk (and in the case of Iron Maiden combined it with Progressive Rock), increased the tempo, and adopted a 'tougher' sound, taking a harder approach to its music."[29] Their song "**The Wicker Man** (has been) released as the first single from their album *Brave New World* in April 2000."[30]

[25] http://en.wikipedia.org/w/index.php?title=Heavy_metal_music&oldid=604129678.

[26] http://en.wikipedia.org/w/index.php?title=Black_Sabbath&oldid=604377063.

[27] http://en.wikipedia.org/w/index.php?title=Paranoid_(Black_Sabbath_song)&oldid=602863818.

[28] http://en.wikipedia.org/w/index.php?title=Iron_Maiden&oldid=604408503.

[29] http://en.wikipedia.org/w/index.php?title=New_Wave_of_British_Heavy_Metal&oldid=603189299.

[30] http://en.wikipedia.org/w/index.php?title=The_Wicker_Man_(song)&oldid=578894246.

Kiss

"Kiss (more often styled as KISS) is an American hard rock band formed in New York City in January 1973."[31]

I Was Made for Lovin' You (Album Version) was "originally released on their 1979 album *Dynasty*. (...) (It) draws heavily from the disco style that was popular in late-1970s United States."[32]

Metallica

> Metallica is an American heavy metal band from Los Angeles, California. The band's fast tempos, instrumentals, and aggressive musicianship placed them as one of the founding 'big four' of thrash metal,[33]
> a sub-genre of heavy metal that is characterized most typically by its fast tempo and aggression. Thrash metal songs typically use fast percussive beats and fast, low-register guitar riffs, overlaid with shredding-style lead work. Lyrically, thrash metal songs often deal with social issues and reproach for The Establishment, often using direct and denunciatory language.[34]

Metallica's song **Master of Puppets (Explicit Version)** is in the 1986 album *Master Of Puppets*. "The song, as lead singer James Hetfield explained, 'deals pretty much with drugs. How things get switched around, instead of you controlling what you're taking and doing, it's drugs controlling you'"[35] and "notable for its extensive use of downpicking and its long instrumental section, beginning at three minutes thirty-four seconds into the song."[35]

A.4 Rap

"Rap music (...) refers to 'spoken or chanted rhyming lyrics' (and) can be broken down into (...) 'content', 'flow' (rhythm and rhyme), and 'delivery'. Rapping is distinct from spoken word poetry in that it is performed in time to a beat. Rapping is often associated with and a primary ingredient of hip hop music,"[36] "a music genre consisting of a stylized rhythmic music."[37]

---

[31] http://en.wikipedia.org/w/index.php?title=Kiss_(band)&oldid=604403205.

[32] https://en.wikipedia.org/w/index.php?title=I_Was_Made_for_Lovin%27_You&oldid=600873385.

[33] http://en.wikipedia.org/w/index.php?title=Metallica&oldid=603818559.

[34] http://en.wikipedia.org/w/index.php?title=Thrash_metal&oldid=603973708.

[35] http://en.wikipedia.org/w/index.php?title=Master_of_Puppets_(song)&oldid=600107813.

[36] http://en.wikipedia.org/w/index.php?title=Rapping&oldid=604005540.

[37] http://en.wikipedia.org/w/index.php?title=Hip_hop_music&oldid=604234652.

Busta Rhymes

"Busta Rhymes is an American rapper, producer, and actor from Brooklyn. (...) Early in his career, he was known for his wild style and fashion, and today is best known for his intricate rapping technique, which involves rapping at a fast rate with lots of internal rhyme and half rhyme."[38] "**Put Your Hands Where My Eyes Could See (Explicit LP Version)** is a 1997 song (...) from his second studio album *When Disaster Strikes*."[39]

Jay-Z

"Jay-Z is an American rapper, record producer, and entrepreneur."[40] Regarding his rap style, he is referred to as "a master of the flow," who uses "'rests' to provide structure to a verse," "'partial linking' to add more rhymes to a verse," and "smoothness and clever wordplay to keep the audience interested and entertained."[41] The track **Big Pimpin' (Album Version)** "is the fifth and final single from (his) fourth album *Vol. 3... Life and Times of S. Carter*."[42]

Snoop Dogg

"Snoop Dogg (...) is an American rapper, singer-songwriter, and actor."[43] "Snoop is known to use syncopation in his flow to give it a laidback quality, as well as 'linking with rhythm' in his compound rhymes, using alliteration, and employing a 'sparse' flow with good use of pauses."[41, cited in 43] His song **Drop It Like It's Hot (feat. Pharrell—Explicit Album Version)** was released on the album *R&G (Rhythm & Gangsta): The Masterpiece* and features "American singer and producer Pharrell Williams. Snoop Dogg performs the chorus and the second and third verses (of three), while Pharrell performs the first verse. (...) It gained some critical attention for its very sparse production, which is essentially just tongue clicks,

---

[38] http://en.wikipedia.org/w/index.php?title=Busta_Rhymes&oldid=604366186.

[39] http://en.wikipedia.org/w/index.php?title=Put_Your_Hands_Where_My_Eyes_Could_See&oldid=597025149.

[40] http://en.wikipedia.org/w/index.php?title=Jay-Z&oldid=602817936.

[41] Edwards, P.: How to Rap: The Art and Science of the Hip-Hop MC. Chicago Review Press (2009). Quotes for Jay-Z on pages 65, 129, 91, and 292, resp.

[42] http://en.wikipedia.org/w/index.php?title=Big_Pimpin%27&oldid=720585452.

[43] http://en.wikipedia.org/w/index.php?title=Snoop_Dogg&oldid=603740234.

keyboards and a drum machine beat which, compared to much early 2000s rap, was very minimalist."[44]

Snow

"Snow is a Canadian reggae musician," who blends "dancehall and reggae with rock and popular music to create his own unique style of music."[45] "He is best known for his 1992 single **Informer (LP Version)**"[45] which was included in the album *12 Inches of Snow*.

A.5 Pop

> Pop music (a term that originally derives from an abbreviation of 'popular') is a genre of popular music which originated in its modern form in the 1950s, deriving from rock and roll. The terms 'popular music' and 'pop music' are often used interchangeably, even though the former is a description of music which is popular (and can include any style). As a genre, pop music is very eclectic, often borrowing elements from other styles including urban, dance, rock, Latin and country; nonetheless, there are core elements which define pop. Such include generally short-to-medium length songs, written in a basic format (often the verse-chorus structure), as well as the common employment of repeated choruses, melodic tunes, and catchy hooks.[46]

Kylie Minogue

"Kylie Ann Minogue, OBE, often known simply as Kylie, is an Australian singer, recording artist, songwriter and actress, working and living in London. (...) Minogue has been known for her soft Soprano vocal range. (...) In musical terms, Minogue has worked with many genres in pop and dance music. However, her signature or suitable music has been contemporary disco music."[47]

"**Get Outta My Way** (was) released as the second single from her eleventh studio album, *Aphrodite*. (...) Musically, Get Outta My Way is a mid-tempo dance-oriented song that draws influences from disco and synthpop music."[48]

---

[44] http://en.wikipedia.org/w/index.php?title=Drop_It_Like_It%27s_Hot&oldid=596500407.

[45] http://en.wikipedia.org/w/index.php?title=Snow_(musician)&oldid=602522684.

[46] http://en.wikipedia.org/w/index.php?title=Pop_music&oldid=604274871.

[47] http://en.wikipedia.org/w/index.php?title=Kylie_Minogue&oldid=604305641.

[48] http://en.wikipedia.org/w/index.php?title=Get_Outta_My_Way&oldid=601448772.

Madonna

"Madonna Louise Ciccone is an American singer-songwriter, actress, and businesswoman. One of the most prominent cultural icons for over three decades, she has achieved an unprecedented level of power and control for a woman in the entertainment industry."[49]

Hung Up

> was written and produced in collaboration with Stuart Price, and released as the first single from her tenth studio album, *Confessions on a Dance Floor* (2005). (...) 'Hung Up' prominently features a sample of pop group ABBA's hit single 'Gimme! Gimme! Gimme! (A Man After Midnight)'. (...) Musically the song is influenced by 1980s pop, with a chugging groove and chorus and a background element of a ticking clock that suggests the fear of wasting time. Lyrically the song is written as a traditional dance number about a strong, independent woman who has relationship troubles.[50]

Maroon 5

"Maroon 5 is an American rock band that originated in Los Angeles, California. (...) The band's songs tend to be very guitar-heavy, often accompanied by piano or synthesizer. The theme in all of their songs is love, frequently lost love."[51]

Their song **Moves like Jagger (feat. Christina Aguilera)** is the bonus track on the rerelease of their album *Hands All Over*. "Moves like Jagger is a dance-pop and electropop song and is backed by synths and electronic drums, the lyrics refer to a male's ability to impress a female with his dance moves, which he compares to Mick Jagger, frontman of the English band The Rolling Stones."[52]

Lady Gaga

"Lady Gaga is an American recording artist, activist, record producer, businesswoman, fashion designer, philanthropist, and actress. (...) Influenced by David Bowie, Michael Jackson, Madonna, and Queen, Gaga is recognized for her flamboyant, diverse, and outré contributions to the music industry through her fashion, performances, and music videos. (...) The structure of her music is said to echo classic 1980s pop and 1990s Europop."[53]

[49] http://en.wikipedia.org/w/index.php?title=Madonna_(entertainer)&oldid=604340142.

[50] http://en.wikipedia.org/w/index.php?title=Hung_Up&oldid=603479571.

[51] http://en.wikipedia.org/w/index.php?title=Maroon_5&oldid=604256717.

[52] http://en.wikipedia.org/w/index.php?title=Moves_like_Jagger&oldid=604018838.

[53] http://en.wikipedia.org/w/index.php?title=Lady_Gaga&oldid=604410320.

"**Bad Romance** is a song (. . .) from her third extended play, *The Fame Monster*. (. . .) Musically, Bad Romance features a spoken bridge, a full-throated chorus and sung lyrics about being in love with one's best friend. The song, which is imbued with elements of German-esque house and techno, as well music from 1980s and the 1990s, was touted by Gaga as an experimental pop record. The song contains a few lines in French."[54]

[54]http://en.wikipedia.org/w/index.php?title=Bad_Romance&oldid=604433877.

References

1. Abeßer, J., Lukashevich, H., Bräuer, P.: Classification of music genres based on repetitive basslines. J. N. Music Res. **41**(3), 239–257 (2012)
2. Adomavicius, G., Mobasher, B., Ricci, F., Tuzhilin, A.: Context-aware recommender systems. AI Mag. **32**, 67–80 (2011)
3. Adomavicius, G., Tuzhilin, A.: Context-aware recommender systems. In: Ricci, F., Rokach, L., Shapira, B., Kantor, P.B. (eds.) Recommender Systems Handbook, pp. 217–253. Springer, New York (2011)
4. Alghoniemy, M., Tewfik, A.: A network flow model for playlist generation. In: Proceedings of the IEEE International Conference on Multimedia and Expo (ICME), Tokyo (2001)
5. Anderson, C.: The Long Tail: Why the Future of Business Is Selling Less of More. Hyperion, New York (2006)
6. Andersen, K., Grote, F.: GiantSteps: semi-structured conversations with musicians. In: Proceedings of the 33rd Annual ACM SIGCHI Conference on Human Factors in Computing Systems Extended Abstracts (CHI EA), Seoul (2015)
7. Anglade, A., Ramirez, R., Dixon, S.: Genre classification using harmony rules induced from automatic chord transcriptions. In: Proceedings of the 10th International Society for Music Information Retrieval Conference (ISMIR), Kobe (2009)
8. Arzt, A., Widmer, G., Dixon, S.: Automatic page turning for musicians via real-time machine listening. In: Proceedings of the 18th European Conference on Artificial Intelligence (ECAI), Patras (2008)
9. Arzt, A., Widmer, G., Dixon, S.: Adaptive distance normalization for real-time music tracking. In: Proceedings of the 20th European Signal Processing Conference (EUSIPCO), Bucharest (2012)
10. Arzt, A., Widmer, G., Sonnleitner, R.: Tempo- and transposition-invariant identification of piece and score position. In: Proceedings of the 15th International Society for Music Information Retrieval Conference (ISMIR), Taipei (2014)
11. Aucouturier, J.J.: Ten experiments on the modelling of polyphonic timbre. Ph.D. thesis, University of Paris 6 (2006)
12. Aucouturier, J.J., Pachet, F.: Music similarity measures: What's the use? In: Proceedings of the 3rd International Conference on Music Information Retrieval (ISMIR), Paris (2002)
13. Aucouturier, J.J., Pachet, F.: Scaling up music playlist generation. In: Proceedings of the IEEE International Conference on Multimedia and Expo (ICME), Lausanne (2002)
14. Aucouturier, J.J., Pachet, F.: Representing musical genre: a state of the art. J. N. Music Res. **32**(1), 83–93 (2003)
15. Aucouturier, J.J., Pachet, F.: Improving timbre similarity: How high is the sky? J. Negat. Results Speech Audio Sci. **1**(1), 1–13 (2004)

P. Knees, M. Schedl, *Music Similarity and Retrieval*, The Information Retrieval Series 36, DOI 10.1007/978-3-662-49722-7

16. Aucouturier, J.J., Pachet, F.: Ringomatic: a real-time interactive drummer using constraint-satisfaction and drum sound descriptors. In: Proceedings of the 6th International Conference on Music Information Retrieval (ISMIR), London (2005)
17. Aucouturier, J.J., Pachet, F., Roy, P., Beurivé, A.: Signal + context = better classification. In: Proceedings of the 8th International Conference on Music Information Retrieval (ISMIR), Vienna (2007)
18. Baccigalupo, C., Plaza, E., Donaldson, J.: Uncovering affinity of artists to multiple genres from social behaviour data. In: Proceedings of the 9th International Conference on Music Information Retrieval (ISMIR), Philadelphia (2008)
19. Bade, K., Nürnberger, A., Stober, S., Garbers, J., Wiering, F.: Supporting folk-song research by automatic metric learning and ranking. In: Proceedings of the 10th International Society for Music Information Retrieval Conference (ISMIR), Kobe (2009)
20. Baeza-Yates, R., Ribeiro-Neto, B.: Modern Information Retrieval – The Concepts and Technology Behind Search, 2nd edn. Addison-Wesley, Pearson (2011)
21. Bainbridge, D., Bell, T.: The challenge of optical music recognition. Comput. Hum. **35**(2), 95–121 (2001)
22. Baltrunas, L., Kaminskas, M., Ludwig, B., Moling, O., Ricci, F., Lüke, K.H., Schwaiger, R.: InCarMusic: context-aware music recommendations in a car. In: International Conference on Electronic Commerce and Web Technologies (EC-Web), Toulouse (2011)
23. Bandari, R., Asur, S., Huberman, B.A.: The pulse of news in social media: forecasting popularity. In: Proceedings of the 6th International AAAI Conference on Weblogs and Social Media (ICWSM), Dublin (2012)
24. Barrington, L., Yazdani, M., Turnbull, D., Lanckriet, G.: Combining feature kernels for semantic music retrieval. In: Proceedings of the 9th International Conference on Music Information Retrieval (ISMIR), Philadelphia (2008)
25. Bauer, C.: A framework for conceptualizing context for intelligent systems (CCFIS). J. Ambient Intell. Smart Environ. **6**(4), 403–417 (2014)
26. Baumann, S.: Artificial listening systems: modellierung und approximation der subjektiven perzeption von musikähnlichkeit. Dissertation, Technische Universität Kaiserslautern (2005)
27. Baumann, S., Hummel, O.: Using cultural metadata for artist recommendation. In: Proceedings of the 3rd International Conference on Web Delivering of Music (WEDELMUSIC), Leeds (2003)
28. Baumann, S., Klüter, A., Norlien, M.: Using natural language input and audio analysis for a human-oriented mir system. In: Proceedings of the 2nd International Conference on Web Delivering of Music (WEDELMUSIC), Darmstadt (2002)
29. Baur, D., Steinmayr, B., Butz, A.: SongWords: exploring music collections through lyrics. In: Proceedings of the 11th International Society for Music Information Retrieval Conference (ISMIR), Utrecht (2010)
30. Bello, J.P.: Measuring structural similarity in music. IEEE Trans. Audio Speech Lang. Process. **19**(7), 2013–2025 (2011)
31. Benetos, E., Dixon, S., Giannoulis, D., Kirchhoff, H., Klapuri, A.: Automatic music transcription: challenges and future directions. J. Intell. Inf. Syst. **41**(3), 407–434 (2013)
32. Bengio, Y.: Learning deep architectures for AI. Found. Trends Mach. Learn. **2**(1), 1–127 (2009)
33. Berenzweig, A., Logan, B., Ellis, D.P., Whitman, B.: A large-scale evaluation of acoustic and subjective music similarity measures. In: Proceedings of the 4th International Conference on Music Information Retrieval (ISMIR), Baltimore (2003)
34. Berkovsky, S., Kuflik, T., Ricci, F.: Entertainment personalization mechanism through cross-domain user modeling. In: Maybury, M., Stock, O., Wahlster, W. (eds.) Intelligent Technologies for Interactive Entertainment. Lecture Notes in Computer Science, vol. 3814. Springer, Heidelberg (2005)
35. Bertin-Mahieux, T., Eck, D., Maillet, F., Lamere, P.: Autotagger: a model for predicting social tags from acoustic features on large music databases. J. N. Music Res. **37**(2), 115–135 (2008)

36. Bertin-Mahieux, T., Ellis, D.P., Whitman, B., Lamere, P.: The million song dataset. In: Proceedings of the 12th International Society for Music Information Retrieval Conference (ISMIR), Miami (2011)
37. Biehl, J.T., Adamczyk, P.D., Bailey, B.P.: DJogger: a mobile dynamic music device. In: Proceedings of the 24th Annual ACM SIGCHI Conference on Human Factors in Computing Systems Extended Abstracts (CHI EA), Montréal (2006)
38. Bishop, C.M.: Pattern Recognition and Machine Learning. Springer, Heidelberg (2006)
39. Blei, D.M.: Probabilistic topic models. Commun. ACM **55**(4), 77–84 (2012)
40. Blei, D.M., Lafferty, J.D.: Topic models. In: Text Mining: Classification, Clustering, and Applications, pp. 71–93. Chapman & Hall/CRC, Boca Raton (2009)
41. Blei, D.M., Ng, A.Y., Jordan, M.I.: Latent Dirichlet allocation. J. Mach. Learn. Res. **3**, 993–1022 (2003)
42. Bogdanov, D.: From music similarity to music recommendation: computational approaches based in audio features and metadata. Ph.D. thesis, Universitat Pompeu Fabra, Barcelona (2013)
43. Bogdanov, D., Serrà, J., Wack, N., Herrera, P., Serra, X.: Unifying low-level and high-level music similarity measures. IEEE Trans. Multimedia **13**(4), 687–701 (2011)
44. Bogdanov, D., Haro, M., Fuhrmann, F., Xambó, A., Gómez, E., Herrera, P.: Semantic audio content-based music recommendation and visualization based on user preference examples. Inf. Process. Manag. **49**(1), 13–33 (2013)
45. Bosch, J.J., Janer, J., Fuhrmann, F., Herrera, P.: A comparison of sound segregation techniques for predominant instrument recognition in musical audio signals. In: Proceedings of the 13th International Society for Music Information Retrieval Conference (ISMIR), Porto (2012)
46. Breiman, L.: Random forests. Mach. Learn. **45**, 5–32 (2001)
47. Breitschopf, G.: Personalized, context-aware music playlist generation on mobile devices. Master's thesis, Johannes Kepler University Linz (2013)
48. Brill, E.: A simple rule-based part of speech tagger. In: Proceedings of the 3rd Conference on Applied Natural Language Processing (ANLC), Trento (1992)
49. Bu, J., Tan, S., Chen, C., Wang, C., Wu, H., Zhang, L., He, X.: Music recommendation by unified hypergraph: combining social media information and music content. In: Proceedings of the 18th ACM International Conference on Multimedia (MM), Firenze (2010)
50. Burke, R.: Hybrid recommender systems: survey and experiments. User Model. User Adap. Inter. **12**(4), 331–370 (2002)
51. Burke, R.: Hybrid web recommender systems. In: Brusilovsky, P., Kobsa, A., Nejdl, W. (eds.) The Adaptive Web: Methods and Strategies of Web Personalization, pp. 377–408. Springer, Heidelberg (2007)
52. Burred, J.J., Lerch, A.: A hierarchical approach to automatic musical genre classification. In: Proceedings of the 6th International Conference on Digital Audio Effects (DAFx), London (2003)
53. Büttcher, S., Clarke, C.L.A., Cormack, G.V.: Information Retrieval: Implementing and Evaluating Search Engines. MIT Press, Cambridge (2010)
54. Byrd, D., Fingerhut, M.: The history of ISMIR – a short happy tale. D-Lib Mag. **8**(11) (2002)
55. Byrd, D., Schindele, M.: Prospects for improving OMR with multiple recognizers. In: Proceedings of the 7th International Conference on Music Information Retrieval (ISMIR), Victoria (2006)
56. Caetano, M., Burred, J.J., Rodet, X.: Automatic segmentation of the temporal evolution of isolated acoustic musical instrument sounds using spectro-temporal cues. In: Proceedings of the 13th International Conference on Digital Audio Effects (DAFx), Graz (2010)
57. Cai, R., Zhang, C., Wang, C., Zhang, L., Ma, W.Y.: MusicSense: contextual music recommendation using emotional allocation modeling. In: Proceedings of the 15th ACM International Conference on Multimedia (MM), Augsburg (2007)
58. Calvo, R.A., D'Mello, S., Gratch, J., Kappas, A. (eds.): The Oxford Handbook of Affective Computing. Oxford University Press, Oxford (2014)

59. Cano, P., Kaltenbrunner, M., Gouyon, F., Batlle, E.: On the use of fastmap for audio retrieval and browsing. In: Proceedings of the 3rd International Conference on Music Information Retrieval (ISMIR), Paris (2002)
60. Cebrián, T., Planagumà, M., Villegas, P., Amatriain, X.: Music recommendations with temporal context awareness. In: Proceedings of the 4th ACM Conference on Recommender Systems (RecSys), Barcelona (2010)
61. Celma, O.: Music recommendation and discovery in the long tail. Ph.D. thesis, Universitat Pompeu Fabra, Barcelona (2008)
62. Celma, O.: Music Recommendation and Discovery – The Long Tail, Long Fail, and Long Play in the Digital Music Space. Springer, Berlin/Heidelberg (2010)
63. Celma, Ò., Herrera, P.: A new approach to evaluating novel recommendations. In: Proceedings of the 2nd ACM Conference on Recommender Systems (RecSys), Lausanne (2008)
64. Celma, O., Nunes, M.: GeoMuzik: a geographic interface for large music collections. In: ISMIR 2008: 9th International Conference on Music Information Retrieval – Late-Breaking/Demo Session, Philadelphia (2008)
65. Celma, O., Cano, P., Herrera, P.: SearchSounds: an audio crawler focused on weblogs. In: Proceedings of the 7th International Conference on Music Information Retrieval (ISMIR), Victoria (2006)
66. Cemgil, A.T., Kappen, H.J., Barber, D.: A generative model for music transcription. IEEE Trans. Audio Speech Lang. Process. **14**(2), 679–694 (2006)
67. Chai, W.: Semantic segmentation and summarization of music. IEEE Signal Process. Mag. **23**(2) (2006)
68. Chakrabarti, S.: Mining the Web: Analysis of Hypertext and Semi Structured Data. Morgan Kaufmann, San Francisco (2002)
69. Chakrabarti, S., van den Berg, M., Dom, B.: Focused crawling: a new approach to topic-specific web resource discovery. Comput. Netw. **31**(11–16), 1623–1640 (1999)
70. Charbuillet, C., Tardieu, D., Peeters, G.: GMM-supervector for content based music similarity. In: International Conference on Digital Audio Effects (DAFx), Paris (2011)
71. Charniak, E.: Statistical techniques for natural language parsing. AI Mag. **18**, 33–44 (1997)
72. Chen, S., Moore, J., Turnbull, D., Joachims, T.: Playlist prediction via metric embedding. In: Proceedings of the 18th ACM SIGKDD International Conference on Knowledge Discovery and Data Mining (KDD), Beijing (2012)
73. Cheng, T., Dixon, S., Mauch, M.: A deterministic annealing em algorithm for automatic music transcription. In: Proceedings of the 14th International Society for Music Information Retrieval Conference (ISMIR), Curitiba (2013)
74. Cheng, W., Hüllermeier, E.: Learning similarity functions from qualitative feedback. In: Althoff, K.D., Bergmann, R., Minor, M., Hanft, A. (eds.) Advances in Case-Based Reasoning. Lecture Notes in Computer Science, vol. 5239. Springer, Heidelberg (2008)
75. Cheng, Z., Shen, J.: Just-for-me: an adaptive personalization system for location-aware social music recommendation. In: Proceedings of the 4th ACM International Conference on Multimedia Retrieval (ICMR), Glasgow (2014)
76. Cleverdon, C.W.: The significance of the cranfield tests on index languages. In: Proceedings of the 14th Annual International ACM SIGIR Conference on Research and Development in Information Retrieval (SIGIR), Chicago (1991)
77. Codognet, P., Diaz, D.: Yet another local search method for constraint solving. In: Steinhöfel, K. (ed.) Proceedings of the International Symposium on Stochastic Algorithms (SAGA): Foundations and Applications. Lecture Notes in Computer Science, vol. 2264. Springer, Heidelberg (2001)
78. Cohen, W.W., Fan, W.: Web-collaborative filtering: recommending music by crawling the web. Comput. Netw. **33**(1–6), 685–698 (2000)
79. Collins, T., Böck, S., Krebs, F., Widmer, G.: Bridging the audio-symbolic gap: the discovery of repeated note content directly from polyphonic music audio. In: Proceedings of the 53rd International AES Conference: Semantic Audio, London (2014)

80. Conklin, D.: Representation and discovery of vertical patterns in music. In: Anagnostopoulou, C., Ferrand, M., Smaill, A. (eds.) Music and Artificial Intelligence. Lecture Notes in Computer Science, vol. 2445. Springer, Heidelberg (2002)
81. Cont, A.: Realtime audio to score alignment for polyphonic music instruments, using sparse non-negative constraints and hierarchical HMMs. In: Proceedings of the IEEE International Conference on Acoustics, Speech and Signal Processing (ICASSP), Toulouse (2006)
82. Cont, A.: ANTESCOFO: anticipatory synchronization and control of interactive parameters in computer music. In: Proceedings of the International Computer Music Conference (ICMC), Belfast (2008)
83. Cooley, J.W., Tukey, J.W.: An algorithm for the machine calculation of complex Fourier series. Math. Comput. **19**(90), 297–301 (1965)
84. Cooper, M.L., Foote, J.: Automatic music summarization via similarity analysis. In: Proceedings of the 3rd International Conference on Music Information Retrieval (ISMIR), Paris (2002)
85. Corpet, F.: Multiple sequence alignment with hierarchical clustering. Nucleic Acids Res. **16**(22), 10881–10890 (1988)
86. Cover, T.M., Hart, P.E.: Nearest neighbor pattern classification. IEEE Trans. Inf. Theory **13**(1), 21–27 (1967)
87. Coviello, E., Chan, A.B., Lanckriet, G.: Time series models for semantic music annotation. IEEE Trans. Audio Speech Lang. Process. **19**(5), 1343–1359 (2011)
88. Cox, T.F., Cox, M.A.A.: Multidimensional Scaling. Chapman & Hall, Boca Raton (1994)
89. Cristianini, N., Shawe-Taylor, J.: An Introduction to Support Vector Machines and Other Kernel-Based Learning Methods. Cambridge University Press, Cambridge (2000)
90. Cunningham, S.J., Reeves, N., Britland, M.: An ethnographic study of music information seeking: implications for the design of a music digital library. In: Proceedings of the 3rd ACM/IEEE-CS Joint Conference on Digital Libraries (JCDL), Houston (2003)
91. Cunningham, S.J., Bainbridge, D., Falconer, A.: 'More of an art than a science': supporting the creation of playlists and mixes. In: Proceedings of the 7th International Conference on Music Information Retrieval (ISMIR), Victoria (2006)
92. Cunningham, S., Caulder, S., Grout, V.: Saturday night or fever? Context-aware music playlists. In: Proceedings of the 3rd International Audio Mostly Conference: Sound in Motion, Piteå (2008)
93. Dannenberg, R.B.: An on line algorithm for real-time accompaniment. In: Proceedings of the International Computer Music Conference (ICMC), Paris (1984)
94. Davis, J.V., Kulis, B., Jain, P., Sra, S., Dhillon, I.S.: Information-theoretic metric learning. In: Proceedings of the 24th International Conference on Machine Learning (ICML), Corvalis (2007)
95. Debole, F., Sebastiani, F.: Supervised term weighting for automated text categorization. In: Proceedings the 18th ACM Symposium on Applied Computing (SAC), Melbourne (2003)
96. Deerwester, S., Dumais, S.T., Furnas, G.W., Landauer, T.K., Harshman, R.: Indexing by latent semantic analysis. J. Am. Soc. Inf. Sci. **41**, 391–407 (1990)
97. de Haas, W.B.: Music information retrieval based on tonal harmony. Ph.D. thesis, Universiteit Utrecht (2012)
98. Deng, L., Yu, D.: Deep learning: methods and applications. Found. Trends Signal Process. **7**(3–4), 197–387 (2013)
99. de Oliveira, R., Oliver, N.: TripleBeat: enhancing exercise performance with persuasion. In: Proceedings of the 10th International Conference on Human Computer Interaction with Mobile Devices and Services (Mobile CHI), Amsterdam (2008)
100. Desrosiers, C., Karypis, G.: A comprehensive survey of neighborhood-based recommendation methods. In: Ricci, F., Rokach, L., Shapira, B., Kantor, P.B. (eds.) Recommender Systems Handbook, pp. 107–144. Springer, New York (2011)
101. Diakopoulos, D., Vallis, O., Hochenbaum, J., Murphy, J., Kapur, A.: 21st century electronica: MIR techniques for classification and performance. In: Proceedings of the 10th International Society for Music Information Retrieval Conference (ISMIR'09), Kobe (2009)

102. Dieleman, S., Brakel, P., Schrauwen, B.: Audio-based music classification with a pretrained convolutional network. In: Proceedings of the 12th International Society for Music Information Retrieval Conference (ISMIR), Miami (2011)
103. Dittenbach, M., Merkl, D., Rauber, A.: Hierarchical clustering of document archives with the growing hierarchical self-organizing map. In: Proceedings of the International Conference on Artificial Neural Networks (ICANN), Vienna (2001)
104. Dittmar, C., Cano, E., Abeßer, J., Grollmisch, S.: Music information retrieval meets music education. In: Müller, M., Goto, M., Schedl, M. (eds.) Multimodal Music Processing. Dagstuhl Follow-Ups, vol. 3. Schloss Dagstuhl–Leibniz-Zentrum für Informatik, Wadern (2012)
105. Dixon, S.: Onset detection revisited. In: Proceedings of the 9th International Conference on Digital Audio Effects (DAFx), Montréal (2006)
106. Dixon, S., Goebl, W., Widmer, G.: The performance worm: real time visualisation of expression based on langner's tempo-loudness animation. In: Proceedings of the International Computer Music Conference (ICMC), Baltimore (2002)
107. Dixon, S., Goebl, W., Widmer, G.: The "air worm": an interface for real-time manipulation of expressive music performance. In: Proceedings of the International Computer Music Conference (ICMC), Barcelona (2005)
108. Donaldson, J.: A hybrid social-acoustic recommendation system for popular music. In: Proceedings of the ACM Conference on Recommender Systems (RecSys), Minneapolis (2007)
109. Dornbush, S., English, J., Oates, T., Segall, Z., Joshi, A.: XPod: a human activity aware learning mobile music player. In: 20th International Joint Conference on Artificial Intelligence (IJCAI): Proceedings of the 2nd Workshop on Artificial Intelligence Techniques for Ambient Intelligence, Hyderabad (2007)
110. Downie, J.S.: The scientific evaluation of music information retrieval systems: foundations and future. Comput. Music J. **28**, 12–23 (2004)
111. Downie, J.S., Byrd, D., Crawford, T.: Ten years of ISMIR: reflections on challenges and opportunities. In: Proceedings of the 10th International Society for Music Information Retrieval Conference (ISMIR), Kobe (2009)
112. Dror, G., Koenigstein, N., Koren, Y.: Yahoo! music recommendations: modeling music ratings with temporal dynamics and item taxonomy. In: Proceedings of the 5th ACM Conference on Recommender Systems (RecSys), Chicago (2011)
113. Dror, G., Koenigstein, N., Koren, Y., Weimer, M.: The Yahoo! Music dataset and KDD-Cup'11. J. Mach. Learn. Res. **8**, 3–18 (2011). Proceedings of KDD-Cup 2011 Competition
114. Drott, E.: The end(s) of genre. J. Music Theory **57**(1), 1–45 (2013)
115. Eck, D., Lamere, P., Bertin-Mahieux, T., Green, S.: Automatic generation of social tags for music recommendation. In: Platt, J., Koller, D., Singer, Y., Roweis, S. (eds.) Advances in Neural Information Processing Systems 20 (NIPS). Curran Associates, Inc., Vancouver (2008)
116. Eerola, T., Lartillot, O., Toiviainen, P.: Prediction of multidimensional emotional ratings in music from audio using multivariate regression models. In: Proceedings of the 10th International Society for Music Information Retrieval Conference (ISMIR), Kobe (2009)
117. Eghbal-zadeh, H., Lehner, B., Schedl, M., Widmer, G.: I-Vectors for timbre-based music similarity and music artist classification. In: Proceedings of the 16th International Society for Music Information Retrieval Conference (ISMIR), Málaga (2015)
118. Ekman, P.: Emotion in the Human Face: Guidelines for Research and an Integration of Findings. Pergamon, New York (1972)
119. Ekman, P.: Basic emotions. In: Dalgleish, T., Power, M. (eds.) Handbook of Cognition and Emotion, pp. 45–60. Wiley, New York (1999)
120. Elliott, G.T., Tomlinson, B.: PersonalSoundtrack: context-aware playlists that adapt to user pace. In: Proceedings of the 24th Annual ACM SIGCHI Conference on Human Factors in Computing Systems Extended Abstracts (CHI EA), Montréal (2006)

121. Ellis, D.P., Whitman, B., Berenzweig, A., Lawrence, S.: The quest for ground truth in musical artist similarity. In: Proceedings of 3rd International Conference on Music Information Retrieval (ISMIR), Paris (2002)
122. Eyben, F., Böck, S., Schuller, B., Graves, A.: Universal onset detection with bidirectional long short-term memory neural networks. In: Proceedings of the 11th International Society for Music Information Retrieval Conference (ISMIR), Utrecht (2010)
123. Fabbri, F.: A theory of musical genres: two applications. Pop. Music Perspect. **1**, 52–81 (1981)
124. Fastl, H., Zwicker, E.: Psychoacoustics, 3rd edn. Springer, Heidelberg (2007)
125. Fiebrink, R.A.: Real-time human interaction with supervised learning algorithms for music composition and performance. Ph.D. thesis, Princeton University (2011)
126. Fields, B., Casey, M., Jacobson, K., Sandler, M.: Do you sound like your friends? Exploring artist similarity via artist social network relationships and audio signal processing. In: Proceedings of the 2008 International Computer Music Conference (ICMC), Belfast (2008)
127. Flexer, A.: On inter-rater agreement in audio music similarity. In: Proceedings of the 15th International Society for Music Information Retrieval Conference (ISMIR), Taipei (2014)
128. Flexer, A., Schnitzer, D., Gasser, M., Widmer, G.: Playlist generation using start and end songs. In: Proceedings of the 9th International Conference on Music Information Retrieval (ISMIR), Philadelphia (2008)
129. Flexer, A., Schnitzer, D., Schlüter, J.: A MIREX meta-analysis of hubness in audio music similarity. In: Proceedings of the 13th International Society for Music Information Retrieval Conference (ISMIR), Porto (2012)
130. Flossmann, S., Goebl, W., Grachten, M., Niedermayer, B., Widmer, G.: The magaloff project: an interim report. J. N. Music Res. **39**(4), 363–377 (2010)
131. Foote, J.T.: Content-based retrieval of music and audio. In: Kuo, C.C.J., Chang, S.F., Gudivada, V.N. (eds.) Proceedings of SPIE: Multimedia Storage and Archiving Systems II, vol. 3229. SPIE, Dallas (1997)
132. Foote, J.T.: Visualizing music and audio using self-similarity. In: Proceedings of the 7th ACM International Conference on Multimedia (MM), Orlando (1999)
133. Freund, Y., Schapire, R.E.: Experiments with a new boosting algorithm. In: Proceedings of the 13th International Conference on Machine Learning (ICML), Bari (1996)
134. Friedman, M.: The use of ranks to avoid the assumption of normality implicit in the analysis of variance. J. Am. Stat. Assoc. **32**(200), 675–701 (1937)
135. Fuhrmann, F.: Automatic musical instrument recognition from polyphonic music audio signals. Ph.D. thesis, Universitat Pompeu Fabra, Barcelona (2012)
136. Fujinaga, I.: Optical music recognition using projections. Master's thesis, McGill University, Montréal (1990)
137. Fujishima, T.: Realtime chord recognition of musical sound: a system using common lisp music. In: Proceedings of the International Computer Music Conference (ICMC), Beijing (1999)
138. Futrelle, J., Downie, J.S.: Interdisciplinary research issues in music information retrieval: ISMIR 2000–2002. J. N. Music Res. **32**(2), 121–131 (2003)
139. Gängler, T.: Semantic federation of musical and music-related information for establishing a personal music knowledge base. Master's thesis, Dresden University of Technology (2011)
140. Gasser, M., Flexer, A.: FM4 soundpark: audio-based music recommendation in everyday use. In: Proceedings of the 6th Sound and Music Computing Conference (SMC), Porto (2009)
141. Geleijnse, G., Korst, J.: Tool play live: dealing with ambiguity in artist similarity mining from the web. In: Proceedings of the 8th International Conference on Music Information Retrieval (ISMIR), Vienna (2007)
142. Geleijnse, G., Schedl, M., Knees, P.: The quest for ground truth in musical artist tagging in the social web era. In: Proceedings of the 8th International Conference on Music Information Retrieval (ISMIR), Vienna (2007)

143. Ghias, A., Logan, J., Chamberlin, D., Smith, B.C.: Query by humming: musical information retrieval in an audio database. In: Proceedings of the 3rd ACM International Conference on Multimedia (MM), San Francisco (1995)
144. Gillhofer, M., Schedl, M.: Iron maiden while jogging, debussy for dinner? - An analysis of music listening behavior in context. In: Proceedings of the 21st International Conference on MultiMedia Modeling (MMM), Sydney (2015)
145. Gjerdingen, R.O., Perrott, D.: Scanning the dial: the rapid recognition of music genres. J. N. Music Res. **37**(2), 93–100 (2008)
146. Goldberg, D., Nichols, D., Oki, B.M., Terry, D.: Using collaborative filtering to weave an information tapestry. Commun. ACM **35**(12), 61–70 (1992)
147. Gómez, E.: Tonal description of music audio signals. Ph.D. thesis, Universitat Pompeu Fabra, Barcelona (2006)
148. Gómez, E.: Tonal description of polyphonic audio for music content processing. INFORMS J. Comput. **18**, 294–304 (2006). Special Cluster on Computation in Music
149. Gómez, E., Grachten, M., Hanjalic, A., Janer, J., Jorda, S., Julia, C.F., Liem, C.C., Martorell, A., Schedl, M., Widmer, G.: PHENICX: performances as highly enriched and interactive concert experiences. In: Proceedings of the Sound and Music Computing Conference (SMC), Stockholm (2013)
150. Gospodnetić, O., Hatcher, E.: Lucene in Action. Manning, Greenwich (2005)
151. Goto, M.: SmartMusicKIOSK: music listening station with chorus-search function. In: Proceedings of the 16th Annual ACM Symposium on User Interface Software and Technology (UIST), Vancouver (2003)
152. Goto, M.: Grand challenges in music information research. In: Müller, M., Goto, M., Schedl, M. (eds.) Multimodal Music Processing. Dagstuhl Follow-Ups, vol. 3. Schloss Dagstuhl–Leibniz-Zentrum für Informatik, Wadern (2012)
153. Goto, M., Goto, T.: Musicream: new music playback interface for streaming, sticking, sorting, and recalling musical pieces. In: Proceedings of the 6th International Conference on Music Information Retrieval (ISMIR), London (2005)
154. Goto, M., Goto, T.: Musicream: integrated music-listening interface for active, flexible, and unexpected encounters with musical pieces. Inf. Process. Soc. Jpn. **50**(12), 2923–2936 (2009)
155. Gouyon, F., Dixon, S.: A review of automatic rhythm description systems. Comput. Music J. **29**(1), 34–54 (2005)
156. Gouyon, F., Pachet, F., Delerue, O.: On the use of zero-crossing rate for an application of classification of percussive sounds. In: Proceedings of the COST-G6 Conference on Digital Audio Effects (DAFx), Verona (2000)
157. Gouyon, F., Klapuri, A., Dixon, S., Alonso, M., Tzanetakis, G., Uhle, C., Cano, P.: An experimental comparison of audio tempo induction algorithms. IEEE Trans. Audio Speech Lang. Process. **14**(5), 1832–1844 (2006)
158. Govaerts, S., Duval, E.: A web-based approach to determine the origin of an artist. In: Proceedings of the 10th International Society for Music Information Retrieval Conference (ISMIR), Kobe (2009)
159. Grace, J., Gruhl, D., Haas, K., Nagarajan, M., Robson, C., Sahoo, N.: Artist ranking through analysis of on-line community comments. In: Proceedings of the 17th International World Wide Web Conference (WWW), Beijing (2008)
160. Grey, J.M., Gordon, J.W.: Perceptual effects of spectral modifications on musical timbres. J. Acoust. Soc. Am. **63**(5), 1493–1500 (1978)
161. Griffiths, T.L., Steyvers, M.: Finding scientific topics. Proc. Natl. Acad. Sci. **101**(Suppl. 1), 5228–5235 (2004)
162. Hall, M., Frank, E., Holmes, G., Pfahringer, B., Reutemann, P., Witten, I.H.: The WEKA data mining software: an update. SIGKDD Explor. Newsl. **11**, 10–18 (2009)
163. Hamasaki, M., Goto, M., Nakano, T.: Songrium: a music browsing assistance service with interactive visualization and exploration of a web of music. In: Proceedings of the 23rd International Conference on World Wide Web (WWW) Companion, Seoul (2014)

164. Hamel, P., Eck, D.: Learning features from music audio with deep belief networks. In: Proceedings of the 11th International Society for Music Information Retrieval Conference (ISMIR), Utrecht (2010)
165. Hart, P., Nilsson, N., Raphael, B.: A formal basis for the heuristic determination of minimum cost paths. IEEE Trans. Syst. Sci. Cybern. **4**(2), 100–107 (1968)
166. Hastie, T., Tibshirani, R., Friedman, J.: The Elements of Statistical Learning: Data Mining, Inference, and Prediction, 2nd edn. Springer, New York (2011)
167. Hauger, D., Schedl, M.: Exploring geospatial music listening patterns in microblog data. In: Nürnberger, A., Stober, S., Larsen, B., Detyniecki, M. (eds.) Adaptive Multimedia Retrieval: Semantics, Context, and Adaptation. Lecture Notes in Computer Science, vol. 8382. Springer, Heidelberg (2014)
168. Hauger, D., Schedl, M., Košir, A., Tkalčič, M.: The million musical tweets dataset: what can we learn from microblogs. In: Proceedings of the 14th International Society for Music Information Retrieval Conference (ISMIR), Curitiba (2013)
169. He, X., King, O., Ma, W.Y., Li, M., Zhang, H.J.: Learning a semantic space from user's relevance feedback for image retrieval. IEEE Trans. Circuits Syst. Video Technol. **13**(1), 39–48 (2003)
170. Heinrich, G.: Parameter estimation for text analysis. Tech. rep., University of Leipzig (2008)
171. Herlocker, J.L., Konstan, J.A., Borchers, A., Riedl, J.: An algorithmic framework for performing collaborative filtering. In: Proceedings of the 22nd Annual International ACM SIGIR Conference on Research and Development in Information Retrieval (SIGIR), Berkeley (1999)
172. Herlocker, J.L., Konstan, J.A., Terveen, L.G., Riedl, J.T.: Evaluating collaborative filtering recommender systems. ACM Trans. Inf. Syst. **22**(1), 5–53 (2004)
173. Herrera, P., Amatriain, X., Battle, E., Serra, X.: Instrument segmentation for music content description: a critical review of instrument classification techniques. In: Proceedings of the International Symposium on Music Information Retrieval (ISMIR), Plymouth (2000)
174. Hevner, K.: Expression in music: a discussion of experimental studies and theories. Psychol. Rev. **42** (1935)
175. Hidaka, I., Goto, M., Muraoka, Y.: An automatic jazz accompaniment system reacting to solo. In: Proceedings of the International Computer Music Conference (ICMC), Banff (1995)
176. Hinton, G., Deng, L., Yu, D., Dahl, G.E., Mohamed, A.R., Jaitly, N., Senior, A., Vanhoucke, V., Nguyen, P., Sainath, T.N., Kingsbury, B.: Deep neural networks for acoustic modeling in speech recognition. IEEE Signal Process. Mag. **29**(6), 82–97 (2012)
177. Hinton, G., Deng, L., Yu, D., Dahl, G.E., Mohamed, A.R., Jaitly, N., Senior, A., Vanhoucke, V., Nguyen, P., Sainath, T.N., Kingsbury, B.: Deep neural networks for acoustic modeling in speech recognition: the shared views of four research groups. IEEE Signal Process. Mag. **29**(6), 82–97 (2012)
178. Hirjee, H., Brown, D.G.: Automatic detection of internal and imperfect rhymes in rap lyrics. In: Proceedings of the 10th International Society for Music Information Retrieval Conference (ISMIR), Kobe (2009)
179. Hirjee, H., Brown, D.G.: Rhyme analyzer: an analysis tool for rap lyrics. In: Proceedings of the 11th International Society for Music Information Retrieval Conference (ISMIR), Utrecht (2010)
180. Hitchner, S., Murdoch, J., Tzanetakis, G.: Music browsing using a tabletop display. In: Proceedings of the 8th International Conference on Music Information Retrieval (ISMIR), Vienna (2007)
181. Hochenbaum, J., Vallis, O.: Bricktable: a musical tangible multi-touch interface. In: Proceedings of Berlin Open Conference (BOC), Berlin (2009)
182. Hoffman, M.D., Blei, D.M., Cook, P.R.: Easy as CBA: a simple probabilistic model for tagging music. In: Proceedings of the 10th International Society for Music Information Retrieval Conference (ISMIR), Kobe (2009)
183. Hofmann, T.: Probabilistic latent semantic analysis. In: Proceedings of the 15th Conference on Uncertainty in Artificial Intelligence (UAI), Stockholm (1999)

184. Hofmann, T.: Latent semantic models for collaborative filtering. ACM Trans. Inf. Syst. **22**(1), 89–115 (2004)
185. Homburg, H., Mierswa, I., Möller, B., Morik, K., Wurst, M.: A benchmark dataset for audio classification and clustering. In: Proceedings of the 6th International Conference on Music Information Retrieval (ISMIR), London (2005)
186. Hörschläger, F., Vogl, R., Böck, S., Knees, P.: Addressing tempo estimation octave errors in electronic music by incorporating style information extracted from Wikipedia. In: Proceedings of the 12th Sound and Music Conference (SMC), Maynooth (2015)
187. Hotelling, H.: Analysis of a complex of statistical variables into principal components. J. Educ. Psychol. **24**, 417–441, 498–520 (1933)
188. Hu, X., Downie, J.S.: Exploring mood metadata: relationships with genre, artist and usage metadata. In: Proceedings of the 8th International Conference on Music Information Retrieval (ISMIR), Vienna (2007)
189. Hu, X., Downie, J.S.: When lyrics outperform audio for music mood classification: a feature analysis. In: Proceedings of the 11th International Society for Music Information Retrieval Conference (ISMIR), Utrecht (2010)
190. Hu, X., Downie, J.S., Laurier, C., Bay, M., Ehmann, A.F.: The 2007 MIREX audio mood classification task: lessons learned. In: Proceedings of the 9th International Conference on Music Information Retrieval (ISMIR), Philadelphia (2008)
191. Hu, X., Downie, J.S., West, K., Ehmann, A.: Mining music reviews: promising preliminary results. In: Proceedings of the 6th International Conference on Music Information Retrieval (ISMIR), London (2005)
192. Hu, Y., Ogihara, M.: NextOne player: a music recommendation system based on user behavior. In: Proceedings of the 12th International Society for Music Information Retrieval Conference (ISMIR), Miami (2011)
193. Hu, Y., Koren, Y., Volinsky, C.: Collaborative filtering for implicit feedback datasets. In: Proceedings of the 8th IEEE International Conference on Data Mining (ICDM), Pisa (2008)
194. Huber, S., Schedl, M., Knees, P.: nepDroid: an intelligent mobile music player. In: Proceedings of the 2nd ACM International Conference on Multimedia Retrieval (ICMR), Hong Kong (2012)
195. Huq, A., Bello, J., Rowe, R.: Automated music emotion recognition: a systematic evaluation. J. N. Music Res. **39**(3), 227–244 (2010)
196. Inskip, C., Wiering, F.: In their own words: using text analysis to identify musicologists' attitudes towards technology. In: Proceedings of the 16th International Society for Music Information Retrieval Conference (ISMIR), Málaga (2015)
197. Izmirli, Ö.: Tonal similarity from audio using a template based attractor model. In: Proceedings of the 6th International Conference on Music Information Retrieval (ISMIR), London (2005)
198. Jacobson, K., Fields, B., Sandler, M.: Using audio analysis and network structure to identify communities in on-line social networks of artists. In: Proceedings of the 9th International Conference on Music Information Retrieval (ISMIR), Philadelphia (2008)
199. Joachims, T.: A support vector method for multivariate performance measures. In: Proceedings of the 22nd International Conference on Machine Learning (ICML), Bonn (2005)
200. Johnson, C.C.: Logistic matrix factorization for implicit feedback data. In: NIPS 2014 Workshop on Distributed Machine Learning and Matrix Computations, Montréal (2014)
201. Jordà, S., Kaltenbrunner, M., Geiger, G., Bencina, R.: The reacTable. In: Proceedings of the International Computer Music Conference (ICMC), Barcelona (2005)
202. Julià, C.F., Jordà, S.: SongExplorer: a tabletop application for exploring large collections of songs. In: Proceedings of the 10th International Society for Music Information Retrieval Conference (ISMIR), Kobe (2009)
203. Juslin, P.N.: What does music express? Basic emotions and beyond. Front. Psychol. **4**(596) (2013)
204. Juslin, P.N., Laukka, P.: Expression, perception, and induction of musical emotions: a review and a questionnaire study of everyday listening. J. N. Music Res. **33**(3), 217–238 (2004)

205. Kendall, M.: A new measure of rank correlation. Biometrika **30**, 81–93 (1938)
206. Kim, J.H., Tomasik, B., Turnbull, D.: Using artist similarity to propagate semantic information. In: Proceedings of the 10th International Society for Music Information Retrieval Conference (ISMIR), Kobe (2009)
207. Kim, J.Y., Belkin, N.J.: Categories of music description and search terms and phrases used by non-music experts. In: Proceedings of the 3rd International Conference on Music Information Retrieval (ISMIR), Paris (2002)
208. Kim, Y.E., Schmidt, E.M., Emelle, L.: MoodSwings: a collaborative game for music mood label collection. In: Proceedings of the 9th International Society for Music Information Retrieval Conference (ISMIR), Philadelphia (2008)
209. Kirke, A., Miranda, E.R. (eds.): Guide to Computing for Expressive Music Performance. Springer, London (2013)
210. Klapuri, A.: Sound onset detection by applying psychoacoustic knowledge. In: Proceedings of the IEEE International Conference on Acoustics, Speech and Signal Processing (ICASSP) (1999)
211. Klapuri, A.: Multipitch estimation and sound separation by the spectral smoothness principle. In: Proceedings of the IEEE International Conference on Acoustics, Speech, and Signal Processing (ICASSP), Salt Lake City (2001)
212. Klapuri, A.: Automatic music transcription as we know it today. J. N. Music Res. **33**(3), 269–282 (2004)
213. Klapuri, A., Virtanen, T., Heittola, T.: Sound source separation in monaural music signals using excitation-filter model and EM algorithm. In: Proceedings of the IEEE International Conference on Acoustics, Speech and Signal Processing (ICASSP), Dallas (2010)
214. Kleedorfer, F., Knees, P., Pohle, T.: Oh Oh Oh Whoah! Towards automatic topic detection in song lyrics. In: Proceedings of the 9th International Conference on Music Information Retrieval (ISMIR), Philadelphia (2008)
215. Kleinberg, J.M.: Authoritative sources in a hyperlinked environment. J. ACM **46**(5) (1999)
216. Knees, P.: Text-based description of music for indexing, retrieval, and browsing. Dissertation, Johannes Kepler University Linz (2010)
217. Knees, P., Schedl, M.: Towards semantic music information extraction from the web using rule patterns and supervised learning. In: Anglade, A., Celma, O., Fields, B., Lamere, P., McFee, B. (eds.) Proceedings of the 2nd Workshop on Music Recommendation and Discovery (WOMRAD), vol. 793. CEUR-WS, Chicago (2011)
218. Knees, P., Widmer, G.: Searching for music using natural language queries and relevance feedback. In: Boujemaa, N., Detyniecki, M., Nürnberger, A. (eds.) Adaptive Multimedia Retrieval: Retrieval, User, and Semantics. Lecture Notes in Computer Science, vol. 4918. Springer, Heidelberg (2008)
219. Knees, P., Pampalk, E., Widmer, G.: Artist classification with web-based data. In: Proceedings of the 5th International Symposium on Music Information Retrieval (ISMIR), Barcelona (2004)
220. Knees, P., Schedl, M., Widmer, G.: Multiple lyrics alignment: automatic retrieval of song lyrics. In: Proceedings of 6th International Conference on Music Information Retrieval (ISMIR), London (2005)
221. Knees, P., Pohle, T., Schedl, M., Widmer, G.: Combining audio-based similarity with web-based data to accelerate automatic music playlist generation. In: Proceedings of the 8th ACM SIGMM International Workshop on Multimedia Information Retrieval (MIR), Santa Barbara (2006)
222. Knees, P., Schedl, M., Pohle, T., Widmer, G.: An innovative three-dimensional user interface for exploring music collections enriched with meta-information from the web. In: Proceedings of the 14th ACM International Conference on Multimedia (MM), Santa Barbara (2006)
223. Knees, P., Pohle, T., Schedl, M., Widmer, G.: A music search engine built upon audio-based and web-based similarity measures. In: Proceedings of the 30th Annual International ACM SIGIR Conference on Research and Development in Information Retrieval (SIGIR), Amsterdam (2007)

224. Knees, P., Pohle, T., Schedl, M., Schnitzer, D., Seyerlehner, K.: A document-centered approach to a natural language music search engine. In: Proceedings of the 30th European Conference on Information Retrieval (ECIR), Glasgow (2008)
225. Knees, P., Schedl, M., Pohle, T.: A deeper look into web-based classification of music artists. In: Proceedings of 2nd Workshop on Learning the Semantics of Audio Signals (LSAS), Paris (2008)
226. Knees, P., Schedl, M., Pohle, T., Seyerlehner, K., Widmer, G.: Supervised and unsupervised web document filtering techniques to improve text-based music retrieval. In: Proceedings of the 11th International Society for Music Information Retrieval Conference (ISMIR), Utrecht (2010)
227. Knees, P., Schnitzer, D., Flexer, A.: Improving neighborhood-based collaborative filtering by reducing hubness. In: Proceedings of the 4th ACM International Conference on Multimedia Retrieval (ICMR), Glasgow (2014)
228. Knees, P., Andersen, K., Tkalčič, M.: "I'd like it to do the opposite": music-making between recommendation and obstruction. In: Ge, M., Ricci, F. (eds.) Proceedings of the 2nd International Workshop on Decision Making and Recommender Systems (DMRS), vol. 1533. CEUR-WS, Bolzano (2015)
229. Knopke, I.: AROOOGA: an audio search engine for the world wide web. In: Proceedings of the 2004 International Computer Music Conference (ICMC), Miami (2004)
230. Koduri, G.K., Miron, M., Serrà, J., Serra, X.: Computational approaches for the understanding of melody in carnatic music. In: Proceedings of the 12th International Society for Music Information Retrieval Conference (ISMIR), Miami (2011)
231. Koenigstein, N., Shavitt, Y.: Song ranking based on piracy in peer-to-peer networks. In: Proceedings of the 10th International Society for Music Information Retrieval Conference (ISMIR), Kobe (2009)
232. Koenigstein, N., Shavitt, Y., Tankel, T.: Spotting out emerging artists using geo-aware analysis of P2P query strings. In: Proceedings of the 14th ACM SIGKDD International Conference on Knowledge Discovery and Data Mining (KDD), Las Vegas (2008)
233. Kohonen, T.: Self-organizing formation of topologically correct feature maps. Biol. Cybern. **43**, 59–69 (1982)
234. Kohonen, T.: New developments of learning vector quantization and the self-organizing map. In: Symposium on Neural Networks; Alliances and Perspectives in Senri (SYNAPSE), Osaka (1992)
235. Kohonen, T.: Self-organizing Maps, vol. 30, 3rd edn. Springer, New York (2001)
236. Konstan, J.A., Miller, B.N., Maltz, D., Herlocker, J.L., Gordon, L.R., Riedl, J.: GroupLens: applying collaborative filtering to usenet news. Commun. ACM **40**(3), 77–87 (1997)
237. Koren, Y.: Factorization meets the neighborhood: a multifaceted collaborative filtering model. In: Proceedings of the 14th ACM SIGKDD International Conference on Knowledge Discovery and Data Mining (KDD), Las Vegas (2008)
238. Koren, Y.: Collaborative filtering with temporal dynamics. In: Proceedings of the 15th ACM SIGKDD International Conference on Knowledge Discovery and Data Mining (KDD), Paris (2009)
239. Koren, Y., Bell, R., Volinsky, C.: Matrix factorization techniques for recommender systems. IEEE Comput. **42**(8), 30–37 (2009)
240. Kornstädt, A.: Themefinder: a web-based melodic search tool. Comput. Musicol. **11**, 231–236 (1998)
241. Korst, J., Geleijnse, G.: Efficient lyrics retrieval and alignment. In: Verhaegh, W., Aarts, E., ten Kate, W., Korst, J., Pauws, S. (eds.) Proceedings of the 3rd Philips Symposium on Intelligent Algorithms (SOIA), Eindhoven (2006)
242. Korzeniowski, F., Böck, S., Widmer, G.: Probabilistic extraction of beat positions from a beat activation function. In: Proceedings of the 15th International Society for Music Information Retrieval Conference (ISMIR), Taipei (2014)
243. Krenmair, A.: Musikspezifische Informationsextraktion aus Webdokumenten. Master's thesis, Johannes Kepler University Linz (2013)

244. Krizhevsky, A., Sutskever, I., Hinton, G.: ImageNet classification with deep convolutional neural networks. In: Pereira, F., Burges, C., Bottou, L., Weinberger, K. (eds.) Advances in Neural Information Processing Systems 25 (NIPS). Curran Associates, Inc., New York (2012)
245. Krumhansl, C.L.: Plink: "Thin Slices" of music. Music Percept. Interdiscip. J. **27**(5), 337–354 (2010)
246. Kulis, B.: Metric learning: a survey. Found. Trends Mach. Learn. **5**(4), 287–364 (2012)
247. Kurth, F., Müller, M., Fremerey, C., Chang, Y.h., Clausen, M.: Automated synchronization of scanned sheet music with audio recordings. In: Proceedings of the 8th International Conference on Music Information Retrieval (ISMIR), Vienna (2007)
248. Lamere, P.: Social tagging and music information retrieval. J. N. Music Res. **37**(2), 101–114 (2008). Special Issue: From Genres to Tags – Music Information Retrieval in the Age of Social Tagging
249. Lamere, P., Celma, O.: Music recommendation. In: ISMIR 2007 Tutorial, Vienna (2007) http://www.dtic.upf.edu/~ocelma/MusicRecommendationTutorial-ISMIR2007
250. Lamere, P., Eck, D.: Using 3D visualizations to explore and discover music. In: Proceedings of the 8th International Conference on Music Information Retrieval (ISMIR), Vienna (2007)
251. Lathauwer, L.D., Moor, B.D., Vandewalle, J.: A multilinear singular value decomposition. SIAM J. Matrix Anal. Appl. **21**(4), 1253–1278 (2000)
252. Laurier, C., Grivolla, J., Herrera, P.: Multimodal music mood classification using audio and lyrics. In: Proceedings of the 7th International Conference on Machine Learning and Applications (ICMLA), San Diego (2008)
253. Laurier, C., Sordo, M., Serrà, J., Herrera, P.: Music mood representations from social tags. In: Proceedings of the 10th International Society for Music Information Retrieval Conference (ISMIR), Kobe (2009)
254. Lavner, Y., Ruinskiy, D.: A decision-tree-based algorithm for speech/music classification and segmentation. EURASIP J. Audio Speech Music Process. **2009** (2009)
255. Law, E., von Ahn, L.: Input-agreement: a new mechanism for collecting data using human computation games. In: Proceedings of the 27th Annual ACM SIGCHI Conference on Human Factors in Computing Systems (CHI), Boston (2009)
256. Law, E., von Ahn, L., Dannenberg, R., Crawford, M.: Tagatune: a game for music and sound annotation. In: Proceedings of the 8th International Conference on Music Information Retrieval (ISMIR), Vienna (2007)
257. Lee, D.D., Seung, H.S.: Learning the parts of objects by non-negative matrix factorization. Nature **401**(6755), 788–791 (1999)
258. Lee, J.H., Downie, J.S.: Survey of music information needs, uses, and seeking behaviours: preliminary findings. In: Proceedings of the 5th International Conference on Music Information Retrieval (ISMIR), Barcelona (2004)
259. Lee, J.H., Price, R.: Understanding users of commercial music services through personas: design implications. In: Proceedings of the 16th International Society for Music Information Retrieval Conference (ISMIR), Málaga (2015)
260. Lee, J.S., Lee, J.C.: Context awareness by case-based reasoning in a music recommendation system. In: Ichikawa, H., Cho, W.D., Satoh, I., Youn, H. (eds.) Ubiquitous Computing Systems. Lecture Notes in Computer Science, vol. 4836. Springer, Heidelberg (2007)
261. Leitich, S., Topf, M.: Globe of music - music library visualization using GeoSOM. In: Proceedings of the 8th International Conference on Music Information Retrieval (ISMIR), Vienna (2007)
262. Lerch, A.: An Introduction to Audio Content Analysis: Applications in Signal Processing and Music Informatics. Wiley–IEEE Press, Hoboken (2012)
263. Levy, M., Sandler, M.: Learning latent semantic models for music from social tags. J. N. Music Res. **37**(2), 137–150 (2008)
264. Lew, M.S., Sebe, N., Djeraba, C., Jain, R.: Content-based multimedia information retrieval: state of the art and challenges. ACM Trans. Multimedia Comput. Commun. Appl. **2**(1), 1–19 (2006)

265. Li, Q., Myaeng, S.H., Kim, B.M.: A probabilistic music recommender considering user opinions and audio features. Inf. Process. Manag. **43**(2), 473–487 (2007)
266. Li, T., Ogihara, M., Li, Q.: A comparative study on content-based music genre classification. In: Proceedings of the 26th Annual International ACM SIGIR Conference on Research and Development in Informaion Retrieval (SIGIR), Toronto (2003)
267. Li, T., Ogihara, M., Tzanetakis, G. (eds.): Music Data Mining. CRC Press/Chapman Hall, Boca Raton (2011)
268. Li, T., Tzanetakis, G.: Factors in automatic musical genre classification of audio signals. In: Proceedings of the IEEE Workshop on Applications of Signal Processing to Audio and Acoustics, New Paltz (2003)
269. Lidy, T., Mayer, R., Rauber, A., Ponce de León, P.J., Pertusa, A., Iñesta, J.M.: A cartesian ensemble of feature subspace classifiers for music categorization. In: Proceedings of the 11th International Society for Music Information Retrieval Conference (ISMIR), Utrecht (2010)
270. Lidy, T., Rauber, A.: Evaluation of feature extractors and psycho-acoustic transformations for music genre classification. In: Proceedings of the 6th International Conference on Music Information Retrieval (ISMIR), London (2005)
271. Liem, C.C., Müller, M., Eck, D., Tzanetakis, G., Hanjalic, A.: The need for music information retrieval with user-centered and multimodal strategies. In: Proceedings of the International ACM Workshop on Music Information Retrieval with User-centered and Multimodal Strategies (MIRUM), Scottsdale (2011)
272. Lillie, A.: MusicBox: navigating the space of your music. Master's thesis, Massachusetts Institute of Technology, Cambridge, MA (2008)
273. Lim, D., McFee, B., Lanckriet, G.R.: Robust structural metric learning. In: Proceedings of the 30th International Conference on Machine Learning (ICML), Atlanta (2013)
274. Linden, G., Smith, B., York, J.: Amazon.com recommendations: item-to-item collaborative filtering. IEEE Internet Comput. **4**(1), 76–80 (2003)
275. Lindosland: Equal-loudness contours. Wikimedia Commons. https://commons.wikimedia.org/wiki/File:Lindos1.svg (2005). Accessed 26 Jan 2016
276. Lippens, S., Martens, J.P., De Mulder, T., Tzanetakis, G.: A comparison of human and automatic musical genre classification. In: Proceedings of the IEEE International Conference on Acoustics, Speech and Signal Processing (ICASSP) (2004)
277. Liu, B.: Web Data Mining – Exploring Hyperlinks, Contents and Usage Data. Springer, Berlin/Heidelberg (2007)
278. Liu, B., Zhang, L.: A survey of opinion mining and sentiment analysis. In: Aggarwal, C.C., Zhai, C. (eds.) Mining Text Data, pp. 415–463. Springer, New York (2012)
279. Liu, H., Hu, J., Rauterberg, M.: Music playlist recommendation based on user heartbeat and music preference. In: Proceedings of the 4th International Conference on Computer Technology and Development (ICCTD), Bangkok (2009)
280. Livshin, A., Rodet, X.: Musical instrument identification in continuous recordings. In: Proceedings of the 7th International Conference on Digital Audio Effects (DAFx), Naples (2004)
281. Logan, B.: Mel frequency cepstral coefficients for music modeling. In: Proceedings of the International Symposium on Music Information Retrieval (ISMIR), Plymouth (2000)
282. Logan, B.: Content-based playlist generation: exploratory experiments. In: Proceedings of the 3rd International Symposium on Music Information Retrieval (ISMIR), Paris (2002)
283. Logan, B., Salomon, A.: A music similarity function based on signal analysis. In: Proceedings of the IEEE International Conference on Multimedia and Expo (ICME), Tokyo (2001)
284. Logan, B., Ellis, D.P., Berenzweig, A.: Toward evaluation techniques for music similarity. In: Proceedings of the 26th Annual International ACM SIGIR Conference on Research and Development in Information Retrieval (SIGIR): Workshop on the Evaluation of Music Information Retrieval Systems, Toronto (2003)
285. Logan, B., Kositsky, A., Moreno, P.: Semantic analysis of song lyrics. In: Proceedings of the IEEE International Conference on Multimedia and Expo (ICME), Taipei (2004)

286. Lu, C.C., Tseng, V.S.: A novel method for personalized music recommendation. Expert Syst. Appl. **36**(6), 10035–10044 (2009)
287. Lübbers, D., Jarke, M.: Adaptive multimodal exploration of music collections. In: Proceedings of the 10th International Society for Music Information Retrieval Conference (ISMIR), Kobe (2009)
288. MacQueen, J.: Some methods for classification and analysis of multivariate observations. In: Le Cam, L.M., Neyman, J. (eds.) Proceedings of the 5th Berkeley Symposium on Mathematical Statistics and Probability, vol. 1. Statistics, Berkeley (1967)
289. Maddage, N.C., Li, H., Kankanhalli, M.S.: Music structure based vector space retrieval. In: Proceedings of the 29th Annual International ACM SIGIR Conference on Research and Development in Information Retrieval (SIGIR), Seattle (2006)
290. Madden, J., Feth, L.: Temporal resolution in normal-hearing and hearing-impaired listeners using frequency-modulated stimuli. J. Speech Hear. Res. **35**, 436–442 (1992)
291. Mahedero, J.P.G., Martínez, A., Cano, P., Koppenberger, M., Gouyon, F.: Natural language processing of lyrics. In: Proceedings of the 13th ACM International Conference on Multimedia (MM), Singapore (2005)
292. Mandel, M.I.: SVM-based audio classification, tagging, and similarity submission. In: Extended Abstract to the Annual Music Information Retrieval Evaluation eXchange (MIREX) (2010)
293. Mandel, M.I., Ellis, D.P.: Song-level features and support vector machines for music classification. In: Proceedings of the 6th International Conference on Music Information Retrieval (ISMIR), London (2005)
294. Mandel, M.I., Ellis, D.P.: A web-based game for collecting music metadata. In: Proceedings of the 8th International Conference on Music Information Retrieval (ISMIR), Vienna (2007)
295. Mandel, M.I., Ellis, D.P.W.: A web-based game for collecting music metadata. J. N. Music Res. **37**(2), 151–165 (2008)
296. Mandel, M.I., Pascanu, R., Eck, D., Bengio, Y., Aiello, L.M., Schifanella, R., Menczer, F.: Contextual tag inference. ACM Trans. Multimedia Comput. Commun. Appl. **7S**(1), 32:1–32:18 (2011)
297. Manning, C.D., Raghavan, P., Schütze, H.: Introduction to Information Retrieval. Cambridge University Press, New York (2008)
298. Markov, Z., Larose, D.T.: Data Mining the Web: Uncovering Patterns in Web Content, Structure, and Usage. Wiley, Hoboken (2007)
299. Marques, G., Langlois, T., Gouyon, F., Lopes, M., Sordo, M.: Short-term feature space and music genre classification. J. N. Music Res. **40**(2), 127–137 (2011)
300. Marsden, A.: Music similarity. In: Workshop on Music Similarity: Concepts, Cognition and Computation. Lorentz Center, Leiden. http://www.lorentzcenter.nl/lc/web/2015/669/presentations/Marsden.pptx (2015). Accessed 26 Jan 2016
301. Mayer, R., Lidy, T., Rauber, A.: The map of Mozart. In: Proceedings of the 7th International Conference on Music Information Retrieval (ISMIR), Victoria (2006)
302. Mayer, R., Neumayer, R., Rauber, A.: Rhyme and style features for musical genre classification by song lyrics. In: Proceedings of the 9th International Conference on Music Information Retrieval (ISMIR), Philadelphia (2008)
303. McFee, B., Barrington, L., Lanckriet, G.: Learning content similarity for music recommendation. IEEE Trans. Audio Speech Lang. Process. **20**(8), 2207–2218 (2012)
304. McFee, B., Bertin-Mahieux, T., Ellis, D.P., Lanckriet, G.R.: The million song dataset challenge. In: Proceedings of the 21st International Conference on World Wide Web (WWW) Companion, Lyon (2012)
305. McFee, B., Lanckriet, G.: Heterogeneous embedding for subjective artist similarity. In: Proceedings of the 10th International Society for Music Information Retrieval Conference (ISMIR), Kobe (2009)
306. McFee, B., Lanckriet, G.R.: Metric learning to rank. In: Proceedings of the 27th International Conference on Machine Learning (ICML), Haifa (2010)

307. McFee, B., Lanckriet, G.R.: Learning multi-modal similarity. J. Mach. Learn. Res. **12**, 491–523 (2011)
308. McFee, B., Lanckriet, G.: Hypergraph models of playlist dialects. In: Proceedings of the 13th International Society for Music Information Retrieval Conference (ISMIR), Porto (2012)
309. McKay, C.: Automatic music classification with jMIR. Ph.D. thesis, McGill University, Montréal (2010)
310. McKay, C., Fujinaga, I.: Musical genre classification: Is it worth pursuing and how can it be improved? In: Proceedings of the 7th International Conference on Music Information Retrieval (ISMIR), Victoria (2006)
311. McKay, C., Fujinaga, I.: Combining features extracted from audio, symbolic and cultural sources. In: Proceedings of the 9th International Conference on Music Information Retrieval (ISMIR), Philadelphia (2008)
312. McKinney, M.F., Breebaart, J.: Features for audio and music classification. In: Proceedings of the 4th International Conference on Music Information Retrieval (ISMIR), Baltimore (2003)
313. Mehrabian, A.: Basic Dimensions for a General Psychological Theory: Implications for Personality, Social, Environmental, and Developmental Studies. Oelgeschlager, Gunn & Hain, Cambridge (1980)
314. Meredith, D. (ed.): Computational Music Analysis. Springer, Berlin (2015)
315. Meredith, D., Lemström, K., Wiggins, G.A.: Algorithms for discovering repeated patterns in multidimensional representations of polyphonic music. J. N. Music Res. **31**(4), 321–345 (2002)
316. Mesnage, C.S., Rafiq, A., Dixon, S., Brixtel, R.P.: Music discovery with social networks. In: Anglade, A., Celma, O., Fields, B., Lamere, P., McFee, B. (eds.) Proceedings of the 2nd Workshop on Music Recommendation and Discovery (WOMRAD), vol. 793. CEUR-WS, Chicago (2011)
317. Miotto, R., Barrington, L., Lanckriet, G.: Improving auto-tagging by modeling semantic co-occurrences. In: Proceedings of the 11th International Society for Music Information Retrieval Conference (ISMIR), Utrecht (2010)
318. Mitchell, T.M.: Machine Learning. McGraw-Hill, New York (1997)
319. Mobasher, B., Burke, R., Bhaumik, R., Williams, C.: Toward trustworthy recommender systems: an analysis of attack models and algorithm robustness. ACM Trans. Internet Technol. **7**(4) (2007)
320. Moens, B., van Noorden, L., Leman, M.: D-jogger: syncing music with walking. In: Proceedings of the 7th Sound and Music Computing Conference (SMC), Barcelona (2010)
321. Moh, Y., Orbanz, P., Buhmann, J.M.: Music preference learning with partial information. In: Proceedings of the IEEE International Conference on Acoustics, Speech and Signal Processing (ICASSP), Las Vegas (2008)
322. Moore, A.F.: Categorical conventions in music discourse: style and genre. Music Lett. **82**, 432–442 (2001)
323. Mörchen, F., Ultsch, A., Nöcker, M., Stamm, C.: Databionic visualization of music collections according to perceptual distance. In: Proceedings of the 6th International Conference on Music Information Retrieval (ISMIR), London (2005)
324. Moreno, P.J., Ho, P.P., Vasconcelos, N.: A Kullback-Leibler divergence based kernel for SVM classification in multimedia applications. In: Thrun, S., Saul, L., Schölkopf, B. (eds.) Advances in Neural Information Processing Systems 16 (NIPS). MIT Press, Cambridge (2004)
325. Moshfeghi, Y., Jose, J.M.: Role of emotion in information retrieval for entertainment (position paper). In: Elsweiler, D., Wilson, M.L., Harvey, M. (eds.) Proceedings of the "Searching 4 Fun!" Workshop (S4F), vol. 836. CEUR-WS, Barcelona (2012)
326. Müller, M.: Fundamentals of Music Processing: Audio, Analysis, Algorithms, Applications. Springer, Cham (2015)
327. Müller, M., Goto, M., Schedl, M.: Frontmatter, table of contents, preface, list of authors. In: Müller, M., Goto, M., Schedl, M. (eds.) Multimodal Music Processing. Dagstuhl Follow-Ups, vol. 3. Schloss Dagstuhl–Leibniz-Zentrum für Informatik, Wadern (2012)

328. Nanopoulos, A., Radovanović, M., Ivanović, M.: How does high dimensionality affect collaborative filtering? In: Proceedings of the 3rd ACM Conference on Recommender Systems (RecSys), New York (2009)
329. Nanopoulos, A., Rafailidis, D., Symeonidis, P., Manolopoulos, Y.: Musicbox: personalized music recommendation based on cubic analysis of social tags. IEEE Trans. Audio Speech Lang. Process. **18**(2), 407–412 (2010)
330. Needleman, S.B., Wunsch, C.D.: A general method applicable to the search for similarities in the amino acid sequence of two proteins. J. Mol. Biol. **48**(3), 443–453 (1970)
331. Nelder, J.A., Mead, R.: A simplex method for function minimization. Comput. J. **7**(4), 308–313 (1965)
332. Neumayer, R., Dittenbach, M., Rauber, A.: PlaySOM and PocketSOMPlayer, alternative interfaces to large music collections. In: Proceedings of the 6th International Conference on Music Information Retrieval (ISMIR), London (2005)
333. Niedermayer, B.: Accurate audio-to-score alignment – data acquisition in the context of computational musicology. Dissertation, Johannes Kepler University Linz (2012)
334. Novello, A., McKinney, M.F., Kohlrausch, A.: Perceptual evaluation of music similarity. In: Proceedings of the 7th International Conference on Music Information Retrieval (ISMIR), Victoria (2006)
335. Ong, B.S.: Structural analysis and segmentation of music signals. Ph.D. thesis, Universitat Pompeu Fabra, Barcelona (2006)
336. Oramas, S., Sordo, M., Espinosa-Anke, L., Serra, X.: A semantic-based approach for artist similarity. In: Proceedings of the 16th International Society for Music Information Retrieval Conference (ISMIR), Málaga (2015)
337. Pachet, F.: The continuator: musical interaction with style. J. N. Music Res. **32**(3), 333–341 (2003)
338. Pachet, F., Cazaly, D.: A taxonomy of musical genre. In: Proceedings of Content-Based Multimedia Information Access (RIAO) Conference, Paris (2000)
339. Pachet, F., Roy, P.: Hit song science is not yet a science. In: Proceedings of the 9th International Conference on Music Information Retrieval (ISMIR), Philadelphia (2008)
340. Pachet, F., Westerman, G., Laigre, D.: Musical data mining for electronic music distribution. In: Proceedings of the International Conference on Web Delivering of Music (WEDELMUSIC), Florence (2001)
341. Page, L., Brin, S., Motwani, R., Winograd, T.: The PageRank Citation Ranking: Bringing Order to the Web. Tech. Rep. 1999-66, Stanford InfoLab (1999)
342. Pampalk, E.: Islands of music: analysis, organization, and visualization of music archives. Master's thesis, Vienna University of Technology (2001)
343. Pampalk, E.: Computational models of music similarity and their application to music information retrieval. Dissertation, Vienna University of Technology (2006)
344. Pampalk, E., Goto, M.: MusicRainbow: a new user interface to discover artists using audio-based similarity and web-based labeling. In: Proceedings of the 7th International Conference on Music Information Retrieval (ISMIR), Victoria (2006)
345. Pampalk, E., Goto, M.: MusicSun: a new approach to artist recommendation. In: Proceedings of the 8th International Conference on Music Information Retrieval (ISMIR), Vienna (2007)
346. Pampalk, E., Rauber, A., Merkl, D.: Content-based organization and visualization of music archives. In: Proceedings of the 10th ACM International Conference on Multimedia (MM), Juan les Pins (2002)
347. Pampalk, E., Rauber, A., Merkl, D.: Using smoothed data histograms for cluster visualization in self-organizing maps. In: Proceedings of the International Conference on Artificial Neural Networks (ICANN), Madrid (2002)
348. Pampalk, E., Dixon, S., Widmer, G.: Exploring music collections by browsing different views. Comput. Music J. **28**(3) (2004)
349. Pampalk, E., Flexer, A., Widmer, G.: Hierarchical organization and description of music collections at the artist level. In: Proceedings of the 9th European Conference on Research and Advanced Technology for Digital Libraries (ECDL), Vienna (2005)

350. Pampalk, E., Pohle, T., Widmer, G.: Dynamic playlist generation based on skipping behavior. In: Proceedings of the 6th International Conference on Music Information Retrieval (ISMIR), London (2005)
351. Pant, G., Srinivasan, P.: Link contexts in classifier-guided topical crawlers. IEEE Trans. Knowl. Data Eng. **18**(1), 107–122 (2006)
352. Park, H.S., Yoo, J.O., Cho, S.B.: A context-aware music recommendation system using fuzzy Bayesian networks with utility theory. In: Proceedings of the 3rd International Conference on Fuzzy Systems and Knowledge Discovery (FSKD), Xi'an (2006)
353. Parsons, D.: The Directory of Tunes and Musical Themes. Brown, S., Cambridge (1975)
354. Paterek, A.: Improving regularized singular value decomposition for collaborative filtering. In: Proceedings of the KDD Cup Workshop at 13th ACM SIGKDD International Conference on Knowledge Discovery and Data Mining (KDD), San Jose (2007)
355. Paulus, J., Müller, M., Klapuri, A.: State of the art report: audio-based music structure analysis. In: Proceedings of the 11th International Society for Music Information Retrieval Conference (ISMIR), Utrecht (2010)
356. Peeters, G., Burthe, A.L., Rodet, X.: Toward automatic music audio summary generation from signal analysis. In: Proceedings of the 3rd International Conference on Music Information Retrieval (ISMIR), Paris (2002)
357. Penny, W.: Kullback-Liebler Divergences of Normal, Gamma, Dirichlet and Wishart Densities. Tech. rep., Wellcome Department of Cognitive Neurology, University College London, London (2001)
358. Picard, R.W.: Affective Computing. MIT Press, Cambridge (2000)
359. Piszczalski, M., Galler, B.A.: Automatic music transcription. Comput. Music J. **1**(4), 24–31 (1977)
360. Plack, C.J.: The Sense of Hearing. Lawrence Erlbaum Associates, Mahwah (2005)
361. Plumbley, M.D., Abdallah, S.A., Bello, J.P., Davies, M.E., Monti, G., Sandler, M.B.: Automatic music transcription and audio source separation. Cybern. Syst. **33**(6), 603–627 (2002)
362. Plutchik, R.: The nature of emotions. Am. Sci. **89**(4), 344–350 (2001)
363. Pohle, T., Knees, P., Schedl, M., Pampalk, E., Widmer, G.: "Reinventing the wheel": a novel approach to music player interfaces. IEEE Trans. Multimedia **9**, 567–575 (2007)
364. Pohle, T., Knees, P., Schedl, M., Widmer, G.: Automatically adapting the structure of audio similarity spaces. In: Proceedings of Workshop on Learning the Semantics of Audio Signals (LSAS), Athens (2006)
365. Pohle, T., Knees, P., Schedl, M., Widmer, G.: Building an interactive next-generation artist recommender based on automatically derived high-level concepts. In: Proceedings of the 5th International Workshop on Content-Based Multimedia Indexing (CBMI), Bordeaux (2007)
366. Pohle, T., Schnitzer, D., Schedl, M., Knees, P., Widmer, G.: On rhythm and general music similarity. In: Proceedings of the 10th International Society for Music Information Retrieval Conference (ISMIR), Kobe (2009)
367. Poli, G.D.: Methodologies for expressiveness modelling of and for music performance. J. N. Music Res. **33**(3), 189–202 (2004)
368. Ponce de León, P.J., Iñesta, J.M.: Musical style classification from symbolic data: a two-styles case study. In: Wiil, U.K. (ed.) Computer Music Modeling and Retrieval. Lecture Notes in Computer Science, vol. 2771. Springer, Heidelberg (2004)
369. Ponceleón, D., Slaney, M.: Multimedia information retrieval. In: Baeza-Yates, R., Ribeiro-Neto, B. (eds.) Modern Information Retrieval – The Concepts and Technology Behind Search, Chap. 14, 2nd edn., pp. 587–639. Addison-Wesley, Pearson (2011)
370. Prockup, M., Ehmann, A.F., Gouyon, F., Schmidt, E., Celma, Ò., Kim, Y.E.: Modeling genre with the music genome project: comparing human-labeled attributes and audio features. In: Proceedings of the 16th International Society for Music Information Retrieval Conference (ISMIR), Málaga (2015)
371. Quinlan, J.R.: C4.5: Programs for Machine Learning. Morgan Kaufmann, San Francisco (1993)

372. Radovanović, M., Nanopoulos, A., Ivanović, M.: Hubs in space: popular nearest neighbors in high-dimensional data. J. Mach. Learn. Res. **11**, 2487–2531 (2010)
373. Raimond, Y.: A distributed music information system. Ph.D. thesis, Queen Mary University of London (2008)
374. Raimond, Y., Abdallah, S., Sandler, M., Giasson, F.: The music ontology. In: Proceedings of the 8th International Conference on Music Information Retrieval (ISMIR), Vienna (2007)
375. Raimond, Y., Sutton, C., Sandler, M.: Interlinking music-related data on the web. IEEE MultiMedia **16**(2), 52–63 (2009)
376. Raphael, C.: Synthesizing musical accompaniments with Bayesian belief networks. J. N. Music Res. **30**(1), 59–67 (2001)
377. Raphael, C.: Automatic transcription of piano music. In: Proceedings of 3rd International Conference on Music Information Retrieval (ISMIR), Paris (2002)
378. Raphael, C.: Aligning music audio with symbolic scores using a hybrid graphical model. Mach. Learn. **65**(2–3), 389–409 (2006)
379. Ras, Z.W., Wieczorkowska, A.A. (eds.): Advances in Music Information Retrieval. Springer, Heidelberg (2010)
380. Rauber, A., Frühwirth, M.: Automatically analyzing and organizing music archives. In: Proceedings of the 5th European Conference on Research and Advanced Technology for Digital Libraries (ECDL), Darmstadt (2001)
381. Rebelo, A., Fujinaga, I., Paszkiewicz, F., Marcal, A.R., Guedes, C., Cardoso, J.S.: Optical music recognition: state-of-the-art and open issues. Int. J. Multimedia Inf. Retr. **1**(3), 173–190 (2012)
382. Reed, J., Lee, C.: Preference music ratings prediction using tokenization and minimum classification error training. IEEE Trans. Audio Speech Lang. Process. **19**(8), 2294–2303 (2011)
383. Ricci, F., Rokach, L., Shapira, B., Kantor, P.B. (eds.): Recommender Systems Handbook, 2nd edn. Springer, New York (2015)
384. Ripeanu, M.: Peer-to-peer architecture case study: gnutella network. In: Proceedings of the IEEE International Conference on Peer-to-Peer Computing (P2P), Linköping (2001)
385. Rocchio, J.J.: Relevance feedback in information retrieval. In: Salton, G. (ed.) The SMART Retrieval System - Experiments in Automatic Document Processing, pp. 313–323. Prentice-Hall, Englewood Cliffs (1971)
386. Rubner, Y., Tomasi, C., Guibas, L.J.: A metric for distributions with applications to image databases. In: Proceedings of the 6th IEEE International Conference on Computer Vision (ICCV), Bombay (1998)
387. Russell, J.A.: A circumplex model of affect. J. Pers. Soc. Psychol. **39**(6), 1161–1178 (1980)
388. Russell, S.J., Norvig, P.: Artificial Intelligence: A Modern Approach, 2nd edn. Prentice Hall, Englewood Cliffs (2003)
389. Sadjadi, S.O., Hansen, J.H.L.: Unsupervised speech activity detection using voicing measures and perceptual spectral flux. IEEE Signal Process. Lett. **20**(3), 197–200 (2013)
390. Sako, S., Yamamoto, R., Kitamura, T.: Ryry: a real-time score-following automatic accompaniment playback system capable of real performances with errors, repeats and jumps. In: Ślęzak, D., Schaefer, G., Vuong, S.T., Kim, Y.S. (eds.) Active Media Technology. Lecture Notes in Computer Science, vol. 8610. Springer, Heidelberg (2014)
391. Salamon, J., Serrà, J., Gómez, E.: Tonal representations for music retrieval: from version identification to query-by-humming. Int. J. Multimedia Inf. Retr. **2**(1), 45–58 (2013)
392. Salton, G., Buckley, C.: Term-weighting approaches in automatic text retrieval. Inf. Process. Manag. **24**(5), 513–523 (1988)
393. Sammon, J.W.: A nonlinear mapping for data structure analysis. IEEE Trans. Comput. **18**, 401–409 (1969)
394. Sapp, C.: Hybrid numeric/rank similarity metrics for musical performance analysis. In: Proceedings of the 9th International Conference on Music Information Retrieval (ISMIR), Philadelphia (2008)

395. Sapp, C.S.: Computational methods for the analysis of musical structure. Ph.D. thesis, Stanford University (2011)
396. Sarwar, B., Karypis, G., Konstan, J., Reidl, J.: Item-based collaborative filtering recommendation algorithms. In: Proceedings of 10th International Conference on World Wide Web (WWW), Hong Kong (2001)
397. Sarwar, B., Karypis, G., Konstan, J., Riedl, J.: Incremental singular value decomposition algorithms for highly scalable recommender systems. In: Proceedings of the 5th International Conference on Computer and Information Technology (ICCIT), Dhaka (2002)
398. Saunders, C., Hardoon, D.R., Shawe-Taylor, J., Widmer, G.: Using string kernels to identify famous performers from their playing style. In: Proceedings of the 15th European Conference on Machine Learning (ECML), Pisa (2004)
399. Scaringella, N., Zoia, G., Mlynek, D.: Automatic genre classification of music content: a survey. IEEE Signal Process. Mag. **23**(2), 133–141 (2006)
400. Schedl, M.: An explorative, hierarchical user interface to structured music repositories. Master's thesis, Vienna University of Technology, Wien (2003)
401. Schedl, M.: Automatically extracting, analyzing, and visualizing information on music artists from the world wide web. Dissertation, Johannes Kepler University Linz (2008)
402. Schedl, M.: Leveraging microblogs for spatiotemporal music information retrieval. In: Proceedings of the 35th European Conference on Information Retrieval (ECIR), Moscow (2013)
403. Schedl, M., Pohle, T.: Enlightening the sun: a user interface to explore music artists via multimedia content. Multimedia Tools Appl. **49**(1), 101–118 (2010)
404. Schedl, M., Tkalčič, M.: Genre-based analysis of social media data on music listening behavior. In: Proceedings of the ACM International Workshop on Internet-Scale Multimedia Management (ISMM), Orlando (2014)
405. Schedl, M., Widmer, G.: Automatically detecting members and instrumentation of music bands via web content mining. In: Boujemaa, N., Detyniecki, M., Nürnberger, A. (eds.) Adaptive Multimedia Retrieval: Retrieval, User, and Semantics. Lecture Notes in Computer Science, vol. 4918. Springer, Heidelberg (2008)
406. Schedl, M., Zhou, F.: Fusing web and audio predictors to localize the origin of music pieces for geospatial retrieval. In: Proceedings of the 38th European Conference on Information Retrieval (ECIR), Padua (2016)
407. Schedl, M., Knees, P., Widmer, G.: A web-based approach to assessing artist similarity using co-occurrences. In: Proceedings of the 4th International Workshop on Content-Based Multimedia Indexing (CBMI), Riga (2005)
408. Schedl, M., Knees, P., Widmer, G.: Discovering and visualizing prototypical artists by web-based co-occurrence analysis. In: Proceedings of the 6th International Conference on Music Information Retrieval (ISMIR), London (2005)
409. Schedl, M., Knees, P., Widmer, G.: Improving prototypical artist detection by penalizing exorbitant popularity. In: Proceedings of the 3rd International Symposium on Computer Music Modeling and Retrieval (CMMR), Pisa (2005)
410. Schedl, M., Knees, P., Pohle, T., Widmer, G.: Towards automatic retrieval of album covers. In: Proceedings of the 28th European Conference on Information Retrieval (ECIR), London (2006)
411. Schedl, M., Knees, P., Widmer, G., Seyerlehner, K., Pohle, T.: Browsing the web using stacked three-dimensional sunbursts to visualize term co-occurrences and multimedia content. In: Proceedings of the 18th IEEE Visualization 2007 Conference (Vis), Sacramento (2007)
412. Schedl, M., Pohle, T., Koenigstein, N., Knees, P.: What's hot? Estimating country-specific artist popularity. In: Proceedings of the 11th International Society for Music Information Retrieval Conference (ISMIR), Utrecht (2010)
413. Schedl, M., Seyerlehner, K., Schnitzer, D., Widmer, G., Schiketanz, C.: Three web-based heuristics to determine a person's or institution's country of origin. In: Proceedings of the 33th Annual International ACM SIGIR Conference on Research and Development in Information Retrieval (SIGIR), Geneva (2010)

414. Schedl, M., Höglinger, C., Knees, P.: Large-scale music exploration in hierarchically organized landscapes using prototypicality information. In: Proceedings of the ACM International Conference on Multimedia Retrieval (ICMR), Trento (2011)
415. Schedl, M., Knees, P., Böck, S.: Investigating the similarity space of music artists on the micro-blogosphere. In: Proceedings of the 12th International Society for Music Information Retrieval Conference (ISMIR), Miami (2011)
416. Schedl, M., Pohle, T., Knees, P., Widmer, G.: Exploring the music similarity space on the web. ACM Trans. Inf. Syst. **29**(3) (2011)
417. Schedl, M., Widmer, G., Knees, P., Pohle, T.: A music information system automatically generated via web content mining techniques. Inf. Process. Manag. **47**, 426–439 (2011)
418. Schedl, M., Hauger, D., Schnitzer, D.: A model for serendipitous music retrieval. In: Proceedings of the 2nd Workshop on Context-awareness in Retrieval and Recommendation (CaRR), Lisbon (2012)
419. Schedl, M., Stober, S., Gómez, E., Orio, N., Liem, C.C.: User-aware music retrieval. In: Müller, M., Goto, M., Schedl, M. (eds.) Multimodal Music Processing. Dagstuhl Follow-Ups, vol. 3. Schloss Dagstuhl–Leibniz-Zentrum für Informatik, Wadern (2012)
420. Schedl, M., Flexer, A., Urbano, J.: The neglected user in music information retrieval research. J. Intell. Inf. Syst. **41**, 523–539 (2013)
421. Schedl, M., Breitschopf, G., Ionescu, B.: Mobile music genius: reggae at the beach, metal on a Friday night? In: Proceedings of the 4th ACM International Conference on Multimedia Retrieval (ICMR), Glasgow (2014)
422. Schedl, M., Gómez, E., Urbano, J.: Evaluation in music information retrieval. Found. Trends Inf. Retr. **8**(2–3), 127–261 (2014)
423. Schedl, M., Hauger, D., Urbano, J.: Harvesting microblogs for contextual music similarity estimation – a co-occurrence-based framework. Multimedia Syst. **20**, 693–705 (2014)
424. Schedl, M., Knees, P., McFee, B., Bogdanov, D., Kaminskas, M.: Music recommender systems. In: Ricci, F., Rokach, L., Shapira, B., Kantor, P.B. (eds.) Recommender Systems Handbook, Chap. 13, 2nd edn., pp. 453–492. Springer, New York (2015)
425. Scherer, K.: What are emotions? And how can they be measured? Soc. Sci. Inf. **44**(4), 693–727 (2005)
426. Schlüter, J., Osendorfer, C.: Music similarity estimation with the mean-covariance restricted Boltzmann machine. In: Proceedings of the 10th International Conference on Machine Learning and Applications and Workshops (ICMLA), Honolulu (2011)
427. Schmidt, E.M., Kim, Y.E.: Projection of acoustic features to continuous valence-arousal mood labels via regression. In: Proceedings of the 10th International Society for Music Information Retrieval Conference (ISMIR), Kobe (2009)
428. Schnitzer, D., Flexer, A., Schedl, M., Widmer, G.: Local and global scaling reduce hubs in space. J. Mach. Learn. Res. **13**, 2871–2902 (2012)
429. Schnitzer, D., Pohle, T., Knees, P., Widmer, G.: One-touch access to music on mobile devices. In: Proceedings of the 6th International Conference on Mobile and Ubiquitous Multimedia (MUM), Oulu (2007)
430. Schubert, E., Wolfe, J., Tarnopolsky, A.: Spectral centroid and timbre in complex, multiple instrumental textures. In: Proceedings of the 8th International Conference on Music Perception and Cognition (ICMPC), Evanston (2002)
431. Schultz, M., Joachims, T.: Learning a distance metric from relative comparisons. In: Thrun, S., Saul, L., Schölkopf, B. (eds.) Advances in Neural Information Processing Systems 16 (NIPS). MIT Press, Cambridge (2004)
432. Şentürk, S., Holzapfel, A., Serra, X.: Linking scores and audio recordings in makam music of Turkey. J. N. Music Res. **43**(1), 34–52 (2014)
433. Serrà, J., Gómez, E., Herrera, P., Serra, X.: Chroma binary similarity and local alignment applied to cover song identification. IEEE Trans. Audio Speech Lang. Process. **16**(6), 1138–1151 (2008)
434. Serra, X.: Data gathering for a culture specific approach in MIR. In: Proceedings of the 21st International Conference on World Wide Web (WWW) Companion, Lyon (2012)

435. Serra, X., Magas, M., Benetos, E., Chudy, M., Dixon, S., Flexer, A., Gómez, E., Gouyon, F., Herrera, P., Jordà, S., Paytuvi, O., Peeters, G., Schlüter, J., Vinet, H., Widmer, G.: Roadmap for Music Information ReSearch. MIReS Consortium (2013)
436. Seyerlehner, K.: Inhaltsbasierte Ähnlichkeitsmetriken zur Navigation in Musiksammlungen. Master's thesis, Johannes Kepler University Linz (2006)
437. Seyerlehner, K.: Content-based music recommender systems: beyond simple frame-level audio similarity. Dissertation, Johannes Kepler University Linz (2010)
438. Seyerlehner, K., Widmer, G., Knees, P.: Frame level audio similarity – a codebook approach. In: Proceedings of the 11th International Conference on Digital Audio Effects (DAFx), Espoo (2008)
439. Seyerlehner, K., Widmer, G., Pohle, T.: Fusing block-level features for music similarity estimation. In: Proceedings of the 13th International Conference on Digital Audio Effects (DAFx), Graz (2010)
440. Seyerlehner, K., Widmer, G., Schedl, M., Knees, P.: Automatic music tag classification based on block-level features. In: Proceedings of the 7th Sound and Music Computing Conference (SMC), Barcelona (2010)
441. Seyerlehner, K., Schedl, M., Knees, P., Sonnleitner, R.: A refined block-level feature set for classification, similarity and tag prediction. In: Extended Abstract to the Annual Music Information Retrieval Evaluation eXchange (MIREX), Miami (2011)
442. Seyerlehner, K., Widmer, G., Knees, P.: A comparison of human, automatic and collaborative music genre classification and user centric evaluation of genre classification systems. In: Detyniecki, M., Knees, P., Nürnberger, A., Schedl, M., Stober, S. (eds.) Adaptive Multimedia Retrieval: Context, Exploration, and Fusion. Lecture Notes in Computer Science, vol. 6817. Springer, Heidelberg (2011)
443. Shannon, C.: Communication in the presence of noise. Proc. Inst. Radio Eng. **37**(1), 10–21 (1949)
444. Shao, B., Wang, D., Li, T., Ogihara, M.: Music recommendation based on acoustic features and user access patterns. IEEE Trans. Audio Speech Lang. Process. **17**(8), 1602–1611 (2009)
445. Shavitt, Y., Weinsberg, U.: Songs clustering using peer-to-peer co-occurrences. In: Proceedings of the 11th IEEE International Symposium on Multimedia (ISM), San Diego (2009)
446. Shavitt, Y., Weinsberg, E., Weinsberg, U.: Mining music from large-scale, peer-to-peer networks. IEEE Multimedia **18**(1), 14–23 (2011)
447. Sheskin, D.J.: Handbook of Parametric and Nonparametric Statistical Procedures, 3rd edn. Chapman & Hall/CRC, Boca Raton (2004)
448. Simon, I., Morris, D., Basu, S.: MySong: automatic accompaniment generation for vocal melodies. In: Proceedings of the 26th Annual ACM SIGCHI Conference on Human Factors in Computing Systems (CHI), Florence (2008)
449. Singhi, A., Brown, D.: Are poetry and lyrics all that different? In: Proceedings of the 15th International Society for Music Information Retrieval Conference (ISMIR), Taipei (2014)
450. Slaney, M.: Web-scale multimedia analysis: Does content matter? IEEE MultiMedia **18**(2), 12–15 (2011)
451. Slaney, M., White, W.: Similarity based on rating data. In: Proceedings of the 8th International Conference on Music Information Retrieval (ISMIR), Vienna (2007)
452. Slaney, M., Weinberger, K., White, W.: Learning a metric for music similarity. In: Proceedings of the 9th International Conference of Music Information Retrieval (ISMIR), Philadelphia (2008)
453. Smaragdis, P., Fevotte, C., Mysore, G., Mohammadiha, N., Hoffman, M.: Static and dynamic source separation using nonnegative factorizations: a unified view. IEEE Signal Process. Mag. **31**(3), 66–75 (2014)
454. Smith, S.W.: Digital Signal Processing: A Practical Guide for Engineers and Scientists. Newnes, Amsterdam (2003)
455. Smolensky, P.: Information processing in dynamical systems: foundations of harmony theory. In: Rumelhart, D.E., McClelland, J.L. (eds.) Parallel Distributed Processing: Explorations in the Microstructure of Cognition, vol. 1, pp. 194–281. MIT Press, Cambridge (1986)

456. Smyth, B., McClave, P.: Similarity vs. diversity. In: Proceedings of the 4th International Conference on Case-Based Reasoning (ICCBR): Case-Based Reasoning Research and Development, Vancouver (2001)
457. Song, Y., Dixon, S., Pearce, M.: Evaluation of musical features for emotion classification. In: Proceedings of the 13th International Society for Music Information Retrieval Conference (ISMIR), Porto (2012)
458. Sordo, M.: Semantic annotation of music collections: a computational approach. Ph.D. thesis, Universitat Pompeu Fabra, Barcelona (2012)
459. Sordo, M., Laurier, C., Celma, O.: Annotating music collections: how content-based similarity helps to propagate labels. In: Proceedings of the 8th International Conference on Music Information Retrieval (ISMIR), Vienna (2007)
460. Sotiropoulos, D.N., Lampropoulos, A.S., Tsihrintzis, G.A.: MUSIPER: a system for modeling music similarity perception based on objective feature subset selection. User Model. User Adap. Interact. **18**(4), 315–348 (2008)
461. Sparling, E.I., Sen, S.: Rating: How difficult is it? In: Proceedings of the 5th ACM Conference on Recommender Systems (RecSys), Chicago (2011)
462. Spearman, C.: The proof and measurement of association between two things. Am. J. Psychol. **15**, 88–103 (1904)
463. Stasko, J., Zhang, E.: Focus+context display and navigation techniques for enhancing radial, space-filling hierarchy visualizations. In: Proceedings of the 6th IEEE Symposium on Information Visualization (InfoVis), Salt Lake City (2000)
464. Stavness, I., Gluck, J., Vilhan, L., Fels, S.: The MUSICtable: a map-based ubiquitous system for social interaction with a digital music collection. In: Proceedings of the 4th International Conference on Entertainment Computing (ICEC), Sanda (2005)
465. Stenzel, R., Kamps, T.: Improving content-based similarity measures by training a collaborative model. In: Proceedings of the 6th International Conference on Music Information Retrieval (ISMIR), London (2005)
466. Stober, S.: Adaptive distance measures for exploration and structuring of music collections. In: Proceedings of the 42nd International AES Conference: Semantic Audio, Ilmenau (2011)
467. Stober, S., Nürnberger, A.: MusicGalaxy - an adaptive user-interface for exploratory music retrieval. In: Proceedings of 7th Sound and Music Computing Conference (SMC), Barcelona (2010)
468. Stober, S., Nürnberger, A.: An experimental comparison of similarity adaptation approaches. In: Detyniecki, M., García-Serrano, A., Nürnberger, A., Stober, S. (eds.) Adaptive Multimedia Retrieval: Large-Scale Multimedia Retrieval and Evaluation. Lecture Notes in Computer Science, vol. 7836. Springer, Heidelberg (2013)
469. Sturm, B.L.: An analysis of the GTZAN music genre dataset. In: Proceedings of the 2nd International ACM Workshop on Music Information Retrieval with User-centered and Multimodal Strategies (MIRUM), Nara (2012)
470. Sturm, B.L.: Classification accuracy is not enough. J. Intell. Inf. Syst. **41**, 371–406 (2013)
471. Sturm, B.L.: A survey of evaluation in music genre recognition. In: Nürnberger, A., Stober, S., Larsen, B., Detyniecki, M. (eds.) Adaptive Multimedia Retrieval: Semantics, Context, and Adaptation. Lecture Notes in Computer Science, vol. 8382. Springer, Heidelberg (2014)
472. Sturm, B.L.: The state of the art ten years after a state of the art: future research in music information retrieval. J. N. Music Res. **43**(2), 147–172 (2014)
473. Sutcliffe, R., Fox, C., Root, D.L., Hovy, E., Lewis, R.: The C@merata Task at MediaEval 2015: Natural Language Queries on Classical Music Scores. In: Larson, M., Ionescu, B., Sjöberg, M., Anguera, X., Poignant, J., Riegler, M., Eskevich, M., Hauff, C., Sutcliffe, R., Jones, G.J., Yang, Y.H., Soleymani, M., Papadopoulos, S. (eds.) Working Notes Proceedings of the MediaEval 2015 Workshop, vol. 1436. CEUR-WS, Wurzen (2015)
474. Suzuki, I., Hara, K., Shimbo, M., Matsumoto, Y., Saerens, M.: Investigating the effectiveness of Laplacian-based kernels in hub reduction. In: Proceedings of the 26th Conference on Artificial Intelligence (AAAI), Toronto (2012)

475. Tiemann, M., Pauws, S.: Towards ensemble learning for hybrid music recommendation. In: Proceedings of the ACM Conference on Recommender Systems (RecSys), Minneapolis (2007)
476. Tobudic, A., Widmer, G.: Learning to play like the great pianists. In: Proceedings of the 19th International Joint Conference on Artificial Intelligence (IJCAI), Edinburgh (2005)
477. Tomašev, N., Radovanović, M., Mladenić, D., Ivanović, M.: The role of hubness in clustering high-dimensional data. In: Huang, J.Z., Cao, L., Srivastava, J. (eds.) Advances in Knowledge Discovery and Data Mining. Lecture Notes in Computer Science, vol. 6634. Springer, Heidelberg (2011)
478. Tsatsishvili, V.: Automatic subgenre classification of heavy metal music. Master's thesis, University of Jyväskylä (2011)
479. Tsunoo, E., Tzanetakis, G., Ono, N., Sagayama, S.: Audio genre classification using percussive pattern clustering combined with timbral features. In: IEEE International Conference on Multimedia and Expo (ICME) (2009)
480. Turnbull, D., Barrington, L., Torres, D., Lanckriet, G.: Towards musical query-by-semantic-description using the CAL500 data set. In: Proceedings of the 30th Annual International ACM SIGIR Conference on Research and Development in Information Retrieval (SIGIR), Amsterdam (2007)
481. Turnbull, D., Liu, R., Barrington, L., Lanckriet, G.: A game-based approach for collecting semantic annotations of music. In: Proceedings of the 8th International Conference on Music Information Retrieval (ISMIR), Vienna (2007)
482. Turnbull, D., Barrington, L., Lanckriet, G.: Five approaches to collecting tags for music. In: Proceedings of the 9th International Society for Music Information Retrieval Conference (ISMIR), Philadelphia (2008)
483. Turnbull, D., Barrington, L., Torres, D., Lanckriet, G.: Semantic annotation and retrieval of music and sound effects. IEEE Trans. Audio Speech Lang. Process. **16**(2), 467–476 (2008)
484. Turnbull, D., Barrington, L., Yazdani, M., Lanckriet, G.: Combining audio content and social context for semantic music discovery. In: Proceedings of the 32th Annual International ACM SIGIR Conference on Research and Development in Information Retrieval (SIGIR), Boston (2009)
485. Typke, R.: Music retrieval based on melodic similarity. Ph.D. thesis, Universiteit Utrecht (2007)
486. Tzanetakis, G.: Marsyas submissions to MIREX 2010. In: Extended Abstract to the Annual Music Information Retrieval Evaluation eXchange (MIREX), Utrecht (2010)
487. Tzanetakis, G., Cook, P.: Musical genre classification of audio signals. IEEE Trans. Speech Audio Process. **10**(5), 293–302 (2002)
488. Ultsch, A., Siemon, H.P.: Kohonen's self-organizing feature maps for exploratory data analysis. In: Proceedings of the International Neural Network Conference (INNC), Dordrecht, vol. 1 (1990)
489. Urbano, J.: Evaluation in audio music similarity. Ph.D. thesis, Universidad Carlos III de Madrid (2013)
490. Urbano, J., Schedl, M., Serra, X.: Evaluation in music information retrieval. J. Intell. Inf. Syst. **41**, 345–369 (2013)
491. van den Oord, A., Dieleman, S., Schrauwen, B.: Deep content-based music recommendation. In: Burges, C., Bottou, L., Welling, M., Ghahramani, Z., Weinberger, K. (eds.) Advances in Neural Information Processing Systems 26 (NIPS). Curran Associates, Inc., Red Hook (2013)
492. van der Maaten, L., Hinton, G.: Visualizing data using t-SNE. J. Mach. Learn. Res. **9**, 2579–2605 (2008)
493. Vapnik, V.N.: The Nature of Statistical Learning Theory. Springer, New York (1995)
494. Vapnik, V.N.: Statistical Learning Theory. Wiley, Chichester (1998)
495. Vargas, S., Castells, P.: Rank and relevance in novelty and diversity metrics for recommender systems. In: Proceedings of the 5th ACM Conference on Recommender Systems (RecSys), Chicago (2011)

496. Vembu, S., Baumann, S.: A self-organizing map based knowledge discovery for music recommendation systems. In: Proceedings of the 2nd International Symposium on Computer Music Modeling and Retrieval (CMMR), Esbjerg (2004)
497. Vesanto, J.: Using SOM in data mining. Licentiate's thesis, Helsinki University of Technology (2000)
498. Vignoli, F., van Gulik, R., van de Wetering, H.: Mapping music in the palm of your hand, explore and discover your collection. In: Proceedings of the 5th International Symposium on Music Information Retrieval (ISMIR), Barcelona (2004)
499. Vincent, E.: Musical source separation using time-frequency source priors. IEEE Trans. Audio Speech Lang. Process. **14**(1), 91–98 (2006)
500. Virtanen, T.: Monaural sound source separation by nonnegative matrix factorization with temporal continuity and sparseness criteria. IEEE Trans. Audio Speech Lang. Process. **15**(3), 1066–1074 (2007)
501. von Ahn, L., Dabbish, L.: Labeling images with a computer game. In: Proceedings of the 22nd Annual ACM SIGCHI Conference on Human Factors in Computing Systems (CHI), Vienna (2004)
502. Voorhees, E.M., Ellis, A. (eds.): Proceedings of the 23rd Text REtrieval Conference (TREC). National Institute of Standards and Technology (NIST), Gaithersburg, MD (2014)
503. Voorhees, E.M., Harman, D.K. (eds.): TREC: Experiment and Evaluation in Information Retrieval. MIT Press, Cambridge, MA (2005)
504. Wanderley, M., Orio, N.: Evaluation of input devices for musical expression: borrowing tools from HCI. Comput. Music J. **26**(3), 62–76 (2002)
505. Wang, A.: The Shazam music recognition service. Commun. ACM **49**(8), 44–48 (2006)
506. Wang, D., Li, T., Ogihara, M.: Are tags better than audio? The effect of joint use of tags and audio content features for artistic style clustering. In: Proceedings of the 11th International Society for Music Information Retrieval Conference (ISMIR), Utrecht (2010)
507. Wang, F., Wang, X., Shao, B., Li, T., Ogihara, M.: Tag integrated multi-label music style classification with hypergraph. In: Proceedings of the 10th International Society for Music Information Retrieval Conference (ISMIR), Kobe (2009)
508. Wang, X., Rosenblum, D., Wang, Y.: Context-aware mobile music recommendation for daily activities. In: Proceedings of the 20th ACM International Conference on Multimedia (MM), Nara (2012)
509. Webb, B.: Netflix update: try this at home. Simon Funk (2006) http://sifter.org/~simon/journal/20061211.html
510. Weinberger, K.Q., Saul, L.K.: Distance metric learning for large margin nearest neighbor classification. J. Mach. Learn. Res. **10**, 207–244 (2009)
511. Whitman, B., Ellis, D.P.: Automatic record reviews. In: Proceedings of 5th International Conference on Music Information Retrieval (ISMIR), Barcelona (2004)
512. Whitman, B., Lawrence, S.: Inferring descriptions and similarity for music from community metadata. In: Proceedings of the 2002 International Computer Music Conference (ICMC), Göteborg (2002)
513. Whitman, B., Smaragdis, P.: Combining musical and cultural features for intelligent style detection. In: Proceedings of the 3rd International Conference on Music Information Retrieval (ISMIR), Paris (2002)
514. Whitman, B., Flake, G., Lawrence, S.: Artist detection in music with minnowmatch. In: Proceedings of the IEEE Signal Processing Society Workshop on Neural Networks for Signal Processing XI (NNSP), North Falmouth (2001)
515. Widmer, G.: Getting closer to the essence of music: the con espressione manifesto. ACM Trans. Intell. Syst. Technol. (2016)
516. Widmer, G., Goebl, W.: Computational models of expressive music performance: the state of the art. J. N. Music Res. **33**(3), 203–216 (2004)
517. Widmer, G., Zanon, P.: Automatic recognition of famous artists by machine. In: Proceedings of the 16th European Conference on Artificial Intelligence (ECAI), Valencia (2004)
518. Widmer, G., Flossmann, S., Grachten, M.: YQX plays Chopin. AI Mag. **30**(3), 35 (2009)

519. Wiggins, G.A.: Semantic gap?? Schemantic schmap!! Methodological considerations in the scientific study of music. In: Proceedings of the 11th IEEE International Symposium on Multimedia (ISM), San Diego (2009)
520. Wolff, D., Weyde, T.: Adapting metrics for music similarity using comparative ratings. In: Proceedings of the 12th International Society for Music Information Retrieval Conference (ISMIR), Miami (2011)
521. Wu, Y., Takatsuka, M.: Spherical self-organizing map using efficient indexed geodesic data structure. Neural Netw. **19**(6–7), 900–910 (2006)
522. Xu, C., Maddage, N.C., Shao, X., Cao, F., Tian, Q.: Musical genre classification using support vector machines. In: Proceedings of the IEEE International Conference on Acoustics, Speech, and Signal Processing (ICASSP), Hong Kong, vol. 5 (2003)
523. Yang, D., Chen, T., Zhang, W., Lu, Q., Yu, Y.: Local implicit feedback mining for music recommendation. In: Proceedings of the 6th ACM Conference on Recommender Systems (RecSys), Dublin (2012)
524. Yang, Y., Pedersen, J.O.: A comparative study on feature selection in text categorization. In: Proceedings of the 14th International Conference on Machine Learning (ICML), Nashville (1997)
525. Yang, Y.H., Chen, H.H.: Ranking-based emotion recognition for music organization and retrieval. IEEE Trans. Audio Speech Lang. Process. **19**(4), 762–774 (2011)
526. Yang, Y.H., Chen, H.H.: Machine recognition of music emotion: a review. Trans. Intell. Syst. Technol. **3**(3) (2013)
527. Yang, Y.H., Lin, Y.C., Su, Y.F., Chen, H.: A regression approach to music emotion recognition. IEEE Trans. Audio Speech Lang. Process. **16**(2), 448–457 (2008)
528. Yiu, K.K., Mak, M.W., Li, C.K.: Gaussian mixture models and probabilistic decision-based neural networks for pattern classification: a comparative study. Neural Comput. Appl. **8**, 235–245 (1999)
529. Yoshii, K., Goto, M., Komatani, K., Ogata, T., Okuno, H.G.: An efficient hybrid music recommender system using an incrementally trainable probabilistic generative model. IEEE Trans. Audio Speech Lang. Process. **16**(2), 435–447 (2008)
530. Yu, H., Xie, L., Sanner, S.: Twitter-driven YouTube views: beyond individual influencers. In: Proceedings of the 22nd ACM International Conference on Multimedia (MM), Orlando (2014)
531. Zadel, M., Fujinaga, I.: Web services for music information retrieval. In: Proceedings of the 5th International Symposium on Music Information Retrieval (ISMIR), Barcelona (2004)
532. Zangerle, E., Gassler, W., Specht, G.: Exploiting Twitter's collective knowledge for music recommendations. In: Rowe, M., Stankovic, M., Dadzie, A.S. (eds.) Proceedings of the WWW'12 Workshop on Making Sense of Microposts (#MSM). CEUR-WS, Lyon (2012)
533. Zhai, C.: Statistical Language Models for Information Retrieval, Chap. 7, pp. 87–100. Morgan and Claypool, San Rafael (2008)
534. Zhang, B., Xiang, Q., Lu, H., Shen, J., Wang, Y.: Comprehensive query-dependent fusion using regression-on-folksonomies: a case study of multimodal music search. In: Proceedings of the 17th ACM International Conference on Multimedia (MM), Beijing (2009)
535. Zhang, M., Hurley, N.: Avoiding monotony: improving the diversity of recommendation lists. In: Proceedings of the 2nd ACM Conference on Recommender Systems (RecSys), Lausanne (2008)
536. Zhang, Y.C., Seaghdha, D.O., Quercia, D., Jambor, T.: Auralist: introducing serendipity into music recommendation. In: Proceedings of the 5th ACM International Conference on Web Search and Data Mining (WSDM), Seattle (2012)
537. Zheleva, E., Guiver, J., Mendes Rodrigues, E., Milić-Frayling, N.: Statistical models of music-listening sessions in social media. In: Proceedings of the 19th International Conference on World Wide Web (WWW), Raleigh (2010)
538. Zhou, F., Claire, Q., King, R.D.: Predicting the geographical origin of music. In: Proceedings of the IEEE International Conference on Data Mining (ICDM), Shenzhen (2014)
539. Zobel, J., Moffat, A.: Exploring the similarity space. ACM SIGIR Forum **32**(1), 18–34 (1998)

Index

P. Knees, M. Schedl, *Music Similarity and Retrieval*, The Information Retrieval Series 36, DOI 10.1007/978-3-662-49722-7

Printed in the United States
By Bookmasters